Crop Production Science in Horticulture Series

Series Editors: Jeff Atherton, Professor of Tropical Horticulture, University of the West Indies, Barbados, and Alun Rees (retired), former Head of Crop Science, Glasshouse Crops Research Institute, Horticultural Consultant and Editor of the *Journal of Horticultural Science and Biotechnology*.

This series examines economically important horticultural crops selected from the major production systems in temperate, subtropical and tropical climatic areas. Systems represented range from open field and plantation sites to protected plastic and glass houses, growing rooms and laboratories. Emphasis is placed on the scientific principles underlying crop production practices rather than on providing empirical recipes for uncritical acceptance. Scientific understanding provides the key to both reasoned choice of practice and the solution of future problems.

Students and staff at universities and colleges throughout the world involved in courses in horticulture, as well as in agriculture, plant science, food science and applied biology at degree, diploma or certificate level, will welcome this series as a succinct and readable source of information. The books will also be invaluable to progressive growers, advisers and end-product users requiring an authoritative, but brief, scientific introduction to particular crops or systems. Keen gardeners wishing to understand the scientific basis of recommended practices will also find the series very useful.

The authors are all internationally renowned experts with extensive experience of their subjects. Each volume follows a common format covering all aspects of production, from background physiology and breeding, to propagation and planting, through husbandry and crop protection, to harvesting, handling and storage. Selective references are included to direct the reader to further information on specific topics.

Titles Available:

1. **Ornamental Bulbs, Corms and Tubers** A.R. Rees
2. **Citrus** F.S. Davies and L.G. Albrigo
3. **Onions and Other Vegetable Alliums** J.L. Brewster
4. **Ornamental Bedding Plants** A.M. Armitage
5. **Bananas and Plantains** J.C. Robinson
6. **Cucurbits** R.W. Robinson and D.S. Decker-Walters
7. **Tropical Fruits** H.Y. Nakasone and R.E. Paull
8. **Coffee, Cocoa and Tea** K.C. Willson
9. **Lettuce, Endive and Chicory** E.J. Ryder
10. **Carrots and Related Vegetable Umbelliferae** V.E. Rubatzky, C.F. Quiros and P.W. Simon
11. **Strawberries** J.F. Hancock
12. **Peppers: Vegetable and Spice Capsicums** P.W. Bosland and E.J. Votava
13. **Tomatoes** E. Heuvelink
14. **Vegetable Brassicas and Related Crucifers** G. Dixon
15. **Onions and Other Vegetable Alliums, 2nd Edition** J.L. Brewster
16. **Grapes** G.L. Creasy and L.L. Creasy
17. **Tropical Root and Tuber Crops: Cassava, Sweet Potato, Yams and Aroids** V. Lebot
18. **Olives** I. Therios

This book is dedicated to the memory of
Dr Robert M. Pool

GRAPES

G.L. Creasy
Senior Lecturer in Viticulture
Centre for Viticulture and Oenology
Lincoln University
Christchurch, New Zealand

and

L.L. Creasy
Professor Emeritus
Department of Horticulture
Cornell University
Ithaca, New York, USA

CABI is a trading name of CAB International

CABI Head Office
Nosworthy Way
Wallingford
Oxfordshire OX10 8DE
UK

Tel: +44 (0)1491 832111
Fax: +44 (0)1491 833508
E-mail: cabi@cabi.org
Website: www.cabi.org

CABI North American Office
875 Massachusetts Avenue
7th Floor
Cambridge, MA 02139
USA

Tel: +1 617 395 4056
Fax: +1 617 354 6875
E-mail: cabi-nao@cabi.org

A catalogue record for this book is available from the British Library, London, UK.

Library of Congress Cataloging-in-Publication Data

Creasy, G. L. (Glen L.)
Grapes / G.L. Creasy and L.L. Creasy.
p. cm. -- (Crop production science in horticulture series ; 16)
Includes bibliographical references and index.
ISBN 978-1-84593-401-9 (alk. paper)
1. Grapes. 2. Grape industry. I. Creasy, Leroy L. II. Title. III. Series: Crop production science in horticulture ; 16.

SB387.7.C74 2009
634.8--dc22

2008029279

ISBN-13: 978 1 84593 401 9

Typeset by Columns Design, Reading, UK.
Printed and bound in the UK by the MPG Books Group.

The paper used for the text pages in this book is FSC certified. The FSC (Forest Stewardship Council) is an international network to promise responsible management of the world's forests.

CONTENTS

PREFACE xi

1
HISTORY, USES AND PRODUCTION 1

Geographical origins of grapevine species 1
Family, genus, species and related plants 2
Natural growth conditions 4
Historical cultivation 5
Uses 5
Fermented grape products 5
Table grapes 6
Raisins 7
Grape juice 7
Grape-producing countries 8

2
CULTIVARS, ANATOMY AND IMPROVEMENT 11

Main cultivars for various uses 11
Clones 14
Anatomy and physiology 15
Roots 15
Above the soil 18
Photosynthesis 21
Flower and berries 24
Ampelography 25
Breeding and genetics 26
Indicators of quality 27

3
Grapevine Growth and Fruit Development 29
Phenology 29
Vine (vegetative) development 34
Patterns of root growth 34
Tendrils 35
Cane maturation 35
Leaf-fall and abscission 36
Dormancy 36
Flower initiation, fruit development and berry maturation 37
Where do tendrils and flower clusters come from? 37
Factors affecting branching 38
Plant growth regulators 39
Flower development and anthesis 40
Fruit set 40
Berry development and maturation 43
Environmental/climatic influences 59
Trying to quantify climate 60
The vine as a perennial plant (carbohydrate partitioning) 63
Balancing vegetative and reproductive growth 64

4
Climatic Requirements 65
Cold hardiness 65
Growing vines in tropical areas 68
Frost tolerance 69
Heat and light 70
Water use 70
Soils 71
Terroir 71
Climate change 72

5
Vineyard Establishment 75
Site selection 75
Climate 76
Soil 77
Slope and aspect of land 79
Other factors 79
Site planning 81
Preparing the site 82

Shelter from wind 83
Vine planting density 85
Rootstocks 88
Rootstock choice 89
Propagation 91
Layering 91
Cuttings 91
Grafting 92
Other methods 93
Planting 94
Vine establishment 96
Vine shelters 98
Young vine care 99
Second season 100

6
Seasonal Management 105

Pruning and training 105
Types of pruning 109
Pruning decisions 114
Matching pruning to vine capacity 115
Other considerations when pruning 118
Training and trellising 118
Self- or stake-supported 119
Single wire 121
Hedge-type 122
Divided canopies 124
Other trellising systems 126
End assemblies 128
Vineyard floor management 131
Frost management 135
Types of frost event 136
Passive control strategies 136
Active control strategies 136
Other methods 138
Canopy management 139
Goals and tools for canopy management 139
Shoot thinning 140
Shoot positioning 142
Shoot topping and hedging 143
Leaf removal 143
Other methods 145
Things to avoid 146
Quantifying change in the canopy 146

Irrigation 148
Methods of monitoring soil moisture 151
Methods of monitoring plant water status 153
Scheduling water application 154
Other management practices 156
Girdling 156
Fruit thinning 156
Application of plant growth regulators 159
Harvest 159

7
NUTRITION 163
Nutrient analysis and correction strategies 164
Testing and nutrient addition 164
The nutrients 165
Macronutrients 166
Micronutrients 170

8
MECHANIZATION 175
Weed management 176
Canopy management 178
Crop management and harvesting 179
Pruning 181
Environmental monitoring 182
Fully mechanized systems 183

9
GRAPEVINE PESTS, DISEASES AND DISORDERS 185
Diseases 187
Fungal diseases 187
Bacterial diseases 201
Viral diseases 202
Control of grape diseases 203
Control of fungal diseases 204
Control of viral diseases 205
Other disorders 206
Insect pests 206
Phylloxera 206
Leafhoppers 208
Borer insects 209

Mealybugs 209
The grape berry moth 209
Thrips 210
Beetles 210
Mites 210
Chemical control 210
Animal pests 212
Nematodes 212
Vine-grazing pests 213
Berry-eating pests 213
Weeds 216
Pesticide resistance 217
Pest and disease control in organic grape production 218
Pesticide application 219
Maximizing spraying efficiency 219
Pesticide application equipment 222

10
Harvest and Postharvest Processing 225
Table grapes 225
Quality parameters 226
Picking 226
Packing 226
Cooling at the packing house 229
Storage 229
Dried grapes 230
Cultivars 230
Harvest 231
Processing 232
Juice/preserves 233
Uses 234
Processing 234
Wine 235
Harvest 235
Processing 237
Other wine styles 239

References 241

Index 281

The colour plate sections can be found following pages 52, 148 and 212.
Sections supported by the California Table Grape Commission, John Coleman of Plasma Physics Corporation and the Corporate Research Sponsors of L. L. Creasy.

Preface

This book targets advanced plant science students with specific interest in viticulture and those who are producing grapes for the first time. We think it will be of interest to enthusiasts of the vine and its products who want to learn a bit more about how grapes are grown and a lot about the whys behind the methods. There are answers to many questions in the following chapters but, as always seems to be the case, the more you learn the more you discover that there is always more to learn!

Grapes have so many uses and are so unique that no fruit can challenge their superiority. The production of grapes continues to grow: China now produces more table grapes than the next nine major producing countries combined. It seems inevitable that China will also dominate wine grape production at some point, which makes life in the industry terribly interesting and exciting!

In fact, methods of grape production and how the fruit is utilized are changing rapidly. For example, the advent of year-round supplies of table grapes, facilitated through economic means of shipping between the northern and southern hemispheres, has bolstered demand for the product, leading to more innovation and investment in the industry and improvement in our ability to grow and market them. Grapes are being produced in many places where it was not considered possible before, largely due to advances in understanding of how the vine works, as well as to improvements in technology and management skills.

Labour is a traditional limiting factor in grape production, but progress is being made to further mechanize production and harvesting. Mechanical harvesters have become the standard for wine and juice grapes and have even changed the way vines are trained because hand harvesting benefits from large clusters in well-defined areas, while for mechanical harvesters it doesn't matter. Improvements in methods of pest control, with new application technologies and improved chemicals incurring reduced environmental impact, also help the bottom line.

Bigger machines and more complex equipment would suggest a trend to larger producers and the decline of small producers. This has not been the case in the wine industry, at least, as the number of small wineries has increased greatly over the past 25 years, and has become the basis for significant tourism industries. This helps to avoid wine becoming a commodity, and therefore engenders curiosity and enthusiasm by producers and consumers alike.

Many small vineyards are being started with great enthusiasm by people with limited experience in grape growing. We hope this book, with its mix of theory and science, will be useful to them when they are pioneering their own path into the world of the vine.

The grapevine is a plastic and adaptable plant. It will grow in the most unlikely of places, and can be shaped into a myriad of different forms in our quest to extract the highest quality product from these. The fruit of this vine is made into many different products, of varying tastes and aromas. We like grapes and all the things that are made from them. Le retired early from academic life to grow grapes, and knows every day that it was the right thing to do. Glen became interested in grapes at about the same time and now lives, breathes and, importantly, drinks, around the subject area.

LLC: I would like to thank the many individuals who have supported my research on grapes and encouraged me to continue this research to the present. Special thanks go to Bill Wagner and his excellent staff at Wagner Vineyards for making available his vines for research. Special thanks to Bruno and Marcello Ceretto of Alba, Italy for introducing me to the best wine and food in this world.

GLC: I wish to acknowledge the large number of people who have contributed to my experiences with all things grape – growers, students, scientists, teachers, winemakers and all the rest that have guided me to where I am today. With reference to this work, I thank my wife Kirsten and kids Bella and Maddy for putting up with the process of writing a book, Le for the chance to write with you and my mom, Min Creasy, for handling the index. I feel truly lucky to have so many supportive people around me!

A few parting words. A book is frozen in time. It is written and endures (or not). Grape growing and viticulture, however, change daily. A book is intended to bring readers up to a specific moment in time and to stimulate them to continue seeking more information. Principles never change but their applications change greatly, so may the quest for knowledge never stop!

June 2008

1

History, Uses and Production

The grapevine (*Vitis* spp.) is cultivated all over the world and the grape itself is used for a myriad of products, many of which are well known to all of us. Viticulture is one of the major horticultural industries of the world, with the area of grapevines cultivated exceeding 7.9 million ha (OIV, 2006). Most grapes are grown for the production of wine but, when first discovered, its appeal as fresh eating fruit was probably what attracted the first hunter-gatherers. Today the fruit is used in a wide variety of products, ranging through fresh fruit, preserves, juice, wine and raisins.

The grapevine is a vigorously growing plant and in some places is considered an invasive weed (Uva *et al.*, 1997). Fortunately, the fact that the grape is used for a great many products means that it can be considered more of a crop plant than a weed. The fact that the grapevine is a climbing plant lends it an unusual plasticity of form. The viticulturist can manipulate it in many ways and change the manner in which is trained, almost yearly if desired. Few perennial plants have this kind of flexibility, which forms part of its fascination as a food crop.

GEOGRAPHICAL ORIGINS OF GRAPEVINE SPECIES

Grapevines have evolved in several different areas of the world, leading to a great many different species developing. The origin of cultivation of the *V. vinifera* grape, now planted throughout the world, is probably in southern Caucasia, now occupied by north-west Turkey, northern Iraq, Azerbaijan and Georgia) (Mullins *et al.*, 1992). At first, grapes (*V. vinifera* spp. *sativa*) were probably gathered from the wild, with the vines growing up into the trees. The association of grapes with oak, now used in the winemaking process in the form of barrels in which wine is aged, may have begun with the vine using oak trees as support, since *Saccharomyces cerevisiae* (or the winemaking yeast) strains have been isolated from oak trees (Sniegowski *et al.*, 2002). The people living in these areas in ancient times discovered the utility of *V. vinifera* grapes and took the vine with them on their trading routes, to Palestine, Syria, Egypt,

Mesopotamia and then to the Mediterranean. The Greeks and Romans took to the vine readily and also spread it, and methods for its cultivation, throughout Europe and as far north as Britain.

From Europe, *V. vinifera* was taken to North America, Peru and Chile, with the Dutch ensuring it travelled with them to South Africa (1616). The English packed grapevines on the First Fleet to Australia (1788) and on travels to New Zealand.

Species of grapes native to North America are numerous, having originated in many different types of environments, from moist to dry (see Plate 1). The *V. labrusca* grape, native to north-eastern USA, is widely used for juice production in many US states and in South America. Other species are not so commonly used for grape production, including many native to Asia, but they do make an important contribution to the production of grapes, as will be discussed later.

FAMILY, GENUS, SPECIES AND RELATED PLANTS

Botanically, the grapevine is a liana, a climbing vine. As such, it does not invest heavily in something as solid as a tree trunk for support, rather having developed to take advantage of trunks already occurring in forests and bushlands. In the wild, the grapevine starts as an understorey plant, growing rapidly and upward, clinging to other plants to eventually reach the top of their canopies (see Fig. 1.1). Once there it fruits heavily, producing a dark-coloured berry that birds, in particular, will eat, thus disseminating the seed. These fruits are not very appetizing to humans, being strong in flavour, high in acidity and relatively low in sugars. In some cases wild grapes can be palatable, such as *V. amurensis*, the Amur grape, which has its origin in north-eastern Asia. This species has been useful in breeding programmes due to its cold hardiness and resistance to some diseases, but it has also been used to make a still table wine.

The seedling, when it sprouts, shows off two rudimentary leaves, thus indicating that it is a dicotyledon, like most broadleaved plants (and unlike monocots, like maize and other grasses). The fruit is a true berry (botanically speaking), containing the seed within, and thus the grapevine is classified as an angiosperm.

The grapevines belong to the family Vitaceae, which are mostly woody, tree-climbing vines, though a few have a shrubby growth habit. They are characterized by tendrils and inflorescences opposite the leaves. There are 12 genera within the family including *Vitis*, *Ampelocissus*, *Clematicissus*, *Parthenocissus* (Virginia creeper), *Ampelopsis* and *Cissus* (kangaroo vine).

The genus *Vitis* is the part of that family in which the grapevine industry is most interested and it consists of two subgenera, *Euvitis* and *Muscadinia*. One peculiarity of the genus is that the flower petals separate from the bottom of the

Fig. 1.1. *Vitis riparia* smothering a tree in an upstate New York winter.

flowers, not from the top as in most other plants (see Plate 2), hence it is referred to as a cap.

The subgenera are distinct because they have different chromosome numbers (38 for *Euvitis* and 40 for *Muscadinia*) and morphological features. *Muscadinia*, in comparison with *Euvitis*, have differences in seed shape, simple as opposed to branched tendrils, smooth bark, continuous pith inside canes (i.e. there is no interruption at the node position), fewer berries per cluster and berries that abscise from the rachis (Williams 1923; Bailey, 1933; Einset and Pratt, 1975). There are three named species in this group, the most important of which is *M. rotundifolia* – discussed further in Chapter 2, *M. munsoniana* and *M. popenoei*.

Because of their different chromosome number, plants in this subgenus will not naturally interbreed with Euvitis species. However, through the use of tissue culture techniques, crosses have been made (Alleweldt and Possingham, 1988). This may be important from the standpoint of producing grapevines with enhanced disease resistance or other desirable characteristics.

Euvitis includes many species, including *V. vinifera*, the most widely planted grape species in the world, which is used for wine, table consumption, juice and raisin production.

Another well-known species is *V. labrusca*, native to North America. Its advantages over *V. vinifera* are that it is much more resistant to pests and diseases but, for many, the flavour of the fruit, described as foxy, is an acquired taste. Hence, though they are made into wine, the grapes are mostly used for production of juice and preserves, though there is also a significant, but regional, industry producing them for fresh consumption.

Regardless of how the fruit tastes, many of the non-*V. vinifera* species have been vitally important to the commercial development of *V. vinifera* cultivars, in finding a solution to the problem of phylloxera and other soil-related pests and conditions (further discussed in Chapters 5 and 9).

NATURAL GROWTH CONDITIONS

The grape is a hardy perennial plant, meaning that it can grow and survive in areas where the temperature goes well below freezing in the winter season: under the right conditions, some grape species can survive temperatures as low as −40°C (Pierquet *et al.*, 1977) and there are active breeding programmes working to develop more cold-hardy cultivars for both table and wine purposes (such as the University of Minnesota's 'Frontenac'). However, *V. vinifera* is much more tender, and generally cannot withstand temperatures below −15°C without suffering damage (Clore *et al.*, 1974). The above-ground parts of the vine develop a hardy outer covering called a periderm as the growing season ends, which is an indicator that the vine tissues are developing resistance to environmental extremes.

The grapevine evolved in temperate climates, so it grows when conditions are warm enough, but stops when temperatures fall below about 10°C. Unlike treefruits, e.g. apple or pear, which set a terminal bud as winter approaches, grapevines will continue to grow as long as conditions permit.

The vine is highly adaptable to different environments which, in part, is why it is found growing in many and varied climates. Typically, at least in terms of commercial production, it is cultivated between the 10 and 20°C annual global isotherms, although grapes are now grown in many areas outside these boundaries due to identification of suitable mesoclimates and increased knowledge about their cultivation. There would be few areas in the world that would be too hot for grapevines to grow, assuming that other plants can also survive there. However, obtaining a commercial crop from those vines may be the most challenging aspect.

HISTORICAL CULTIVATION

It did not take long for people to recognize the advantages of cultivating crops rather than collecting from the wild. The Egyptians were using grapes from approximately 3000 BC, and pictures showing vines growing on structures date back to around 1500 BC (Singer *et al.*, 1954, cited in Janick, 2002). The Chinese had probably started cultivating *V. vinifera* vines by 2000 BC (Huang, 2000) and native cultivars (based on *V. amurensis*, for example) before that. The plastic nature of the vine lends itself well to manipulation, and doing so with some form of trellis exploits the vine's tendency to fruit heavily. Early trellising systems appeared to be forms of arbours, rather than the more common hedge or overhead trellises seen today. However, there has been little commercial adoption of trellis systems that have been introduced over the past 100 years. Be that as it may, structures for supporting vines can and do take many different forms and are discussed in more detail in Chapter 6.

USES

Just as the grapevine is highly adaptable to where and how it is grown, its fruit is also highly adaptable to different uses. Somewhat unique, the grape attains a high concentration of sugar when ripe, and also (depending on cultivar) pectin, as well as a wide range of aromatic compounds. These factors, in concentration with the presence of relatively high levels of acids (particularly tartaric acid), mean that the fruit is amenable to many different end uses. Grapes are mainly used for wine and related fermented products, with table grapes and raisins a distant second and third, respectively, in ranking (FAO, 2006).

Fermented grape products

There is an association between grapes and various types of yeasts (usually living on the surface of the berry) (Parish and Carroll, 1985; Martini *et al.*, 1996; Cavalieri *et al.*, 2003), so it is likely that fruit that had been picked and stored may have started fermenting naturally. This will have produced an alcoholic mixture that some found enjoyable, and thus people wanted to be able to repeat the process. This was possibly the beginning of winemaking.

Much of the world production of grapes ends up as wine, and it is made into a bewildering array of types and price points. Wine can be thought of as a naturally made storage form of the fruit as it retains characteristics of the grape and, protected from oxidation, can remain palatable for many years. The production of wine from grapes can be simple – with few additions postharvest – to complex, with addition of a variety of substances designed to modify its appearance, aroma and taste.

Wines can be further manipulated in the form of sweetening and the addition of additional alcohol, which produces fortified wine products such as port and sherry, which were even more stable and travelled better on long sea journeys. With distillation, wine can be transformed into products such as brandy, grappa and marc, which often carry the characteristics of the grape variety from which they were made. However, the skill of the grape grower, in producing the starting material for the beverage, is still paramount in the production of a quality wine.

Table grapes

The first use of grapes by man was probably as fresh fruit, and this continues to be an important industry. Table grapes have their own important attributes, such as looking attractive, having good eating properties (e.g. flavour and berry firmness), large and consistently sized berries, a sturdy rachis and strong attachment of the berry to the rachis. Ideally, grapes for fresh consumption should also be resistant to both injury caused by handling and postharvest diseases (discussed in more detail in Chapters 9 and 10).

Major table grape cultivars include 'Sultana' (also known as 'Sultanina', 'Thompson Seedless', 'Oval Kishmish' (Winkler *et al.*, 1974; Plate 3), 'Almeria' ('Ohanez'), 'Cardinal', 'Dattier' ('Waltham Cross'), 'Emperor', 'Malaga', 'Perlette', 'Ribier' ('Alphonse Lavallée'), 'Rish Baba', 'Tokay' and 'Ruby Seedless'. New cultivars are always under development. Seedless French hybrid cultivars developed in New York (see Plate 3), Arkansas and other places have more winter hardiness and therefore can be planted in colder climates than the major cultivars planted in California and most of Europe.

Curiously, 'Zinfandel' (also known as 'Primitivo'), a cultivar now well known as a wine grape in California, was first planted on the eastern coast of the USA as a table grape, before being taken west and used as a wine grape (Sullivan, 2003).

The table grape industry has changed several times in the last 100 years. In North America the development of the iced railroad car signalled the end of the eastern North American table grape industry, as table grapes from the western states were then available in eastern markets. Improved storage methods also greatly contributed to the change. Recently, better surface transportation and storage methods have resulted in a supply of southern hemisphere grapes to the northern hemisphere, primarily from Chile and other South American countries. In 2006/2007 Chile exported 815,000 t of table grapes, Italy 500,000 t, the USA 300,000 t and South Africa 230,000 t (from USDA FAS Production, Supply and Distribution Data for 2007). The popularity of seeded grapes, usually capable of being stored for longer periods of time, has declined in favour of fresh seedless grapes (Alston *et al.*, 1997). More information about the postharvest handling of table grapes is presented in Chapter 10.

Raisins

Raisins, or dried grapes, may have also occurred in nature as berries that had not been eaten by predators and left to dry on the vine to create a very sweet and decay-resistant product. They have been eaten probably since grapes were first found to be edible, but evidence exists to suggest that they have been created on purpose since Egyptian times (Janick, 2002). Today, they are an important facet of grape production but there are relatively few cultivars used for their production, with 'Sultana' being the most important. Grapes can be dried out in the open (see Plate 4) or in forced-air heaters, the latter usually reserved for the more premium market fruit due to the extra cost.

In terms of production, the USA and Turkey lead the way, contributing about 80% of world production, with significant amounts made in China, Chile, South Africa, Greece and Australia (FAS, 2007). The production of raisins is highly dependent on the weather, as the vast majority of grapes are dried in the sun. Because of this vulnerability, production of raisins is decreasing in areas where rain is more likely at the end of the season, and also because of fluctuations in the need for other products made from 'Sultana' grapes (e.g. for fresh market, juice or bulk wine). See Chapter 10 for further information about the production of raisins.

Grape juice

The juice of grapes is appealing to the human palate, through a combination of the sweetness, acidity and flavours (Morris, 1989). Juice can be consumed immediately or further processed (e.g. pasteurized) to create a longer-lasting product. Grapes are also used in the production of jellies and other preserves, and juice is widely used in the bottled drinks industry as a natural sweetener as well as on its own (Winkler *et al.*, 1974; Olien, 1990). However, the amount of world production used for juice and these other products is very small compared with that used for wine or table grapes.

Some cultivars, such as 'Sultana' and 'French Colombard' (another white grape), produce a neutral-flavoured juice well suited to concentration and then for use as a sweetener, whereas other juices, such as that from *V. labrusca* grapes, have a distinctive flavour all their own. Red juices can be made from heat-extracted grapes of 'Grenache'/'Garnacha' (a *V. vinifera* grape), 'Concord' (*V. labrusca*) or other cultivars. In general, however, juice from *V. vinifera* grapes does not process well and has less flavour and acidity than the *V. labrusca* grape juices, which is less palatable for consumers (Morris, 1989). Postharvest processing of juices is discussed in Chapter 10.

A relatively recent use of grapes is for the production of health-related products, as grapes are a natural source of readily extractable phenolics that have antioxidant properties (Teissedre *et al.*, 1996; Nuttall *et al.*, 1998; Yilmaz and Toledo, 2004). In many cases, the phenolic compounds are extracted out of the material left over from primary processing of the grapes, e.g. pomace. These are quite active and can be sold as nutritional supplements or used to bolster the antioxidant capacity of foodstuffs (Bonilla *et al.*, 1999).

GRAPE-PRODUCING COUNTRIES

Many of the world's countries produce at least some grapes. Table 1.1 shows production by country, averaged over 3 years to minimize the effect of seasonal variation in crops. The top ten producers are responsible for 70% of world production, and it is notable that China is now one of the top five countries.

The amount of land given over to grape production in 2005 and the level of change in the 10 years before that are shown in Table 1.2. The European producers have, in general, been pulling out vineyards, whereas there has been huge growth in newer areas such as Namibia, New Zealand, China and Australia. Although most of these countries have been starting from a very small base, it is clear that there is a shift in vineyard development on a global

Table 1.1. Estimated average annual grape production (2001–2003, in 100,000 kg) by country (from OIV, 2003).

Rank	Country	Production	Rank	Country	Production
1	Italy	78,436	22	Yugoslavia[a]	4,437
2	France	67,921	23	Bulgaria	4,253
3	USA	61,933	24	Mexico	4,184
4	Spain	59,896	25	Korea	4,173
5	China	44,450	26	Ukraine	4,000
6	Turkey	34,500	27	Afghanistan	3,650
7	Iran	26,736	28	Croatia	3,544
8	Argentina	23,353	29	Syria	3,461
9	Chile	19,459	30	Iraq	2,950
10	Australia	15,989	31	Morocco	2,739
11	South Africa	14,884	32	Russia	2,669
12	Germany	12,075	33	Austria	2,553
13	Greece	11,793	34	Algeria	2,362
14	India	11,400	35	Japan	2,260
15	Egypt	10,956	36	Macedonia	1,954
16	Romania	10,820	37	Turkmenistan	1,667
17	Brazil	10,794	38	Yemen	1,654
18	Portugal	9,988	39	Georgia	1,467
19	Hungary	6,789	40	Peru	1,364
20	Moldavia	6,036		All others	17,790
21	Usbekistan	4,970		Total	616,309

[a]Yugoslavia was not officially abolished as a political entity until 2003.

Table 1.2. Estimated area of land cultivated to grapes in 2005, by country. The percentage change in area since 1995 is also indicated (from FAO, 2006).

Country	Production area (ha)	Change from 1995 (%)
Spain	949,100	–18
France	851,615	–5
Italy	837,845	–7
Turkey	530,000	–6
China	453,200	+187
USA	380,000	+20
Iran	275,000	+18
Romania	217,006	–13
Portugal	210,000	–18
Argentina	208,000	+1
Chile	178,000	+57
Australia	153,204	+145
Moldova	145,000	–18
Greece	127,000	0
South Africa	123,190	+19
Bulgaria	113,334	+1
Uzbekistan	110,000	+16
Germany	98,000	–5
All others	1,387,118	+5
Of note:		
New Zealand	19,960	+227
Switzerland	15,000	+1
Canada	9,259	+29
Namibia	2000	+233

scale, with most movement concerning grapes grown for wine production. There has been an emphasis in more wine production outside Europe, leading to more coming from places such as the USA, Australia, South Africa, China and Chile. Grape growers in the European Union (EU) have long been in a state of overproduction, with surplus wines being distilled into ethanol (Unwin, 1994). This situation appears to be changing, as there is now rationalization of vineyard area, as well as new marketing initiatives to export grape products outside the EU (Pike and Melewar, 2006).

Some of the most significant vineyard developments have come from countries with centuries of growing tradition, such as Spain and Kazakhstan, for instance, where the high-labour input and low-yielding vineyards are being replaced with modern cultivars and management practices (Smart, 1996). China is well placed to make a large impact on world wine (and table grape) production, exploiting their wide range of climates and soils with the idea of earning export dollars as well as supply growing internal demand.

In North America, California, Washington, Oregon and other states continue to refine where and how they grow grapes, lowering costs of production and increasing quality to remain competitive.

In South America, Chile is the country that seems to have the greatest international presence, with established international markets for wine and table grapes. The national wine industry is becoming more organized, with the formation of Viños de Chile A.G., which represents 90% of the volume of exported wine (http://www.vinasdechile.cl). Argentina is also a significant exporter of grapes, with 60% of their table grape production being shipped to the EU (Ernst, 2007).

South Africa is another country that has a strong focus on the national and export grape product market, with wine making up a large part of it. One sign of this is the developing tourism market, which is catering to a domestic and international clientele (Bruwer, 2003).

Overall, wine producers realize that, in order to be competitive in the world marketplace, they must produce better-quality products more cheaply. As the highest profit margins lie in the more expensive bottles of wine, for example, there will be thinner margins on the basic table wine and more competition to sell fewer bottles of premium product. Grapes make up a significant portion of the cost of a bottle of wine, so the pressure filters down to the growers as well.

For growers of grapes for the fresh market the trend is for increased consumption, but also greater international competition. China is by far the largest producer of grapes for fresh consumption, ahead of Turkey, Italy, Chile and the USA (FAS, 2006), but the vast majority of the fruit is consumed domestically and forecasts are for increasing demand for more imported grape products as well. China will be a country to watch, as it has already had a huge impact on the world apple juice market (Huang and Gale, 2006), and predictions are that it could have a similar impact on the grape market.

Cultivars, Anatomy and Improvement

Grapes are one of the world's most planted horticultural crops, and there is a wide range of cultivars in use, many of them well suited to a particular area or end use. In addition to listing the most used cultivars, this chapter will discuss clones, or selections within a cultivar, which are another important variable in the growing of quality grapes. After a review of the parts of the vine there will be a brief discussion about vine breeding and improvement.

MAIN CULTIVARS FOR VARIOUS USES

Though there are many different kinds and names of grapes, there are relatively few that make up the bulk of those produced. The terms cultivar and variety are used interchangeably in the industry to refer to particular types of grapes. There are something on the order of 24,000 named cultivars (Viala and Vermorel, 1909), but because there is often more than one name for the same grape (e.g. 'Syrah'/'Shiraz' or 'Sultana'/'Sultanina'/'Thompson Seedless'/'Oval Kishmish'/'Kouforrogo'/'Tchekirdeksiz'. For the latter, it was said that the grapevine was originally brought to California named as 'Lady De Coverly' by a William Thompson, but growers referred to the cultivar as 'Thompson Seedless' and the name stuck (Christensen and Peacock, 2000a)). The number of different and distinguishable cultivars is in the order of 5000. As an example, of these the Organisation Internationale de la Vigne et du Vin (OIV) list approximately 250 as being significant to the wine industry (OIV, 2000), so there are many cultivars that are planted in only very small numbers.

Due to this complexity, there are several internet databases available that attempt to maintain clarity in the relationships between species, cultivars and their common names (see the European Network for Grapevine Genetic Resources Conservation and Characterization, http:/www.genres.de/vitis/vitis.htm; Greek Vitis Database, http://gvd.biology.uoc.gr/gvd/; European Cooperative Programme for Plant Genetic Resources, http://www.ecpgr.cgiar.

org/workgroups/vitis/vitis.htm). The International Grape Genome Program (http://www.vitaceae.org) was established to act as a clearing house for genetic information relating to grape, with the aim of aiding research in improving the understanding of grape physiological processes and the efficacy of breeding programmes. The entire genome of *V. vinifera* has now been completely sequenced (Velasco *et al.*, 2007), the second food plant to have reached this milestone, after rice. This achievement will be of assistance in developing new cultivars, introducing resistance to diseases or pests into existing cultivars and understanding the physiology of the grapevine: how it works.

The addition of such tools as DNA fingerprinting is helping to elucidate the genetic relationship between cultivars. For example, Goto-Yamamoto *et al.* (2006) have demonstrated the genetic similarity of a number of grape cultivars through the use of microsatellite DNA markers. Figure 2.1 is a diagrammatic representation of these relationships, where the length of the branches represents the genetic similarity (e.g. 'Chardonnay' and 'Pinot noir' are quite closely related, but *V. rotundifolia* and *V. riparia* are not).

The main cultivars used for grape production on a worldwide basis are changing rapidly. Wine grapes dominate this scene, and traditional cultivars such as 'Airén', grown in large areas of Spain, partly due to the use of low vine density in the very dry areas, are now being pulled out and replaced by more

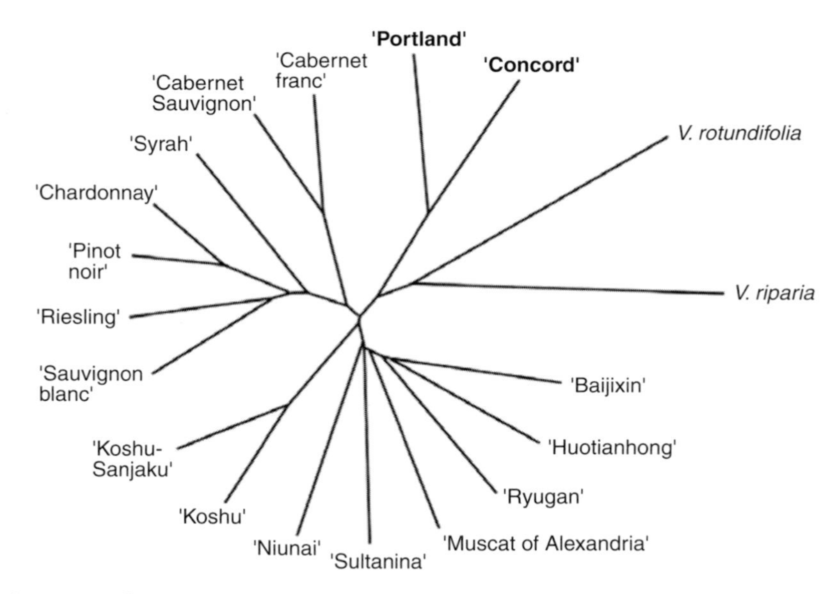

Fig. 2.1. A diagram representing the genetic similarity of grape species and cultivars through the length of the lines between them (reproduced with permission from Goto-Yamamoto *et al.*, 2006).

profitable cultivars such as 'Cabernet Sauvignon', 'Syrah'/'Shiraz' and 'Chardonnay' (Smart, 1996). Similar trends are being seen in many other traditional grape-growing areas – a result of the internationalization of the wine market.

Within individual grape-growing regions, particularly in places that have been producing grapes for many years, there are local cultivars that are not much planted outside those areas. Particularly with the preference for labelling wines by cultivar instead of by region of origin (e.g. 'Pinot noir' instead of Burgundy), the average consumer has come to recognize a small number of cultivar names. This puts pressure on grape growers to plant those cultivars, resulting in traditional cultivars being pulled out: a potential loss in terms of diversity of plant material and different wines from which to choose. The same is true for table grapes as well.

Even though there are selections of cultivars in table, raisin and juice production, they appear not to be as numerous as in the wine industry. This could be due to the added layer of complexity in postharvest processing, as well as to the product being more carefully analysed by the consumer.

Still further cultivars are not used for their fruit but in breeding programmes or as rootstocks, e.g. *V. champini, V. berlandieri, V. rupestris, V. riparia*, etc. Though these are not as well known, they are vitally important to the long-term health of the grape industry, as they represent a reservoir of diversity that can be used to develop disease resistance or cultivars better suited to a particular region or use.

The crossing of *V. vinifera* and other species of grapevine was not widespread until the 19th century, when North American vine diseases and phylloxera (a plant louse) spread throughout Europe. Vine breeders recognized that the best way to combat these problems was to breed new cultivars and rootstocks that were resistant. To do this, they used species of North American vines that had co-evolved with the pests and had resistance to them. And so were developed the French hybrids, which could produce grapes without succumbing completely to the disease and root pests. Specialized rootstocks were also developed by crossing a wide range of *Vitis* species, for example, to develop vines with tolerance to high-pH soils (rootstocks are discussed further in Chapter 5).

Muscadines, the close cousin to the *Euvitis* grapes, are grown primarily in the south-eastern USA, where they support a small but persistent wine, table grape, juice and preserves market. Well-known cultivars include 'Scuppernong', 'Carlos', 'Magnolia' and 'Fry', the latter being the most widely planted (Olien, 1990). Famously, 'Fry' was the cultivar that was most used in Virginia Dare, the best selling wine in the USA before Prohibition set in (Adams, 1985; Olien, 1990).

With each category of use, there are cultivars of grapes that have been specially selected as they perform best for those uses. In table grape and raisin production the market consists of more and more grapes without seeds (Alston *et al.*, 1997), whereas in wine grapes it is desirable to have seeds, which contribute flavour and mouthfeel characteristics to wine, as well as aiding in juice extraction during processing. Table 2.1 shows some of the main cultivars used for different end products. 'Sultana' and other neutral-flavoured grapes do end up being used for multiple purposes, which brings some advantages in terms of finding a market for them.

CLONES

The *Oxford Companion to Wine* defines a clone as a 'population of vines all derived by vegetative propagation from cuttings or buds from a single 'mother vine' by deliberate clonal selection' (Smart and Coombe, 1999). In practical terms, a clone is a selection of a cultivar that has some distinguishing characteristic that someone noticed and thought was significant enough to warrant separate propagation of the vine.

Examples of characteristics that may be used to select a clone are vine growth (vigour), disease resistance, leaf or shoot appearance, cluster shape, fruitfulness, fruit composition, etc. So far, clones have been selected on the basis of morphological differences. It is already possible to distinguish clones using DNA screening (Cervera *et al.*, 1998; Scott *et al.*, 2000), which could eventually be used to help select desirable new clones as well as identify unknown clones already planted.

Since the historical appearance of clones is due to natural variation in the population of vines, they are often the result of a single gene mutation – some part of the DNA not being expressed, being expressed differently or perhaps being over-expressed. These point mutations are relatively common, and many desirable clones (and even cultivars) have arisen by chance through this

Table 2.1. Examples of grape cultivars used for various end products.

End use	Cultivars
Fresh consumption: seeded	'Muscat of Alexandria', 'Red Globe', 'Isabella', 'Emperor', 'Concord', 'Tokay'
Fresh consumption: seedless	'Sultana', 'Flame Seedless', 'Perlette'
Raisin: seeded	'Muscat of Alexandria'
Raisin: seedless	'Black Corinth', 'Sultana', 'Monukka', 'Fiesta'
Juice/preserves	'Sultana', 'Grenache', 'Concord', 'Niagara'
Wine	'Airén', 'Grenache', 'Cabernet Sauvignon', 'Marechal Foch'
Fortified wines	'Palomino', 'Pedro Ximénez', 'Touriga National'
Distillation	'Muscat of Alexandria', 'Gouais blanc', 'Ugni blanc'

pathway. Recently, it was demonstrated that white grapes could have arisen as a mutation of red grapes – there having been no white grapes before this. Two chance mutations that affected the ability of the plant to turn on anthocyanin synthesis was all it took (however unlikely the occurrences were) to create white-fruited grapevines (Walker *et al.*, 2007).

Most cultivars of grape have different clones available for planting, the number of which depends on the natural genetic stability of the cultivar and how carefully someone is looking for new sports of it. Genetic mutants of a vine appear spontaneously in the vineyard, and in some cases these display improved growth or fruiting properties and can be propagated asexually to plant new areas with the new clone (or, in some cases, cultivar). 'Pinot noir', for example, expresses a considerable amount of variability in comparison with other cultivars (Konrad *et al.*, 2003). Over the centuries, it is thought that 'Pinot noir', 'Pinot Meunier', 'Pinot gris' and 'Pinot blanc' all have a common ancestry, and it is still relatively easy to spot variants in the vineyard (Plate 5).

Note that, in some countries, clonal names may differ, i.e. the clone may be the same but the name is different, as happens with cultivars. As well, in some cases, due to natural variation or mix-ups in labelling, the name given to clones in some areas may not be correct. This highlights the importance of being able to reliably distinguish clones, a task that is not yet fully completed through the use of ampelography (discussed later in this chapter), as clones growing in different areas may exhibit slightly different growth habits, which can make consistent typing of a vine difficult. Thus, until genetic testing is available for routine determination of clones, a paper trail is the best means of identifying clones.

ANATOMY AND PHYSIOLOGY

The grapevine structure is very similar to that of many other woody perennials. It has a root system that serves to anchor the plant in the ground, but also gathers and sends water and nutrients to support plant growth and acts as a carbohydrate reservoir for carrying over energy from season to season. There is also a trunk, which serves structural, carbohydrate storage and conduit roles, and the branches (called canes or cordons) that support the shoots, fruit and leaves.

Roots

The root system is made up of the larger arms and branches, down to root tips and root hairs. The latter organ is the workhorse of the root system, where the vast majority of nutrient and water uptake occurs (Pratt, 1974). Basic root structure is shown in Fig. 2.2.

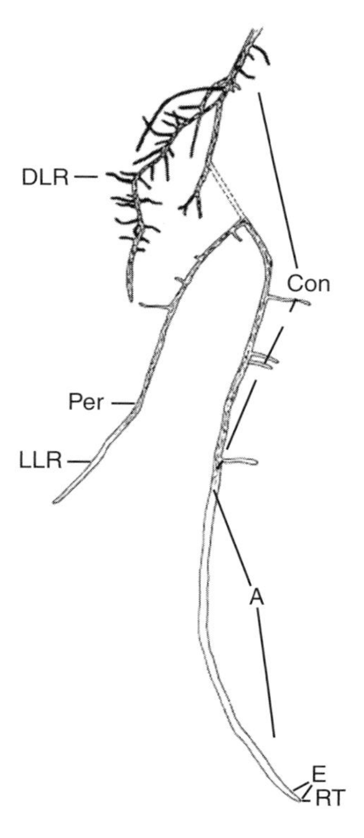

Fig. 2.2. Root of *Vitis vinifera* ('White Riesling') showing actively growing and inactive or dead portions. A, zone of absorption; Con, zone of conduction; DLR, dead lateral root; E, zone of cell elongation; LLR, living lateral root; Per, periderm; RT, root tip (reproduced with permission from Pratt, 1974).

The root system, as does the shoot system, produces plant growth regulators that can modify the growth of other parts of the vine (further discussed in Chapter 3).

The vascular tissues, mainly the xylem and phloem, are in the centre of the root and surrounded by the periderm, as found in many woody perennial roots. It is important for the xylem and phloem to remain protected from the rest of the root and the soil beyond, as this is a key control point for the entry of water and nutrients (Tanton and Crowdy, 1972).

The root tip is the most important part of the root system, in that it is the part that will allow the plant to explore new areas of soil that have the necessary nutrients needed for the vine to survive. The root cap, covering the growing point (root meristem), protects the actively dividing cells in the meristem and also contains cells with starch granules – important for indicating to the root which way is up (Wilkins, 1966).

Root hairs are the specialized organs that grow from the epidermal cells behind the tip and along for about 5 cm (zone of absorption, Fig. 2.2). As these, collectively, have a very large surface area, they are the part of the root that perform most of the water and nutrient gathering, and also interact with the mycorrhizae (see below). As a section of root continues to grow and thicken, the root hairs die back, but are continually replaced by new ones emerging near the tip. The outer layer of cells on the older roots do serve to absorb water and nutrients but, as their relative surface area is very small in comparison with the root hairs, the amount of absorption is low (Steudle and Meshcheryakov, 1996).

Mycorrhizae

These are a group of fungi that form beneficial relationships with most species of plants, including grapevines. In trees, shrubs and woody perennial vines, vesicular–arbuscular mycorrhizae (VAM) are the most common. These are symbiotic relationships, as both the plant and the fungi benefit. VAM are usually most associated with plants in low-phosphorus soils, as the hyphal network is more efficient at gathering this immobile nutrient than native plant roots, however, VAM associations are also known to occur in higher-phosphorus soils (Morin *et al.*, 1994).

Grapevines normally have VAM infections (Possingham and Groot Obbink, 1971; Deal *et al.*, 1972), and this has been shown to have a beneficial effect on the growth of vines (Biricolti *et al.*, 1997; Aguin *et al.*, 2004), with improved root and shoot development (see Fig. 2.3).

Fig. 2.3. The effect of inoculating two different rootstocks in grafted nursery vines with mycorrhizae. Both shoot dry weight and root dry weight were increased through the use of the mycorrhizal fungus *Glomus aggregatum* (reproduced with permission; redrawn from Aguín *et al.*, 2004).

The exploration of soil is very important to the survival of the vine. Some nutrients, such as phosphorus, are immobile in the soil and thus, for the plant to obtain a supply of it, the roots must grow into areas of soil that have those nutrients available (Mullins *et al.*, 1992). This is unlike a nutrient like nitrogen, which is mobile in the soil solution and is brought to the roots by water percolating through the soil profile (see Chapter 7). It is worth noting that roots do not grow toward areas of higher water or nutrient availability, i.e. they cannot pinpoint rich areas and target them. Roots grow in all directions and, if an area of higher nutrient or water availability is grown into, there is a more rapid proliferation of roots there.

Roots exhibit significant turnover, with parts of the root system dying off as other parts grow (Comas *et al.*, 2000; Anderson *et al.*, 2003). Management decisions, such as severity of pruning, also affect root life, with more severe pruning reducing the average lifespan of a root (Comas *et al.*, 2000). Those parts of the root that continue to develop become more permanent branches and are a site for carbohydrate storage.

Above the soil

If we look at a mature vine in the dormant season, the above-ground parts consist of the trunk, which is the portion of the vine from the ground to about the fruiting wire and provides support for canopy growth as well as being a carbohydrate storage site (see Fig. 2.4). Also shown are new canes, which were the previous year's shoots and a non-count cane, which are shoots arising from latent buds.

Along the cane are nodes, separated by internodes. At this point in the season, the nodes are where the following season's shoots will arise. Positioned at alternate sides of the cane are compound buds, so called because they contain three (the primary, secondary and tertiary) pre-formed shoots (see Fig. 2.5). Each of these will have six to nine leaf primordia and, in some cases, flower cluster primordia already formed and indeed, if these dormant buds are dissected they can be seen with a reasonably powerful stereo microscope (Antcliff and Webster, 1955).

The compound bud is designed for overwintering, and is protected by tough bud scales and woolly fibres. Primary buds are less winter hardy than secondary or tertiary buds. If the primary bud is killed by cold temperatures in winter, the secondary bud will grow; if the secondary bud is killed the tertiary bud can grow. The largest clusters are found on the primary buds; in comparison, the secondary bud has inferior clusters and results in a lower yield. Tertiary buds generally do not have clusters, but following a severe winter or early-season frost damage, at least the vine is able to grow to normal size from those.

When environmental conditions become favourable, one of the shoot primordia – usually the primary one – will begin to grow and push out of the

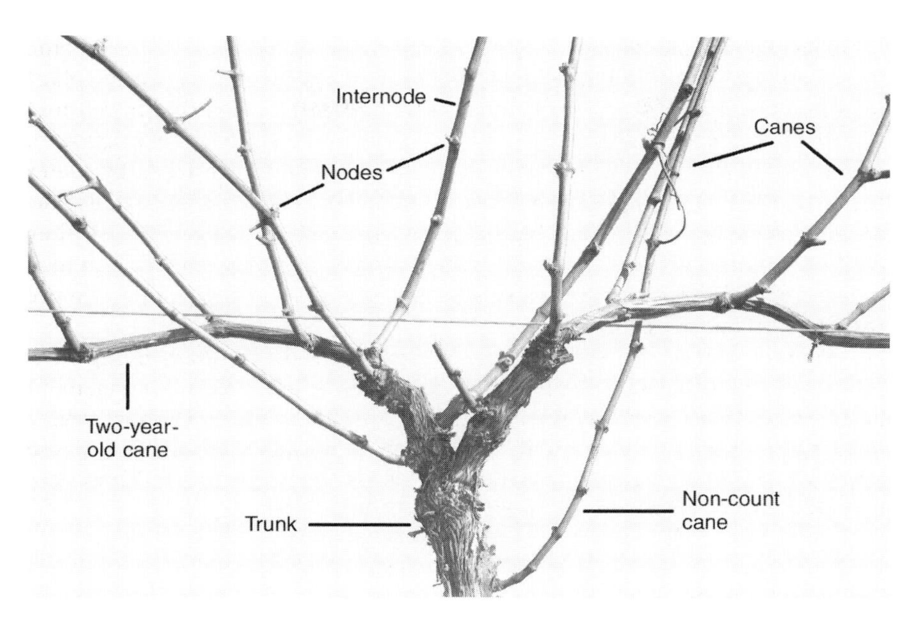

Fig. 2.4. Photograph of a dormant, cane-pruned vine showing the trunk, 2-year old canes, new canes, a non-count cane, nodes and internodes.

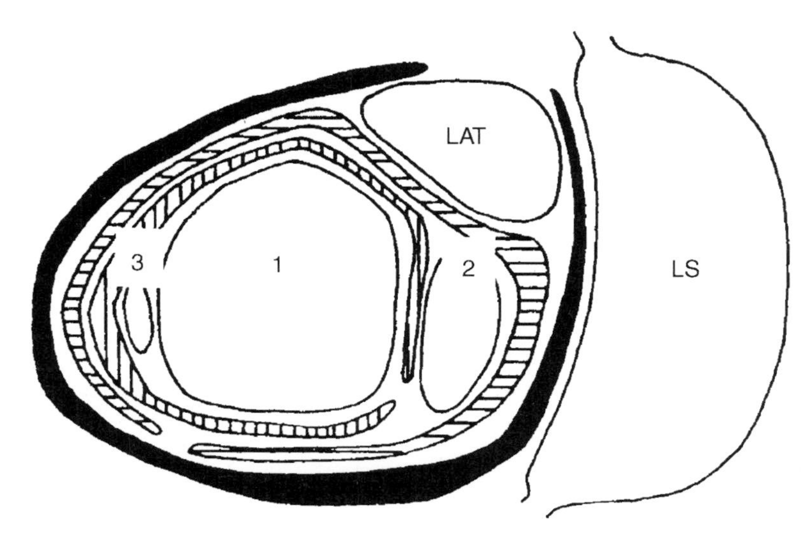

Fig. 2.5. Diagram of a transverse section through a compound bud (eye) of *Vitis labrusca* ('Concord') showing relative positions of leaf scar (LS), lateral shoot (LAT, also known as the axillary bud) and three dormant buds (1, 2 and 3) (reproduced with permission from Pratt, 1974).

bud. Temperature is the main driver of plant growth, and shoots (and any flower clusters that may be on it) develop more rapidly as temperatures rise during the season.

The early-season shoot development is called fixed growth because it is the growth of the pre-formed nodes and leaves and results in the early appearance of leaves and flower clusters. If a shoot is examined closely, a pattern to tendril and leaf appearance emerges: every third node lacks a tendril (Mullins *et al.*, 1992), but this pattern starts only once the first flower cluster or tendril appears at the base of the shoot, which may have anything up to five or more nodes without any tendril or flower cluster appearing.

At each internode and at the base of the leaf petiole (leaf stem) there forms another compound bud and an axillary (or lateral) bud (see Fig. 2.5) The axillary bud will develop into a noticeable lateral shoot if conditions are favourable during the season, and the compound bud is the one that will overwinter and be the source of shoots for the following season. It is worth noting that grapevines do not form adventitious, or spontaneously formed, buds – all shoot growth originates from a previous node position. Since there can be several compressed nodes at the base of every shoot, which are not removed through normal practice, this results in many latent buds that can grow later in the vine's life (see Fig. 2.4); however, shoots will never arise from an internode area.

Further elongation of the shoot beyond the fixed growth is called the free growth, and is the result solely of the environmental conditions of the current season and the shoot tip meristem. Grapevines will grow as long as conditions (primarily temperature) permit, so their growth is called indeterminate.

The previously mentioned nodes are distinguished by being greater in diameter than the internode, much like a finger's knuckle. The internode keeps the leaves spaced apart and acts as a conduit, containing the xylem and phloem as well as having its own meristem, which causes growth in cane diameter through the season. The outer layer of meristem growth is old phloem, but as the cane grows it becomes what we see as the bark of the vine, which covers the older canes and vine trunks. A cross-section of a cane shows the relative positions of the phloem and xylem on either side of the cambial layer, as well as the way in which the inner cortical cells (pith) break down to leave the centre open (see Fig. 2.6).

Internode length influences shoot length as it is variable, probably depending on growing conditions when it is formed and shortly thereafter. Quinlan and Weaver (1970) reported that internodes were most lengthened when gibberellic acid was applied to them when they were less than 1 cm in length rather than later, demonstrating that this is a key time for determination of this parameter.

Arising from the node positions are leaf petioles, tendrils, flower clusters and, as already mentioned, they house the axillary and compound buds. At the distal (furthest from origin, as opposed to basal, closest to the origin) end of the petiole is the leaf blade, which is the main photosynthetic organ of the vine.

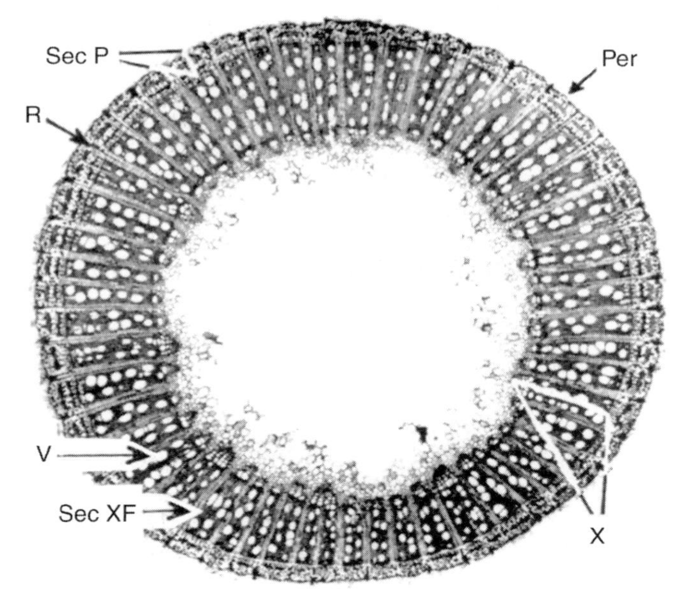

Fig. 2.6. Cross-section of a 1-year-old grapevine shoot of *V. vinifera*. Per, periderm; Sec P, secondary phloem; X, primary xylem; Sec XF, secondary xylem fusiform cell; R, ray; V, xylem vessel. Note that the centre of the shoot is open – the cortical cells have collapsed (reproduced with permission from Pratt, 1974).

Photosynthesis

Photosynthesis is perhaps the most important process on Earth. In basic terms it is the harnessing of light energy to convert carbon dioxide to chemical energy, while releasing oxygen. It is the opposite of respiration, which is the process of using chemical energy by combining it with oxygen, which releases carbon dioxide. Animals survive through the process of respiration, while plants, although they respire as well, produce the chemical compounds that can be respired.

The workhorse of photosynthesis is the chloroplast, which collects light energy using chlorophyll. Chlorophyll appears green because it absorbs most red and blue wavelengths of light. The light energy collected is used to split water molecules (H_2O), adding the hydrogen (H) to carbon dioxide (CO_2) to form carbohydrates (CHO) in the form of sugar (glucose). As a result of this process, oxygen (O_2) is released. Ribulose bis-phosphate carboxylase/oxygenase (Rubisco) is the key enzyme in photosynthesis, with up to half of the protein in the chloroplast being just Rubisco. Demonstrating the importance of plants in the ecosystem, Rubisco is probably the most abundant protein on Earth (Ellis, 1979).

The ingredients that the photosynthetic process needs are light, moderate temperature, CO_2 and water. Temperature also influences respiration, where its rate increases the warmer it gets. Because the process of respiration is happening along with photosynthesis, under certain conditions there will exist a compensation point, where the amount of chemical energy photosynthesis is creating equals the amount being used by respiration. In this case, no net photosynthesis is taking place – the tissue is making energy as quickly as it is using it.

In order for the chloroplast to have access to CO_2 and release the O_2 that it makes, there needs to be a transfer of gases between the inside of the leaf and the outside environment. Because the outside environment is quite hostile (in terms of relative moisture content) to the easily desiccated interior cells of the leaf, there needs to be tight regulation of the movement of gases in and out of the leaf. This is accomplished with a passage through the leaf 's surface, called a stomate (pl. stomata) (see Fig. 2.7). Specialized guard cells form the pore when the cells become turgid, and the pore closes when the cells lose their turgidity. In this way, there is an efficient method for the vine leaf to regulate the passage of gases, including water vapour, in and out of the leaf. When the vine has a good water supply, the guard cells are turgid and the stomatal pores are open; when the vine suffers water stress, the guard cells become flaccid and the pores close. A typical 'Cabernet Sauvignon' leaf may have approximately 180 pores/mm^2, but there will be more in sun-exposed than shaded leaves (Palliotti *et al.*, 2000). There are far more stomata on the abaxial (lower) side of the leaf compared with the adaxial (upper) side. Berries also have stomatal pores, though their frequency is very low (Blanke and Leyhe, 1987).

Wind can significantly affect the rate of water loss through the stomata by reducing the thickness of the boundary layer (area of still air) over the leaf. This increases the strength of the humidity gradient and thus the rate at which water exits the leaf. If the vine cannot supply enough water from the roots to match that lost through the leaves, the leaves and shoots would wilt. To prevent this, the vine shuts the stomata, preventing possibly lethal water loss, but also preventing photosynthesis and the production of carbohydrates. So as wind speed increases, more and more stomata close (Freeman *et al.*, 1982; Campbellclause, 1998). This simple response to wind is the cause of many changes to vine growth and fruit development. Making sure the vine is well supplied with water under hot and windy conditions can minimize stomatal closure, but there can be many instances where the vine roots simply cannot keep up with loss through the leaves, and the vine shuts down.

The critical wind speed where photosynthetic efficiency starts to drop is altered according to numerous variables, but one study suggests that values above 4 m/s (14 km/h) tend to decrease stomatal openings (Dry *et al.*, 1989). In areas with considerable wind, windbreaks will increase the number of hours during the day that the vines can retain open stomata, which results in

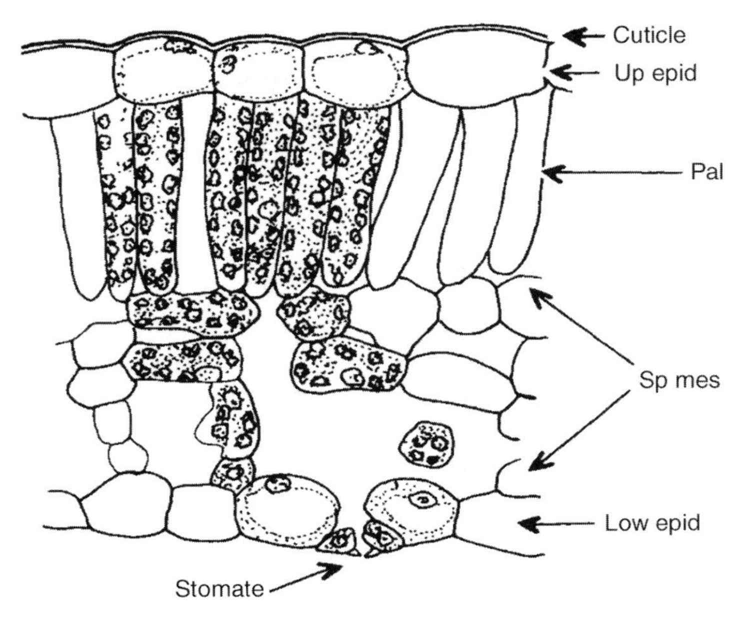

Fig. 2.7. Transverse section of a grapevine leaf (*Vitis vulpina* L.). The bulk of photosynthesis occurs in the palisade and spongy mesophylls. Up epid, upper epidermis; Pal, palisade layer; Sp mes, spongy mesophyll layer; Low epid, lower epidermis (reproduced with permission from Pratt, 1974).

increased photosynthesis (Freeman *et al.*, 1982) and thus more energy available for growth or fruit development.

Wind also moves away heat in the vineyard (Smart and Sinclair, 1976). Since photosynthesis is optimal at temperatures just below 30°C, if wind cools the vine canopy from 28° to 25°C, photosynthesis won't be operating as efficiently as it could be. Conversely, if wind is cooling a hot vineyard, there may be a net benefit. Light, water and temperature thus have a strong influence on the photosynthetic process, and therefore grapevine productivity.

The vine can alter itself to help maximize photosynthesis or minimize the effect of too much sunlight or heat. Leaf blades are able to change their angle relative to the petiole and follow the position of the sun through the day, to obtain the best light capture. However, this ability can also be used to tilt the blade away from the sun to reduce leaf temperature when light intensity is excessively high (Gamon and Pearcy, 1989). In this way, the vine is able to take best advantage under a range of conditions.

Continuing with the characteristics of the leaf, the blade may evidence noticeable hairs, usually on the abaxial side, which can influence insect and disease susceptibility as well as water loss from the vine. The hairs may physically hinder insect access to the epidermis, and they can increase the depth of the boundary layer above the stomata. On young leaves, flower clusters and shoots, there may be clear, round objects, which may look like insect eggs (see Fig. 2.8). These are a normal occurrence and have been called sap balls; their origin and purpose is unknown (Mullins *et al.*, 1992), but they tend to go more opaque as time goes on.

Flower and berries

The grapevine flower is one of a number arranged on a panicle inflorescence (Pratt, 1971). The individual flower, or floret, can be perfect (having both male and female parts), female or male, though on cultivated vines it is always perfect (which allows for self-pollination). Similarly, vines can be dioecious (having separate male and female plants) or monoecious. Many rootstock cultivars are dioecious, unlike their cultivated-for-fruit counterparts.

The fruit of grapevines is a true berry, like blueberry or the cranberry. It develops from the fertilized flower, with the bulk of the flesh being made up of

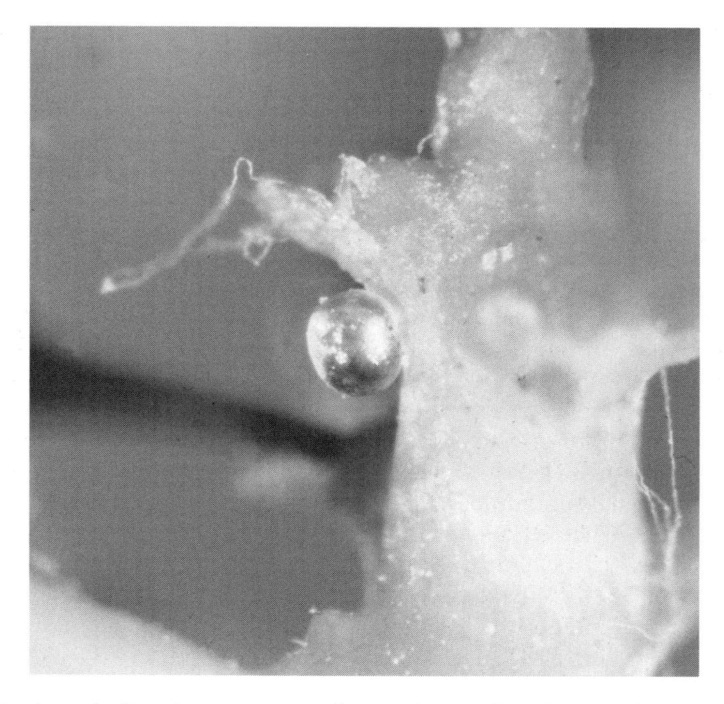

Fig. 2.8. A sap ball, as it appears on a flower cluster. Though it may look like an egg of some kind, it is normal to see these on leaves, shoots and flower clusters.

ovarian tissue. *Euvitis* can have up to four seeds (Winkler and Williams, 1935), while the muscadine grapes can have up to six (Olien, 1990).

The flower clusters are closely related to the tendril (see discussion in Chapter 3) and usually arise from the 3rd or higher node on a shoot. The number of flower clusters per shoot varies with cultivar, management and environmental conditions, but can range from none to five or even more.

The floret is attached to the rachis by the pedicel, and the rachis is attached to the peduncle, or stem, of the cluster. Flowers, because they open from the base rather than the tip, form a cap (called the calyptra), which pops off at flowering (anthesis) (see Plate 2). As with other flowers, there are anthers, which produce the pollen, and a stigma, which receives the pollen.

The berry that develops from the fertilized flower has a waxy outer covering called the cuticle. Later in berry development this will appear as the white or greyish bloom on the skin. The cuticle helps protect the berry against water loss as well as disease invasion (Possingham *et al.*, 1967; Blanke and Leyhe, 1988). The epidermal layer of cells, or skin, of the berry also serves this function, as well as being the source of colour and some flavour compounds. In some cultivars lenticels are visible on the berry surface as well, which help exchange gas through the berry skin. Lenticels may become more visible later in the season as the berries start to senesce (see Plate 6).

Within the berry are seeds, which are high in phenolic content as well as containing the embryo that could develop into another grape plant. If the berry is pulled off the pedicel, the latter often has a portion of the berry left on it, called the brush, which is pulp and vascular tissue that has been pulled from inside the berry.

AMPELOGRAPHY

Ampelography is the science or art of distinguishing grapevine cultivars by examining their appearance. Through the observation of leaves, shoots, shoot tips, clusters and fruit, it is possible to determine the cultivar and, in most cases, the clone of a cultivar.

Some examples of parameters that are examined in ampelography include (i) the number of teeth between major leaf veins; (ii) the form and colour of the shoot tip; (iii) shoot growth habit; (iv) presence and position of hairs on plant parts; (v) cluster openness or shape; (vi) berry shape and skin thickness; (vii) leaf shape and angles between veins (see Fig. 2.9); and (viii) tolerance to phylloxera or botrytis, etc. (OIV, 1983). Clearly, as there is still a measure of subjectivity to identification in this manner and the potential for changes to leaf morphology in different environments, its usefulness is limited, though it is still an important method of cultivar identification. It is expected that, at some point, analysis of plant genetics will be powerful enough readily to distinguish between clones and take some of the art out of the equation.

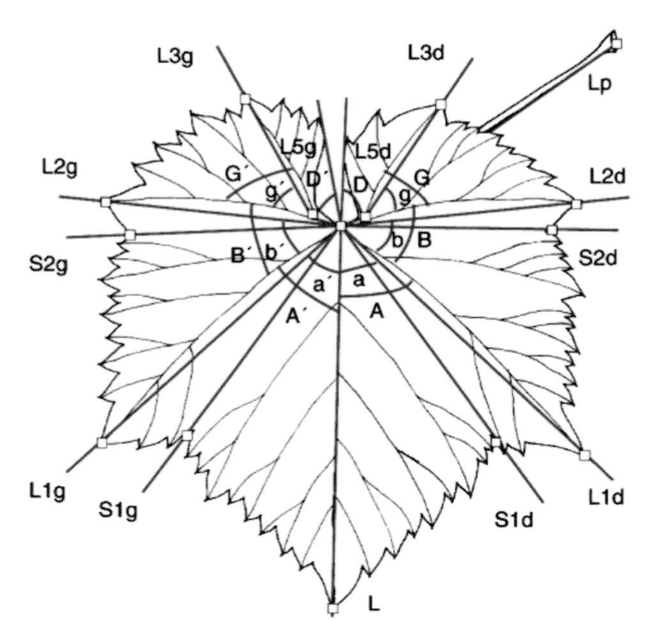

Fig. 2.9. An example of ampelographic measurements that can be made on mature grape leaves. Abbreviations correspond to parameters as set out by the OIV (Office International de la Vigne et du Vin (OIV, 1983)) (reproduced with permission from Santiago *et al.*, 2005).

BREEDING AND GENETICS

In looking at the most-planted grape cultivars (Robinson and Harding, 2006), many, if not all, have been in cultivation for centuries. This is not to say that new cultivars are not being produced. There are several active grape-breeding programmes around the world, usually with the goal of producing cultivars that will produce better end products in a certain region, allowing reduced production costs, greater yields, better quality, etc.

It is unlikely that a grower would commit the vineyard space and time necessary to grow seedlings in search of superior cultivars. The genetics of the genus are complex enough that many thousands of seedlings must be grown for several years to reach fruiting age for only a small chance of finding a superior vine that could be launched as a new cultivar. Luckily, there are many existing cultivars available to accommodate most needs.

Small changes to the genome may result in different clones, as discussed earlier, or they may result in what people consider new cultivars. In conventional breeding programmes, pollen from one grapevine (the male of the cross) is brushed onto the stigma of the other vine (the female parent). This results in a recombination of genes that most typically results in a wide variety of unique individuals, some of which may be very similar (perhaps indistinguishable from) to one of the parents, and some which may be very different. So, a breeding programme requires the harvesting of many seeds associated with a particular set of parents, planting them, growing them for a number of years until they reach fruiting maturity and then evaluating their growth habits as well as the fruit they produce. Grapevines grown from seed will remain juvenile (in a vegetative state) for a time, though treatment with cytokinins has been successful in initiating flower clusters on young seedlings (Srinivasan and Mullins, 1981). There is hope that, through the manipulation of genes regulating fruiting, tools may be made that can minimize the juvenile phase and therefore increase the rate at which seedlings can be evaluated (Pillitteri *et al.*, 2004). However, and at present, the time between creating the cross and being able to examine the fruit is usually many years. To fully evaluate a vine and the fruit it produces takes over a decade due to the need to see how the vine performs in different seasons and, if wine is to be made, there are further variables such as looking at appropriate styles of wine, as well as how the wine ages in the bottle.

Molecular biology is increasingly being used in breeding programmes, in some cases to introduce desired characteristics directly into vines, such as disease resistance (Aguero *et al.*, 2005; Vidal *et al.*, 2006) and, in some cases, to aid in tracking the movement of desirable traits from parent to offspring (Grando *et al.*, 2003). Increasingly, as the debate about the use of genetically modified organisms continues, molecular biology is being used to learn more about how the plants or other organisms work (Giovannoni, 2001), which can lead to more effective management or control techniques (Purcell and Hopkins, 1996).

Changing the ploidy level (number of sets of chromosomes the organism has) in grapes usually has an effect on flower and berry size (Dermen, 1964). The development of tetraploid or higher levels of duplication grapevines from the native diploid state can be achieved through a conventional breeding programme or through the use of colchicine (Dermen, 1954). The Kyoho grape bred in Japan is a tetraploid cultivar that was developed in part for its large fruit (Okamoto, 1994), which can reach 13 g/berry and above (Wang *et al.*, 2001), compared with the typical wine grape berry weight of approximately 1–2 g.

INDICATORS OF QUALITY

Although a vastly important attribute of a product, the trouble with quality is that it is not something that can be readily quantified, especially when it comes to something as complex as some grape products. Examples of basic quality parameters for table grapes are size, shape, sweetness and acidity; however, these are all relative. A small example of a large-berried table grape could still be far larger than a large-berried wine grape. Therefore, factors determining quality should be end use-specific.

It has been very difficult to quantify flavour and aroma compounds, and often we are left with human sensory analysis to determine some aspect of quality. The human perception of aroma and taste can be very sensitive (often more sensitive than many of the analytical methods currently available), but it does vary from person to person (Amerine and Roessler, 1976), and being able to communicate the quality of an aroma (for example, if it smells like strawberries or raspberries), as well as the intensity of that aroma, is problematic (Noble *et al.*, 1984). As a result of this, there have been attempts to standardize the terminology used for aspects of human sensory perception (Noble *et al.*, 1987; Gawel *et al.*, 2000), so that people in different places can use the same criteria to describe these attributes. However, quantifying these attributes still requires trained sensory panels or specialized equipment (e.g. gas chromatography–olfactometry), used in concert with trained panelists.

As indicators of quality, Table 2.2 shows some examples of parameters that are important for those grapes' end use. It is evident that Brix (or percentage soluble solids, which are made up mostly of sugars in grapes) is important in many cases, but for some end uses other parameters come into play.

In some instances, objective measures have been developed for characteristics (e.g. Brix by refractometer or hydrometer, acidity by titration, moisture by weight), but in others there is no standardized method available yet (e.g. tannins relating to astringency or flavour). Because of the importance of these attributes to end use (particularly when it comes to valuing grapes according to quality) there is much research effort invested in finding ways of measuring and quantifying these aspects.

Table 2.2. Examples of grape characteristics that are important for different end uses.

End use	Fruit attributes
Wine grapes	Brix, acidity, pH, colour, tannins, flavour
Table grapes	Brix, acidity, colour, flavour, presence of bloom, Brix:acid ratio
Raisins	Brix, colour, percentage moisture, flavour, presence of bloom
Juice	Brix, acidity, colour, flavour
Preserves	Brix, acidity, pectin, solids content

3

GRAPEVINE GROWTH AND FRUIT DEVELOPMENT

This chapter reviews the growth cycle of the grapevine, along with factors that influence it. The fact that the vine is a perennial plant means that management is quite different to that for annual crops, as well as there being unique aspects to cultivating young versus established vines. An overarching concept, that of vine balance, is introduced as it has implications for decisions made from the very start of vineyard planning as to how the weather affects the vine after harvest. Fruit quality parameters, which comprise a great many things but of which we can only measure a relative few, will also be mentioned, particularly with reference to grape berry maturation.

PHENOLOGY

Phenology is the study of events or growth stages of a plant or animal that recur seasonally and their relations with various climatic factors, including temperature, solar radiation and day length (Mullins *et al.*, 1992). With respect to grapevines, it can be used as a predictive tool, allowing the manager to plan vineyard operations in advance. In combination with other plants' phenological data, it can help to sequence events even though the weather patterns in the current year may be substantially different to those of previous years.

It is useful to track the growth stages of grapevines so that development can be monitored with respect to other seasons (or some other benchmark), and so that pesticide sprays can be applied in a timely fashion. In discussing grapevine phenology, it is useful to use a starting point prior to budbreak.

Before any visible signs of growth, the vine begins to come out of dormancy at about the same time as the soil temperatures starts to rise (due to increased solar radiation and air temperatures). The roots reactivate, and the phloem and xylem tissues start to function again. Visibly, this leads to sap exuding from fresh cuts on the vine, as root pressure builds to fill the empty xylem vessels (see Fig. 3.1; Sperry *et al.*, 1987). The volume of liquid that can exude from the cut end of a cane can be high – Bennett (2002) collected 174 ml in a 24 h period.

Fig. 3.1. Part of a vine trunk cut with a saw late in the dormancy period. Xylem sap, flowing from the roots and the cut vessels, oozes out; it contains low concentrations of sugars (Bennett, 2002) – in the range of 5–6mg/ml, which supports the proliferation of fungi and bacteria, making the sap here appear white.

Actual breaking (or appearance of green tissue through the bud scales) of the dormant buds has usually been said to occur when the air temperature reaches approximately 10°C (Winkler *et al.*, 1974; Williams *et al.*, 1985), though Moncur *et al.* (1989) report no physiological basis for this threshold. Indeed, their experiments on potted cuttings revealed that the base temperature for bud development in grapevines was as low as 0.4°C (depending on cultivar), which more closely matches values for other woody perennials (Anstey, 1966; Richardson *et al.*, 1975).

The progression of the budbreak process is illustrated in Plate 7. Buds are at their most hardy prior to budbreak with respect to both physical and cold temperature damage. The first shoots start to grow powered by energy derived from stored carbohydrate, as there is no photosynthesis yet occurring (Winkler *et al.*, 1974; May, 1986). The pre-formed leaves expand, as do the internodes. Any flower clusters present also develop, rapidly forming individual florets (see Fig. 3.2). Shoots can grow as fast as 2–5cm/day (4 cm/day, Creasy, 1996; 1.8 cm/day, Keller *et al.*, 2005; 5 cm/day, Wolf and Warren, 1995), forming a new internode every 2–3 days (Lovisolo and Schubert, 2000). Photosynthesis occurs as soon as there is green tissue on the shoots; however, due to the high

Fig. 3.2. Young shoot showing the first leaves and two flower clusters. Individual florets have already differentiated by this stage.

metabolic activity and use of stored carbohydrates, there is no net production of photoassimilates until several leaves have fully expanded (Hale and Weaver, 1962; Winkler *et al.*, 1974).

Unlike many other perennial woody crops, grapes flower long after budbreak. As it takes some time for the shoot to develop leaves that are capable of supplying the carbohydrate needs of the rest of the vine, it is critical that enough stored carbohydrate is available to support the development of shoots, roots and flower clusters. If there is not, then it is the flower clusters that suffer the most, as they can drop off the vine due to a lack of available carbohydrate (Ollat, 1992).

When the shoots have approximately 15–17 nodes formed on them (Pratt and Coombe, 1978), the flowers begin to open and the calyptra fall from the rest of the flower (see Plate 8). The flowering and fruit set process in grapes is very weather dependent. Pollination (transfer of the pollen from the anther to the stigma) is mostly by wind, though insects may also contribute. Self-pollination, occurring as the cap comes off the ovary, is the norm (Winkler *et al.*, 1974; Kimura *et al.*, 1998), though some studies have suggested that cross-pollination results in bigger fruit and higher seed counts (Sampson, *et al.* 2001; Milne *et al.*, 2003). Airborne pollen counts have been used as a predictor for fruit set in some grape-growing areas (Cristofolini and Gottardini, 2000; Cunha *et al.*, 2003) as there is a good correlation between pollen in the air and fruit set, but this may be due to more pollen being released when the weather is warmer and dry, which is generally conducive to fruit set.

Fertilization occurs 2–3 days after pollination, as the pollen tube must grow (a highly temperature-dependent process) down through the style and up into the micropyle in a 'J' shape (see Fig. 3.3) in order to reach the embryo. Even relatively brief spells of cool temperatures cause degeneration of embryos and decrease the chance of fruit set (Ebadi *et al.*, 1996), which goes some way to explaining why this is a problem in cool climate grape-growing areas.

Typical success rates for fruit set may be in the 30–45% range. Flower clusters may contain from just a few to more than 1000 flowers (Woodham and Alexander, 1966), depending on conditions at flower cluster initiation and cultivar.

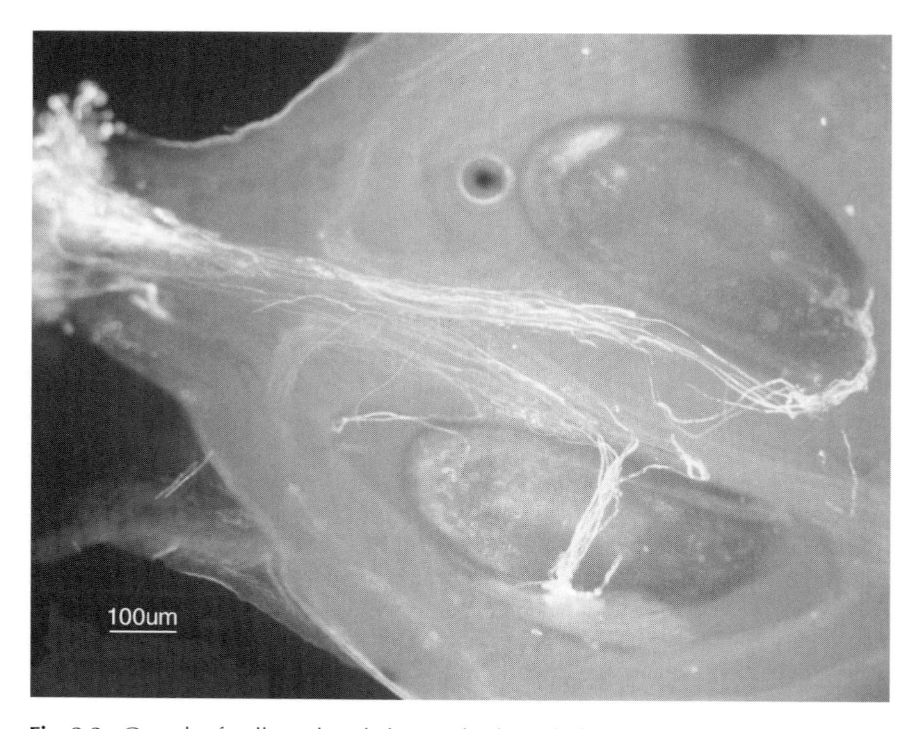

Fig. 3.3. Growth of pollen tubes (lighter paths through the style) from stigmatic surface to the micropyle, shown using a fluorescent dye (image courtesy of M. Longbottom).

Flower cluster initiation occurs about 18 months before the fruit from those flowers is harvested. At about the time of flowering, the flower cluster primordia are being initiated in the compound buds located in the axils of the basal leaf petioles (Winkler *et al.*, 1974). Both temperature and light are important determinants of flower cluster size and number (Buttrose, 1969). As the season progresses, the initiation of flower cluster primordia works its way up the shoot, meaning that along different parts of the shoot (next season's cane) there can be significantly different levels of fruitfulness (this has implications for methods of pruning, see Chapter 6). Individual florets do not form until after budbreak the following season, but the the amount of branching, and thus cluster size and total floret number, is largely determined in the current season.

Lateral shoots arising from axillary buds may also have fruit on them, which is termed second set. Generally, these are quite far behind in terms of development compared with the primary crop and, as well as being smaller in size than the primary crop, are rarely used in production. However, in tropical climates, there would be the opportunity for this crop to be harvested. In most cases this fruit is a nuisance, as hand-pickers can mistake it for ripe fruit, and machine-harvesters harvest all grapes in the canopy, regardless of maturity.

Once fruit is set on the vine, it is unlikely that the vine will lose it. Apple and other tree fruit crops have one or more times of the year when the crop abscises naturally (Westwood, 1993; Bangerth, 2000; Wertheim, 2000); however, with grape the fruit cannot be dropped, and the vine is pretty much committed to bringing it to maturity. As such, after fruit set it becomes the most important destination for the vine's carbohydrate supply, as carbon radio-tracer studies have shown (Hale and Weaver, 1962; Quinlan and Weaver, 1970).

Berry expansion is rapid following fertilization, due to both cell division and cell expansion. The growth in terms of weight, diameter, volume or other measurements can be described as a double sigmoid curve (see Fig. 3.4), particularly for seeded berries – for seedless ones the curve is less pronounced (Coombe, 1960; Winkler *et al.*, 1974; Friend, 2005). The first period of rapid growth is often called Phase I, which is followed by Phase II, a time of relatively little growth as the seed matures and begins to lignify (develop a hard outer coating). At the end of this period the berry undergoes an amazing transformation, softening, becoming translucent, beginning to colour (if a red cultivar) and increasing in size rapidly (Phase III). On the inside, the berry beings to metabolize malic acid, accumulate sugar and develop characteristic flavour and aroma compounds. The French use the term *véraison* to describe the shift in colour, which signals that all these other changes are taking place, too. It is not known exactly what triggers or even regulates this massive and rapid change in berry physiology, but the term engustation (Coombe and McCarthy, 1997) has been proposed to describe the great upheaval of events.

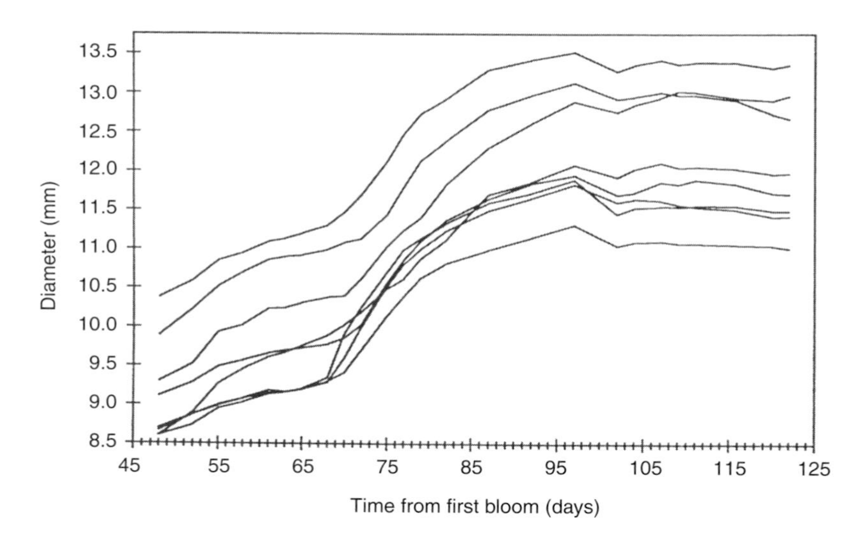

Fig. 3.4. Individual growth curves (by measurement of diameter) for eight grape berries, demonstrating the double-sigmoid shape characteristic of seeded fruit (from Creasy, 1991).

VINE (VEGETATIVE) DEVELOPMENT

Patterns of root growth

The exploration of soil available to a vine is extensive. Vine roots have been found to penetrate to depths of more than 10 m in favourable soils (Winkler *et al.*, 1974), and in general will make as full use as possible of the soil profile. As with any plant, many factors influence root growth, including soil structure, compaction, water availability (including precipitation and water table level), plant genotype, planting density, competing plants, etc. Most roots in a typical soil profile will be found in the top 1 m, but more sparse rooting is found to varying depths and can contribute greatly to water and nutrient uptake.

An important point to make is that vines grown from seed have a taproot, and will explore soil to a greater depth than vines propagated by cuttings (by far the more common method of propagating), which have roots that form adventitiously on the sides of the cane and thus do not grow downwards as much. Even so, different species of vines have varying tendencies to develop deeper root systems, even from cuttings (Perold, 1927), which becomes an important consideration when choosing a rootstock (see Chapter 5). Those roots that tend to explore horizontally are termed spreader roots, and those that tend to grow further down into the soil profile are termed sinker roots. Rootstocks can be classified in terms of their root angles, i.e. the degree to which roots grow away from the vertical plane (Perold, 1927). Thus, a vine with a steep angle tends to grow many sinker roots, whereas one with a shallow angle tends to grow spreader roots. In terms of determining the suitability of rootstock vines to areas with or without a high water table, this information can be very useful.

Like the above-ground parts of the vine, the roots also have a period of quiescence in the dormant season. As the soil warms in the spring, root tips begin to explore the soil, but their development is relatively slow (in terms of dry matter accumulation) compared with the growth and development of the shoots and flower clusters. After fruit set, the full canopy of the vine is able to support more root development but, as the berries ripen toward the end of the season, root growth again slows due to the shift in partitioning of carbohydrates towards the fruit. In areas where there is sufficient warmth to allow photosynthesis after harvest, there may be another flush of root growth before the vine becomes dormant, which appears to be related to the excess photosynthetic capacity on the vine after the fruit is removed (Williams, 1996; Conradie, 2005). In cooler climate areas where there is little or no time between harvest and leaf-fall this cannot, of course, happen.

Tendrils

Most, if not all, vinous plants produce tendrils as a means to support themselves on another structure. It is through this specialized lateral branch (Pratt, 1974) that the vine can invest less energy in developing a solid trunk and more into growing in length, which is in keeping with the vine's evolutionary origins.

To facilitate grasping on to a support, the tendrils grow away from light and curl around or within objects, eventually becoming lignified and woody (see Plate 9). If a tendril does not latch onto something, then it will wither and fall from the shoot. Unlike shoots, tendrils have determinant growth, i.e. they grow to a finite length.

Cane maturation

Similarly to tendrils, shoots will lignify, become more woody and change colour from green to brown (sometimes distinctly red) as they age (see Plate 10). It is this formation of a periderm that is associated with the shoot becoming more resistant to drought and cold – an adaptation that allows the plant to survive below freezing temperatures that are found in many areas where *Vitis* species have developed. The compound buds at each node on the hardened shoot – now called a cane – also go through a similar process, with the bud scales becoming lignified and resistant to drought, temperature and mechanical damage. As shoots mature, there is a build-up of starch and/or soluble sugars within, which is associated with the vine being able to survive below-freezing temperatures (Hamman *et al.*, 1996; Chapter 4) and which provides energy for growth in the following spring.

Leaf-fall and abscission

Towards the end of the growing season environmental signals trigger the vine to accelerate its physiological preparations for the dormant season. Leaves may begin to yellow (or become red if a red-skinned grape cultivar, see Plate 11) and fall from the vines. However, if the vines still have their crop, senescence may be delayed due to the demand for photosynthates from the fruit (Petrie *et al.*, 2000a). The ability of vines to photosynthesize after harvest, and therefore augment storage of photosynthates before the dormant period, is significant in terms of vine performance. Increased carbohydrate status in vines at the end of the season is correlated with increased winter hardiness and fruitfulness in the following season (Howell *et al.*, 1978; Bennett *et al.*, 2005). The coincidence of harvest and leaf-fall in cool climate vineyard areas, and thus limitation on photosynthesis, may be a reason why vine yields are lower in cool climates as opposed to warmer ones (Howell, 2001).

Dormancy

The period of dormancy is integral to the survival of the vine in areas where winter temperatures fall below 0°C, although dormancy is also an important part of the vine cycle in warmer areas, too. Grapevines are grown in tropical areas, and as shoots are not determinant they will grow indefinitely. This can create some production and management problems, however, so vines in these areas are often forced into and back out of dormancy through the use of severe pruning or chemicals such as hydrogen cyanamide (Lavee, 1974; Shulman *et al.*, 1983). It has also been proposed that cyanamide could be used to force shoots to grow even before leaves have been removed or senesce, resulting in evergreen production of grapes (Lin and Wang, 1985).

Naturally occurring dormancy is brought about by low temperatures in the late growing season as well as by shortening day length (Schnabel and Wample, 1987). Non-*V. vinifera* species, such as *V. labrusca* and *V. riparia*, seem to be more sensitive to day length than the *V. vinifera* vines (Fennell and Hoover, 1991; Wake and Fennell, 2000). Once the leaves have fallen the vine is in a state of ecodormancy, which is controlled by the environment (e.g. cold temperatures) as well as in a state of endodormancy (through physiological factors within the plant). Therefore, buds will not develop immediately even if environmental conditions improve. While vines do not have a chilling requirement (i.e. needing a certain amount of time below 10°C in order to emerge from endodormancy) per se, cold temperatures do facilitate the process, decreasing the length of time needed for the plant to start growing again when environmental conditions improve (Schnabel and Wample, 1987).

The major task of pruning occurs during the dormant season (Chapter 6).

FLOWER INITIATION, FRUIT DEVELOPMENT AND BERRY MATURATION

The perennial nature of the grapevine is nicely illustrated through an examination of the vine's reproductive cycle. Here we will see that fruiting potential is initiated over 1 year before the crop is harvested.

The fruitfulness, or how much potential crop it has, of a vine can be determined well before budbreak by dissecting and examining the primordial shoots within, or by forcing the buds to grow out early. Such studies have shown than flower clusters for the next season are initiated about two to three weeks before the current season's flowers are opening (Lavee *et al.*, 1967; Pratt, 1979). An examination of buds at different node positions along 'Thompson Seedless' shoots in Davis, California showed that basal buds had greater fruitfulness than more distal buds at the start of June, but another sample taken in July revealed that buds at all node positions were fruitful. This led to the conclusion that the initiation of flower clusters occurred over time, and in sequence from the bottom of the shoot upwards (Winkler *et al.*, 1974).

Though this indicates when the inflorescences appear to be initiated, it doesn't tell us the whole story. Microscopic examination of developing buds has allowed the sequence of flower cluster initiation to be derived.

Where do tendrils and flower clusters come from?

Earlier, it was mentioned that tendrils and flower clusters are related. This is so because they are thought to be derived from the same uncommitted primordium (called an anlage) produced as another growing point off the primordial shoot's apical meristem. The primordium can become one of three things, two vegetative and one reproductive: a tendril, a shoot or a flower cluster. If it becomes a tendril it will develop, but not branch very much; if it becomes a shoot, then the main shoot will split, forming a fasciation, or bifurcated shoot (see Fig. 3.5). If it is destined to become a flower cluster it will undergo much branching, which will lead to what we see as a flower cluster.

It is of note that tendrils are not found on shoots below flower cluster positions (if present), so tendrils can be thought of as the default form of the anlage, where conversion to a flower cluster happens only under certain conditions. Indeed, the distinction between a flower cluster and a tendril can become blurred, with the former having tendril-like appendages and the latter having their ends populated by groups of flowers, ending up as fruit (see Plate 12).

For the flower cluster to appear as we would expect, it needs to undergo repeated branching. This branching can continue through the season and, by leaf-fall, much of it has been completed, laying the foundation for next season's

Fig. 3.5. A bifuricated shoot (centre), where a new shoot tip arises from a node position opposite a leaf.

potential crop. However, more branching can occur around the time of budbreak after the dormant season, or whenever environmental conditions allow (May, 2000).

Factors affecting branching

Because buds in the varying node positions go through the process of flower cluster initiation at different times, changing environmental factors that affect this process will also affect the potential crop for each bud in the following season.

Temperature is one such factor, as it affects all biological processes. Cool and excessively hot temperatures discourage the initiation process, though absolute values for minimum temperatures necessary for any initiation to occur vary from cultivar to cultivar (Buttrose, 1969). The optimum for maximizing potential crop is thought to be in the range of 30–35°C, but this also appears to vary with cultivar (Buttrose, 1969). In many cool climates, achieving this range of temperature during flowering, when flower cluster initiation is going on, is relatively rare, hence the chance for high crop loads is already limited by this factor.

Light is also important, as it affects the rate at which photosynthesis can occur and hence photoassimilate supply. The carbohydrate supply near bloom is an important factor affecting the number and potential size of the flower clusters being initiated (May, 1965; Sommer *et al.*, 2000), but Bennett *et al.* (2005) have also shown that vine carbohydrate status following fruit set will also affect flower cluster number and size in the following season. Hence, overly vigorous shoots are associated with fewer flower clusters because vigorous vines generally have shadier canopies, and also the growing points of a vigorous shoot are much better at drawing carbohydrates away from the developing flower clusters.

Plant growth regulators

As with many aspects of plant development, plant growth regulators play an important role in flower cluster initiation. The two most involved are gibberellic acid (GA) and cytokinins (CKs).

GA enhances the development of an uncommitted primordium that forms to the side of the shoot apex, encouraging the development of tendrils (Srinivasan and Mullins, 1980) but inhibiting the formation of inflorescences (Srinivasan and Mullins, 1980; Palma and Jackson, 1989). CKs will encourage the formation of flower clusters from the tendrils (Srinivasan and Mullins, 1980), so the net effect of applying GA to a grapevine shoot during the flower cluster initiation period will be to promote the production of tendrils rather than flower clusters. The net effect of applying CKs to the shoot would be more flower clusters.

The role that each of these plant growth regulators plays has been elucidated in part through the use of chlormequat, which has been used as a herbicide. This chemical blocks the synthesis of GA (primarily, though it may have some effect on CK production as well) (Skene, 1968; Lang, 1970).

Cytokines are also important later on in flower development, as their presence promotes the formation of flower primordia on the inflorescence near budbreak in the following spring, and can improve fruit set during flowering as well (for a review, see Srinivasan and Mullins, 1981). Analysis of the bleeding sap from cuts on vines before budbreak show that CKs are being transported from the roots to the upper parts of the vine at this time of year (Skene and

Kerridge, 1967), which demonstrates the potential importance of soil temperature early in the season, as warmer soils will have more root growth and potentially more CK production. Work by Woodham and Alexander (1966) showed that, for each of two 10°C increments (from approximately 10°C), if the root temperature was higher than that of the shoot, there was an approximate increase of 18% in flower number per inflorescence.

Flower development and anthesis

Differentiation of the flowers on the inflorescences starts near the time of budburst (Srinivasan and Mullins, 1981) as soil and air temperatures begin to rise, which also means that the full cropping potential of the vine has been realized.

The pollen sacs in the anthers mature shortly before capfall (Staudt, 1999), as does the receptivity of the style itself. Thus it is thought that the mechanical disruption caused by the movement of the cap off the ovary and style, with probable transfer of pollen from the anthers to the style (a process called pollination) as well as other factors, means that grapes are mostly self-pollinated (Lavee and Nir, 1986).

Capfall has been associated with a certain number of nodes on the flower cluster's shoot, though the number seems to vary between cultivars. However, it does not seem to vary much within a cultivar or between seasons (Pratt and Coombe, 1978), which means that counting the number of nodes on shoots can be a way of predicting flowering date, as long as the number of days it takes to form a node is known. Some research suggests this is on the order of 3 days per node (Lovisolo and Schubert, 2000), though this rate depends on environmental conditions.

The duration of flowering is also highly dependent on the environment at the time. Cool, overcast weather, associated with rainfall, lengthens the flowering period, whereas warm and sunny conditions hasten it. Thus, flowering can occur over a period of a few days to longer than a month. This is thought to contribute to variation found in the fruit through the rest of the season, although there is also considerable variation found within the flower cluster itself (May, 1986; Friend *et al.*, 2003).

Capfall occurs mainly in the morning hours (Staudt, 1999), with the highest rate occurring between 7.00 and 9.00, and the final ones falling by 12.00. This is thought to be a result of changes in turgor pressure within the cells in the calyptra's abscission zone (Swanepoel and Archer, 1988).

Fruit set

Once viable pollen lands on a receptive stigmatic surface (see Plate 13), a germ tube emerges from the pollen grain within a short period of time, as little as

30 min (Staudt, 1982). It then begins to grow down through the style to the micropyle, following a 'J'-shaped path, and on to the nucellus (see Fig. 3.3). There is some thought that the size of the flower, and thus the length the pollen tube must grow, is a factor in the success of fruit set. This is because the speed at which the pollen tube grows is very much dependent on the temperature, with colder temperatures resulting in drastically lower rates of elongation (Staudt, 1982). If the growth is slowed to the point where the ovules degenerate before the pollen tube reaches them, fertilization cannot take place.

Plant growth regulators are also involved in this phase of reproductive growth, as auxins are released from the pollen tube (Taylor and Hepler, 1997) as it grows, which stimulates growth of the ovarian tissues.

If the pollen tube does reach past the micropyle, the generative nucleus in the pollen unites with the synergids in the ovarian embryo sac, causing a rapid increase in metabolic activity (Kassemeyer and Staudt, 1981) that results in further development of the now fertilized embryo. The ovule becomes the seed, and the ovule wall enlarges to become the berry pericarp, or flesh.

Commonly, one or two seeds are found per berry in *V. vinifera* fruit, usually more than two on average for *V. labrusca* (Ebadi *et al.*, 1995), though there is considerable variation between cultivars (Forsline *et al.*, 1983). Berry size seems to be driven initially by the presence or absence of seeds, and then by the mass of seed within the berry rather than just the number of seed (Scienza *et al.*, 1978; Ebadi *et al.*, 1996; Trought and Tannock, 1996; Roby and Matthews, 2004; Friend, 2005). In terms of vine yield, it is most important to have at least one viable seed.

Seedlessness

Evolutionarily speaking, the seeds within the fruit are the most important part of the vine: the vine produces seed to propagate and spread the offspring. So how have seedless berries come about? After all, if no seeds are produced, then how can the vine propagate itself? As with the appearance of clones, random mutations can cause a vine to arise that has seedless berries. In addition, some crosses of plants can result in sterile offspring, so a seeded cultivar pollinating another seeded cultivar may result in a seedless offspring.

In most cases seedless cultivars are not, in fact, really seedless. Many of the popular seedless table grapes are stenospermocarpic. In this case, the flowers are pollinated and the embryos fertilized, but soon after the embryo aborts; however, in the time that the fertilized embryo is developing, it produces enough plant growth regulators to encourage growth of a large berry. If a seedless grape is cut apart, often a small remnant is visible inside, which is the aborted seed (see Fig. 3.6). In some cases there may be something that looks like a viable seed, but is in fact the lignified shell of a seed with no embryo within. 'Einset Seedless' (developed at the NYSAES, Geneva) is one cultivar that is prone to having these crunchy seed remnants.

Parthenocarpy, a process that requires pollination but not fertilization to set fruit, does occur with cultivars like 'Black Corinth' (syn. 'Zante Current',

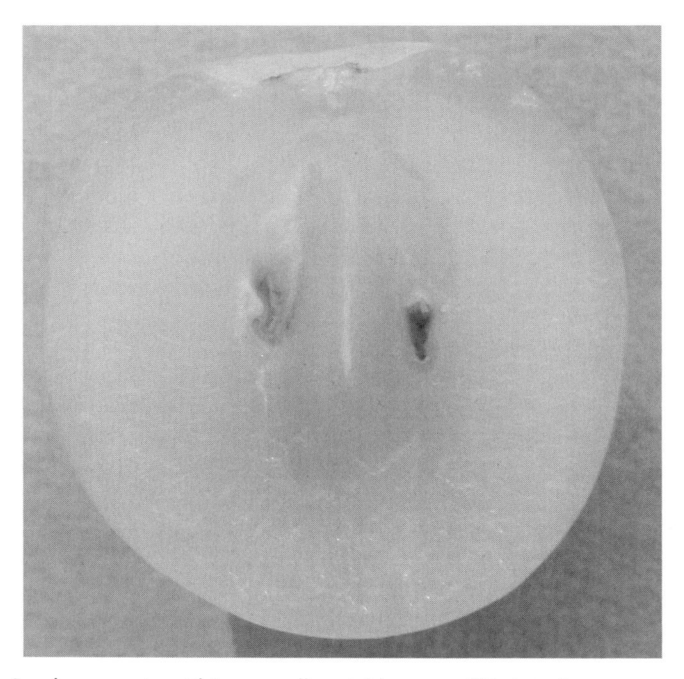

Fig. 3.6. Seed remnants within a seedless table grape ('Perlette').

Plate 14). These vines have very small berries, from a lack of cell division, in comparison with seeded ones (Coombe, 1973), as without the seed present, division and growth of the cells of the ovarian wall are not encouraged (Olmo, 1946).

Factors influencing fruit set

There are many factors that will affect the percentage fruit set in grapevine. Among the most important are the availability of light, moderate temperatures and dry weather. The effects of temperature on pollen tube growth and root-produced plant growth regulators have already been mentioned. However, light is at the core of many vine functions as it is the source of carbohydrate energy for the plant. If conditions are overcast, the vine will not be able to produce enough photosynthates to feed the growth of shoots and all other sinks, and so flower clusters, being a weak sink, will not attract a great deal of them. This correlates with how the grapevine grows in the wild, as the vine is vegetative until it has grown up the tree trunks and emerged into the higher light levels in the canopy above.

As the vine is a perennial plant, the amount of carbohydrate stored is also a factor – if the vine has sufficient reserves in its root and other permanent wood, then it is better able to supply energy to the growing parts of when photosynthesis falters. This is used as evidence to support the notion that older vines (having more permanent wood than younger vines) perform better (Howell, 2001). In addition, since the early-season shoot growth is highly dependent on stored reserves, if reserves are not sufficient then shoot growth will suffer and so, too, will the leaf area available to manufacture photoassimilates.

Recent research has demonstrated a positive link between vine carbohydrate levels pre-budburst and percentage fruit set in a cool-climate growing region (Bennett *et al.*, 2005). This has led to the suspicion that stored carbohydrates are more important than previously thought when it comes to determining how many fruit will set at bloom.

Water must be available to the vine near flowering to allow photosynthesis to proceed at the optimal rate, as water stress has been found to be highly detrimental to fruit set (Hardie and Considine, 1976; Jackson, 1991). However, as flowering occurs early in the season, water is usually available in the soil profile, and stress must be actively avoided in only the driest of regions, or in those areas with soils of very low water-holding capacity.

One physiological disorder that is probably connected to a lack of available carbohydrate is early bunchstem necrosis (syn. inflorescence necrosis). This fruit set problem results in the loss of individual flowers, branches of flower clusters or even entire flower clusters in the weeks leading up to and including the fruit set period (Jackson and Coombe, 1988; Ibacache, 1990; Plate 15). It is thought that this problem is a result of a build-up of toxic ammonium in the flower cluster tissues because of an inability of those tissues to convert ammonium to the amino acid glutamate – a process that requires carbohydrates (Jordan *et al.*, 1993). Its appearance is sporadic and difficult to predict, but can have a significant effect on yields.

Deficiencies in certain nutrients can also reduce fruit set, chief among them being zinc and boron. Zinc is necessary for activation of many enzymes, as well as for the production of the plant growth regulator indole acetic acid (Marschner, 1986), and boron is necessary for the formation of cell walls (Thellier *et al.*, 1979), which is involved in all aspects of plant growth.

In fact, any significant stress the vine is under, be it abiotic (e.g. nutrient or water) or biotic (disease or insect pest damage), can reduce the potential of the vine to set fruit (Creasy, 1991).

Berry development and maturation

Once the embryo is fertilized in grapes, the vine is committed to maturing the fruit, unlike some other fruit crops where there is fruit abscission and thus a chance to reduce what might otherwise be too high a crop. Fertilization results in immediate and rapid cell division (from 1 to 2.5 per cell on average) to move from approximately 200,000 cells at anthesis to a maximum of 600,000 by véraison (Harris *et al.*, 1968). However, the highest activity in terms of cell division occurs before flowering, where about 17 cell divisions are responsible for attaining that figure of 200,000 in the first place.

In Phase 1 of growth there is no significant accumulation of sugars in the berry, as much of the photoassimilate is used for cell division and expansion. However, there is an accumulation of organic acids, primarily malic and tartaric (Hrazdina *et al.*, 1984), the latter of which is relatively unusual in fruit crops (McGovern *et al.*, 2004). The duration of Phase I seems to be similar for most grape cultivars (Nakagawa and Nanjo, 1965; Coombe, 1976), while that for Stage II can vary considerably depending on cultivar, management and environment (Winkler and Williams, 1935; Coombe, 1960). Many characteristics in the grape change in the period leading up to maturity.

Grape composition

The primary component of mature grapes is water, making up about 75–85% of their weight (see Table 3.1). Approximately 15–25% is in the form of sugar, a higher percentage than in many other fresh fruits. The organic acids tartaric, malic and citric make up 0.5–1.0% of the fruit, pectin about 0.25% and there is a long list of other nutritional components. If all of these are totalled they amount to over 99% of the weight of grapes but, if they were the only contributors, we would not be able tell grapes apart from many other fruits.

Fleshy fruits like grapes are made up of living cells and therefore have all the ‘primary metabolites’ of all plants. Primary metabolites are the proteins, amino acids, nucleic acids, fatty acids, cellulose and others found in all plant cells that facilitate respiration, photosynthesis and, to all intents and purposes, life. These primary metabolites make up a higher percentage of the total mass in leaves, buds, flowers and other non-reproductive parts compared with fruit.

Most of the attributes that make grapes desirable to us (and to certain pest species!) accumulate or develop in Phase III of berry development. One of the most important of these is sugar, in the forms of glucose and fructose.

The currency of the phloem, which is the network of conduits that moves photoassimilates from one place to another in the vine, is sucrose, which is not the same as that for photosynthesis or for ripening grapes. Photosynthesis creates glucose, which is freely inter-converted with fructose. One each of these molecules is combined into sucrose prior to their journey through the phloem and then on to the sink (growing point, fruit, etc.), where it’s taken out of the phloem and converted back into glucose and fructose. Therefore, very little sucrose (which is the form of sugar we’re most familiar with, as it is what makes up the granulated sugar we put in much of our food) is found in ripe grapes (0.15% sucrose compared with 7 and 8% by weight glucose and fructose, respectively (see Table 3.1).

Grapes are high in carbohydrates and not a particularly good source of dietary fibre. However, they are a useful source of many minerals and vitamins B_6, C, E and K. They are also a source of antioxidant compounds through the phenolics in their skins and possibly seeds (Yilmaz and Toledo, 2004).

Table 3.1. Raw grape composition, comparing *V. labrusca* with *V. vinifera* (from US Department of Agriculture, Agricultural Research Service, 2005; USDA Nutrient Database for Standard Reference, Release 18, http://www.ars.usda.gov/ba/bhnrc/ndl (used with permission)).

		Composition/100 g	
Nutrient	Units	*V. labrusca*	*V. vinifera*
Water	g	81.3	80.5
Energy	kcal	67.0	69.0
Protein	g	0.63	0.72
Total lipid (fat)	g	0.35	0.16
Ash	g	0.57	0.48
Fibre (total dietary)	g	0.9	0.9
Carbohydrate (by difference)	g	17.1	18.1
Sugars (total)	g	16.2	15.5
Minerals			
Calcium (Ca)	mg	14.0	10.0
Iron (Fe)	mg	0.29	0.36
Magnesium (Mg)	mg	5.0	7.0
Phosphorus (P)	mg	10.0	20.0
Potassium (K)	mg	191	191
Sodium (Na)	mg	2.0	2.0
Zinc (Zn)	mg	0.04	0.07
Copper (Cu)	mg	0.04	0.13
Manganese (Mn)	mg	0.72	0.71
Vitamins			
Vitamin C (total ascorbic acid)	mg	4.0	10.8
Thiamin	mg	0.09	0.07
Riboflavin	mg	0.06	0.07
Niacin	mg	0.30	0.19
Pantothenic acid	mg	0.02	0.05
Vitamin B_6	mg	0.11	0.09
Folate (total)	mcg	4.0	2.0
Vitamin A	IU	100.0	66.0
Vitamin E (alpha-tocopherol)	mg	0.19	0.19
Vitamin K (phylloquinone)	mcg	14.6	14.6
Others			
Amino acids	g	0.50	0.57
Carotenoid pigments	mcg	131	111

What is fruit maturity?

Fruit maturity and ripeness are often-used terms in the grape industry, but it is difficult making the linkage from these to the quality of the grapes. This is because quality is, by its very nature, a qualitative term, not a quantitative one – there are very few absolute measures of quality, as one person's quality may differ from another's. Today, industries talk about suitability for end use to try to get around this elusive concept of quality.

There are several measurable parameters in grapes that relate in some way to quality factors. One of these is some measure of sugar concentration, which usually is accomplished by estimating the amount of dissolved compounds in the juice. As the vast majority of dissolved compounds are sugars (glucose, fructose and sucrose), this ends up being a pretty accurate representation.

There are several different and commonly used measures. Baumé is a convenient term for winemakers as the number corresponds to the approximate alcohol percentage once the grapes have been fermented to dryness. The °Brix value corresponds to the approximate percentage sugar in the solution and is an adaptation of the percentage soluble solids, which is used in many industries to measure sugar in liquids. Specific gravity is an actual measure of the density of a solution, commonly determined through the use of a hydrometer. Its value is expressed with respect to the density of water, as a solution with dissolved sugar will be heavier than water. Oechsle is a derivative of specific gravity where the value of one is subtracted from the value for specific gravity and the result multiplied by 1000. Table 3.2 shows the relationship between these different measures. Note that the specific gravity of a liquid changes with its temperature, so this is a factor that must be held constant or corrected for when comparing units.

Because many parameters associated with quality are linked with changes to sugar in the berry, °Brix (or other measures related to sugar concentration) is a commonly used measure linked to meeting minimum requirements for harvest. For example, 'Concord' grapes harvested for juice production may have to reach a minimum of 15 °Brix to be accepted by the processing plant. Bonuses can be associated with exceeding this target, or there may be deductions from a base pay rate if the crop comes in below the target.

As the degree of sweetness is affected by the amount of acid mixed in with the sugar, and vice versa (McBride and Johnson, 1987), the amount of acid in grapes is also often measured. This is most often performed by titrating a sample

Table 3.2. Equivalent measures of sugar concentration for four scales used in the grape industry. Note that the relationships between the measurements are altered with changes in temperature of the liquid.

Baumé	°Brix	Specific gravity	Oechsle
12.5	22.5	1.094	94

of juice with a standardized basic solution to a certain end point (usually pH 8.2). For wine grapes, the titration is usually done with 0.10 normal sodium hydroxide (NaOH) and the result expressed as titratable acidity (TA):g tartaric acid (the predominant organic acid in grapes)/l solution. In some areas data are expressed on a sulphuric acid basis and sometimes to a slightly different end point, which means that TA measurements are not always directly comparable.

Because the ratio of sugar to acid (usually °Brix:TA) is important to perceived flavour, in table grape production it is particularly important and has been formally recognized as being such since the 1960s (Nelson *et al.*, 1963). As long as sugar levels are high, grapes with relatively high acidity will still be acceptable to the market. At the other end, if acidity is too low in relation to sugars, the grapes can taste overly sweet and insipid (Liu *et al.*, 2007).

Related to, but not the same as titratable acidity, is the concentration of hydrogen ions (H^+) in a solution. This is expressed as pH, and is important for the biological stability of grape juice and wine as well as having an effect on the ionic forms of some molecules, such as anthocyanins, affecting their colour. pH is an important measure that is not directly related to TA as one might think; pH measures the H^+ present in solution at any given time whereas, through the process of titration, H^+ that are bound to organic acids are progressively removed and measured. So, juices with identical TA can be coupled with either low or high pH, depending on the relative amounts of organic acids and other cations present (such as potassium) that can take the place of H^+.

Sometimes it is not what makes up the fruit that is important, but other factors, such as physical damage, presence of disease or non-grape materials (leaves, stems, insects, sand, etc.). These can often be measured objectively. For example, the level of botrytis infection in grapes can be estimated by measuring the amount (or activity) of the enzyme laccase in the juice (Zouari *et al.*, 1988), as this enzyme is not found naturally in grapes but comes from the disease organism.

More recently, quantifying other aspects of quality has been attempted, such as objective measures of (i) colour through spectrophotometry; (ii) tannins through precipitation assays (Harbertson *et al.*, 2002; Herderich and Smith, 2005) or high-performance liquid chromatography (HPLC) (Vrhovsek *et al.*, 2001); and (iii) flavour compounds (Ortega-Heras *et al.*, 2002; Howard *et al.*, 2005). Most of these factors had previously been assessed through sensory means. This process introduces an amount of subjectivity into the picture, which is often incompatible with determination of scales such as pay rates for fruit.

However, it is worth noting that aspects that were once considered purely subjective are now able to be measured objectively (e.g. sugar concentration). It is inevitable that progress will be made in this area.

Content versus concentration

Distinguishing between concentration and content in berries is also of value. Many parameters are measured on a concentration basis (Brix would be an example of this) but, as the size of the berry changes considerably while ripening, it can give a skewed view of what is actually happening. For example, conventional Brix monitoring during ripening may show the solid line in Fig. 3.7, which indicates that Brix increased through ripening, but the rate of increase was less on 31 March. Usually, a decrease in Brix is associated with rainfall, but in this case there was no rain during this period. The cause of the slowing in Brix accumulation is not apparent but, if the sugar concentration per berry is calculated and plotted (dashed line), we see that the sugar content of the berries decreased at this sampling date. However, this is not possible at this stage of berry development, so another reason could be sampling error: a less ripe population of berries was taken on this date, perhaps because a different person did the berry collection. So it is likely that there was not a dip in the Brix at that date, but rather it was the result of an artefact of sampling. Because accurate sampling is crucial for making harvest decisions, tracking the content of sugar is much more reliable than tracking the concentration.

Factors affecting fruit maturity

It may be useful to think of grapes at harvest as being a summation of environmental conditions for the whole of the season. Aspects such as berry size, colour, flavour, their number and how diseased they are relate back to the conditions under which they developed. Decisions that were made before planting have a continuing effect on the grapes, as both macro and mesoclimate determine how the grapes will turn out in a broad sense. For example, the day/night temperature regime can affect aroma and colour in fruit (Kliewer and Torres, 1972; Becker, 1977; Ewart, 1987), so that grapes grown in a continental climate taste, and may look, very different to the same grapes grown in a maritime climate.

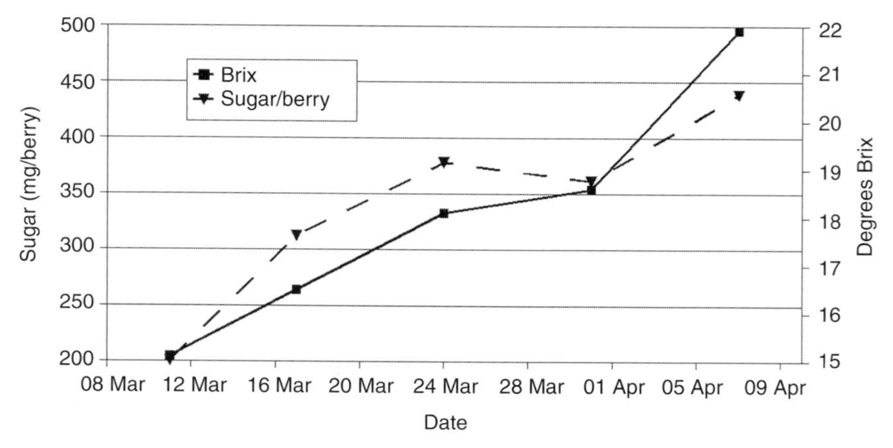

Fig. 3.7. Example of differences in trends in sugar levels when data are expressed on a per berry versus concentration basis.

How the vines have been managed also has a large impact, as row orientation and many canopy management decisions (e.g. shoot positioning, shoot thinning, leaf removal) can change harvest composition dramatically (Jackson and Lombard, 1993).

Some unexpected changes can also occur, e.g. the use of a sterol biosynthesis inhibitor as a spray to control powdery mildew has been found to interfere with terpene biosynthesis in 'Muscat of Alexandria' berries (Aubert *et al.*, 1998). In addition, bird netting – essential for the production of grapes in many areas – can also have an effect, slightly delaying fruit ripening (M.C.T. Trought and G.L. Creasy, 1999, unpublished results), possibly from the slight shading effect, but also from deformation of the canopy, creating shade.

There are very few things that do not change with berry development. A pivotal point of this process is the time near véraison, or between Phases II and III of berry development. At this time there is a dramatic shift towards expansion, which brings with it a number of changes. The cells within the berry lose structural integrity very suddenly, resulting in a mixing of the symplast (intracellular contents) and apoplast (extracellular contents) and causing the berry to soften and become more translucent. Cell walls become thinner (Considine and Knox, 1979) and the berry skin more stretchable (Matthews *et al.*, 1987) as the berry expands and ripens.

Amino acids also change with grape maturity. These are the building blocks of proteins and so are extremely important to plant function. They are also an important grape constituent, particularly for winemaking, as the yeast cells that ferment the sugar into alcohol utilize nitrogen from the must. Stuck fermentations with wine grapes are sometimes attributable to a lack of yeast-assimilable nitrogen (YAN), which is mostly that in the form of amino acids (especially argenine) or free ammonium ions in solution. However, yeast is not so well able to utilize nitrogen from the amino acid proline, which tends to be in higher concentrations in riper grapes (Kliewer, 1968; Bisson and Butzke, 2000). Hence, monitoring YAN in grapes can be an important predictor for possible stuck fermentations. It is notable that ripe *V. labrusca* grapes have much higher argenine than proline concentrations (Kluba *et al.*, 1978), potentially leading to fewer stuck fermentations. However, the amount of wine made from this species is very small compared with that from *V. vinifera* grapes.

Secondary metabolites

The components that make a grape distinctive are loosely grouped together and termed 'secondary metabolites', because they are not core to the survival of the plant (those were mentioned before as being primary metabolites). Many thousands of secondary components have been identified and many more are probably not yet recognized. Some are found in all plants while others in only a single species at a specific time in their life cycle. Their synthesis by plant cells is under genetic control. Fifty years ago the field of chemo-taxonomy was established on the recognition that these secondary metabolites were sometimes very specific. The discovery that native American grapes and *V. vinifera* grapes had different types of anthocyanins (Ribereau-Gayon and Ribereau-Gayon, 1958) was used as a method of distinguishing between the species. Research analysing plants for their content of secondary products was used to help differentiate them and test taxonomic and phylogenetic hypotheses (e.g. Moore and Giannasi, 1994).

Once DNA analysis was discovered, chemo-taxonomy was replaced by direct comparison of the genetic message (e.g. DNA fingerprinting) rather than the products of metabolism. This was because chemo-taxonomy was not directly determined by the genetic code of the plant – there were other factors involved as well (e.g. plant age, condition and environment of the individual plant) that influenced secondary metabolite composition. Ultimately, viticulturists attempt to manipulate the secondary metabolite expression in vines as this enhances the quality (or suitability for purpose) of harvested grapes.

Within the general concept of 'secondary metabolites' are a large group of plant components called phenolic compounds. Phenolic refers to phenol, and the commonality of the group derives from a 6-carbon aromatic ring with at least one hydroxyl group (see Fig. 3.8).

Phenolic compounds are in highest concentrations at around 50 days post-bloom and then fall progressively as berries mature (Ristic and Iland, 2005). In addition, flavour compounds, or their precursors, can be synthesized in the berry at this early stage, leading to what is known as the unripe or green characteristic of developing fruit.

Some phenolic compounds are synthesized from branch pathways of the biosynthetic sequence to phenylalanine. These include the benzoic acids – para-hydroxy benzoic acid, protocatechuic acid, vanillic acid and gallic acid – all of which occur in plants. Plants alone have the full biosynthetic sequence to manufacture the aromatic amino acid phenylalanine, and animals are dependent on plants for a dietary supply of this essential amino acid. Plant reactions convert phenylalanine to thousands of aromatic (in terms of structure, not necessarily aroma) compounds, many of which occur in grapes.

The products of phenylalanine metabolism include the chemical groups of non-flavonoids and flavonoids. Major flavonoid groups in grapes are tannins (e.g. flavanol oligomers and polymers contributing approximately 0.5% by weight), flavonols (0.01%) and anthocyanins (0.1%) (Brossaud *et al.*, 1999). The group's (closely related) structures are shown on the left axis of Figure 3.8.

Anthocyanins are those molecules that contribute the bulk of red colour to grapes. The best-known flavonol is quercetin, found in high concentrations in onions, but also in significant amounts in tea, grapes and wine (Hollman and Arts, 2000). Kaempferol, myricetin and isorhamnetin are other flavonols found commonly in grapes; however their concentration is much lower than that of quercetin (Cheynier and Rigaud, 1986). The tannins are made up of compounds from the flavanol group, and contribute to grape astringency and bitterness.

A-ring and heterocyclic ring → B-ring ↓	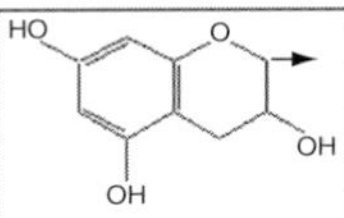 HO O OH OH catechins	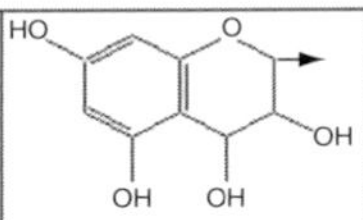 HO O OH OH OH flavan-3,4-diols	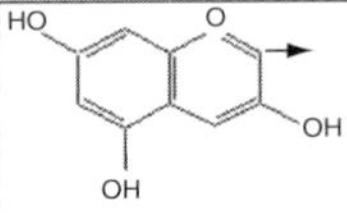 HO O OH OH anthocyanidins	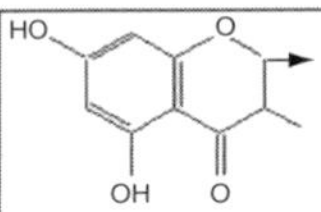 HO O OH O flavonols
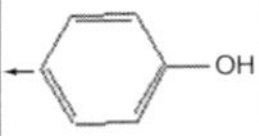 OH				kaempferol
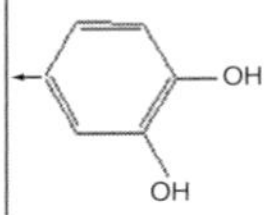 OH OH	catechins, epicatechin	leucocyanidin (as dimers and polymers terminated with catechin	cyanidin	quercetin
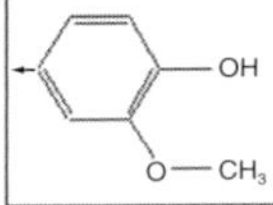 OH O—CH_3			peonidin	isorhamnetin
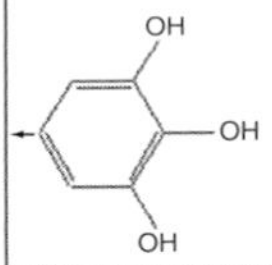 OH OH OH	gallocatechin, epigallocatechin		delphinidin	myricetin
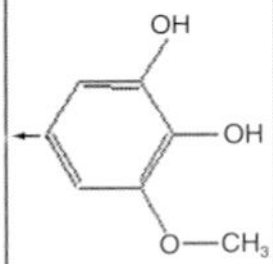 OH OH O—CH_3			petunidin	
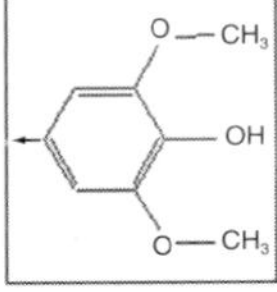 O—CH_3 OH O—CH_3			malvidin	

Fig. 3.8. Major phenolic compounds of mature grape berries. The A ring substitution runs across the top with the trivial name of the components in the crossing boxes.

In addition, flavonols and anthocyanins occur as glycosides for which there may be different sugars, which may also be in different positions on the hydroxyl groups and/or linked together to form polysides. Chemo-taxonomy of anthocyanins in wines was used to trap unscrupulous wineries in France that were diluting traditional varietal wines with higher-producing and cheaper-to-grow hybrid cultivars, through the different mixtures of anthocyanins in the species. This technique has also been used to detect the addition of elderberry juice (which also has different anthocyanins) to wine (Bridle and Garcia-Viguera, 1996).The phenolics are thus a diverse group, contributing to colour and aspects of aroma, bitterness and astringency in grapes as well as participating in many interactions with other molecules.

The non-flavonoids are smaller compounds that often associate with flavonoids, and are also important in browning reactions that occur in juice and wine. They can also contribute to bitterness and aroma, particularly in wines, though their overall effect is much less than that of the flavonoids. Examples of non-flavonoids would be gallic acid, protocatechuic acid, *p*-coumaric acid and ferulic acid. Gallic acid is a simple and easily obtainable phenolic compound, so therefore phenolic compounds are sometimes expressed on the basis of gallic acid equivalents (GAE).

Phenolics are found in all plant tissues, hence the inclusion of leaves, rachides and stems in grapes destined for juice or wine production can alter the character of the resulting product. Within the berry, however, the non-flavonoids are primarily found in the pulp, and therefore are easily extracted from the berry. Of the flavonoids, the anthocyanins are the most readily apparent in red grape cultivars and are usually found only in the berry skin. There are, however, some grape cultivars that have anthocyanins in the pulp, which are known as teinturier type grapes (see Plate 11). These will express red juice if the ripe berries are squeezed, quite a contrast to most red grapes. Examples of these grapes include 'Alicante Ganzin', 'Carmina', 'Kolor' and 'Pontac', which are usually used as a blending agent in wines to increase the intensity of red colour.

The flavonols are found in the skin of grapes, with almost none present in the pulp or seeds (Singleton and Esau, 1969), though green parts of the vine – such as leaves and rachis tissue – contain high amounts (Price *et al.*, 1996). Price *et al.* (1996) found that there was a strong relationship between the amount of flavonols in berries and the degree to which they were exposed to the sun. Spayd *et al.* (2002) showed that if berries were shielded from UV radiation the amount of flavonols produced dropped significantly, but not to the levels of shaded fruit, demonstrating that it is not just light that is responsible for the relationship. The connection between berry exposure and the amount of flavonols formed is such a strong one that the relative amount of quercetin (it being the flavonol in highest concentration) found in the grapes could be used as an indicator of the exposure of those grapes to the sun during the growing season.

Flavan-3-ols are found in both the seeds and skins, with there being qualitative differences between the two sources. Additionally, it is easier to extract (and perceive) those from the skins as opposed to the seeds, as the seeds are enclosed by a hardy covering, reducing the rate of transfer to the liquid juice or wine (Singleton and Draper, 1964). Similarly to flavonols, flavan-3-ols are found in the green parts of the vine as well (Price *et al.*, 1996).

1

2

3

Plate 1. *Vitis riparia* grapes growing wild in the North East of the United States. The berries are densely coloured and strong of flavour.
Plate 2. Grape cluster showing individual florets, some of which have their fused petals (calyptra) separating from the basal part of the flower (top and left).
Plate 3. Examples of Sultana (left; also known as Thompson Seedless. This cluster has not been grown for commercial table grape production and so the berries are smaller than those found on clusters in a shop) and Einset Seedless (right; a French-American hybrid grape) clusters.

Plate section supported by the California Table Grape Association

Plate 4. Raisins drying out in between the vineyard rows (Photograph courtesy of the California Raisin Commission).
Plate 5. Example of a mutation of berries on a Pinot blanc cluster that starts production of anthocyanins in the skins.
Plate 6. Chardonnay berries near harvest, showing the brown spots that are lenticels in the skin.

9

Plate 7. A series of photos showing early bud growth, from dormant, Stage 1 (top left); budswell, Stage 2 (top middle); wooly bud, Stage 3 (top right); green tip, Stage 4 (bottom left); rosette of first leaves visible, Stage 5 (bottom right). Stages of development are labelled according to the Modified E-L System (Coombe 1995).

Plate 8. Anthesis (cap fall) in a flower cluster. This is the first day any of the florets have opened on this flower cluster.

Plate 9. A dormant tendril that had wrapped around a foliage wire during the growing season. Tendrils such as these can be quite woody and difficult to remove at pruning, which demonstrates their role in helping to support the vine.

10

11

Plate 10. Section of shoot late in the season showing the reddish lignified periderm and the still-green tissue of the cluster peduncle.

Plate 11. A cluster and leaf from a teinturier type grape, Grand noir, taken late in the season. Not only is the flesh and juice of the berry filled with anthocyanins, but the stems and leaves also express a lot of colour.

12

14

13

Plate 12. Tendril-like cluster of berries, the result of only a small number of flowers being initiated when the tendril was being formed.

Plate 13. Close up of grapevine flowers, showing a moist stigmatic surface on the open one. Once the tip of the ovary has dried out it is no longer receptive to pollen tube growth.

Plate 14. A Black Corinth / Zante Current cluster showing the small size of the parthenocarpic berries.

15

16

17

Plate 15. Left; an example of part of a cluster that has been badly affected by inflorescence necrosis (also known as early bunchstem necrosis). Right; some clusters affected by this disorder can end up looking like this at harvest.

Plate 16. Cross-sectional cut of dormant grapevine bud following severe cold temperatures. The primary shoot tissue has been killed (left side of bud), but the secondary shoot tissues are still active and should produce a shoot in the following spring (Photograph reproduced with permission, G.O'Rourke).

Plate 17. Left; a vine with poor growth in early spring. Right; when the bark was cut away, it was clear that the phloem and cambial layer had been killed (brown tissue under bark), causing the poor shoot development.

18

19

20

Plate 18. Vine canes being de-hilled after a winter under the soil, which protects them from potentially damaging cold temperatures (Photograph reproduced with permission, K.Ker).
Plate 19. Spring frost-affected grapevine shoots. One (centre), however, seemed to escape damage and is pushing forth.
Plate 20. Yellowing of leaves caused by early season sub-lethal cold temperatures while the leaf tissues were developing in the bud.

21

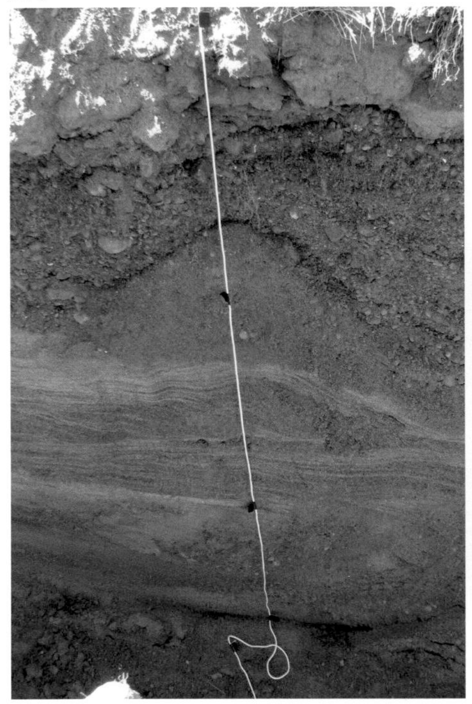

23

22

Plate 21. Soil pit used for evaluating a potential vineyard site. Note that several horizons are visible in the soil.

Plate 22. Grapevine layering, which uses a long cane still attached to a vine as the basis for a new plant. Top left; the vine has been allowed to grow a long shoot in the previous season. Top right; Lateral growth is trimmed off. Bottom left; the cane is left attached to the vine, but buried with the end sticking up out of the ground and staked. Bottom right; admiring the finished product.

Plate 23. Waxed cuttings, ready to be incubated to allow callus to form at the graft union. The cuttings are packed in boxes and kept moist.

Anthocyanidin pigments in grapes include these five: malvidin, cyanidin, delphinidin, petunidin and peonidin. However, they are not found naturally in these forms, but rather as the glycosylated anthocyanins: malvidin-3-glucoside (malvin), cyanidin-3-glucoside, (cyanin), delphinidin-3-glucoside (delphinin), petunidin-3-glucoside (petunin) and peonidin-3-glucoside (peonin). This is because the un-glycosylated form (aglycone) of the pigment is unstable (and less soluble in water), which is true of flavonols as well. In fact, adding on other molecules can further enhance stability, such as attaching another glucose molecule to the A-ring (creating a 3,5-diglucoside) or attaching a non-flavonoid such as caffeic, or *p*-coumaric acids (creating an acylated pigment) (see Figure 3.8). A peculiarity of the wine grape 'Pinot noir' is that its anthocyanins are not acylated (Boss *et al.*, 1996; Gao *et al.*, 1997), which contributes to this cultivar's wines being shy of colour.

Malvin was first isolated from grapes in 1915 (Willstätter and Zollinger, 1915), and is the pigment found in the highest concentrations (usually 40–60% of the total, depending on cultivar) (Boss *et al.*, 1996; González-Neves, *et al.* 2001). Interestingly, red-coloured clones or sports of white grapes (i.e. 'Pinot gris', 'Pink Sultana' and 'Red Chardonnay') usually contain a different primary pigment as well as possessing non-acylated anthocyanins, like 'Pinot noir' (Boss *et al.*, 1996). This probably contributes to the lack of colour stability in products made from them.

Anthocyanin production is affected by the environment, including such aspects as exposure to sunlight, temperature and temperature fluctuations. An elegant study by Spayd *et al.* (2002) in Washington State, USA demonstrated the individual effects that light and heat have on anthocyanin production in 'Merlot' berries. By cooling exposed berries to shaded berry temperature and heating shaded berries to exposed berry temperature, the effects of each could be readily examined. As a result, they found that exposure to temperatures around 35–40°C reduced the production of anthocyanins, but more exposure to light increased colour. Unfortunately, as the study was conducted in a relatively warm climate situation, the effect of cooler temperatures on anthocyanin production could not be determined.

Other studies have shown that temperature is important to the production of anthocyanins. High temperatures of 37/32°C (day/night) inhibited anthocyanin formation as well as reducing their concentration if plants with coloured berries were transferred to that environment (Kliewer, 1977). Yamane *et al.* (2006) have also shown that high temperatures inhibit colour formation in berries, but also that the stage of development at which they are most sensitive to this is just after véraison, when the berries are starting to accumulate colour. Cultivar differences are also apparent, with 'Tokay' preferring cooler night temperatures, but 'Cabernet Sauvignon' and 'Pinot noir' able to manufacture anthocyanins at higher night temperatures (Kliewer and Torres, 1972).

We know that, by changing the degree of exposure of grapes to the sun, we can alter the amount of colour they have. Is improvement in juice or wine colour as simple as increasing the amount of colour in the grapes? Probably not: in the case of wine, while abundant anthocyanins result in a more intensely red product, the susceptibility of the wine's colour to sulphite bleaching (as sulphites are commonly added to grape juices and wines to protect them from oxidation and microbial spoilage) is relatively high (Somers *et al.*, 1983). The wine is susceptible to changes in colour as a result of pH shifts (Somers, 1971) and, as the wine ages, the tendency is for the colour to fade.

We are beginning to discover what makes a deep and long-lasting wine colour. We know that improving fruit exposure also increases levels of both flavonols and flavan-3-ols (Price *et al.*, 1995; Phelps, 1999). Researchers have been getting a handle on more of the interactions happening between these different, but closely related, phenolic compounds.

Flavonols, such as quercetin, seem to be a key factor in improving grape wine and possibly juice colour, through a process called co-pigmentation. Quercetin is only a light yellow pigment, so does not contribute much to colour by itself. However, if a small amount of quercetin is added to a solution of anthocyanin pigment (such as malvidin-3-glucoside), the result is a fairly dramatic increase in colour density of the solution (Mirabel *et al.*, 1999). Figure 3.9 shows the positive effect on absorbance at 520 nm through the addition of caffeic acid, *p*-coumaric acid and the flavonol rutin (quercetin-3-rutinoside). While there is a benefit from adding these acids, the rutin provides the greatest effect in this young wine. In this case, one plus one equals three: by increasing the amount of flavonols in the grapes, more colour is visible from the anthocyanins present. This effect is somewhat short lived in wine, however, in that as time passes, the number of anthocyanin molecules able to participate in co-pigmentation decreases (Somers, 1998).

Other substances have an effect on the colour of anthocyanins. There are several metal ions that associate with anthocyanin molecules, such as Al_3^+, Cu_2^+ and Fe_2^+, which result in a slight colour shift in the pigments (Singleton and Esau, 1969). These associations may be partly responsible for the colour differences between *vinifera* and *labrusca* grapes, the latter of which have a more bluish hue to them (Ingalsbe *et al.*, 1963).

The flavanols are the third group of the most important flavonoid compounds. These include the flavan-3-ols catechin, epicatechin, epicatechin gallate and epigallocatechin as the most common in grapes (Souquet *et al.*, 1996). Their main importance lies in their involvement in the formation of tannins.

Tannins can be an important component of grape quality; they are phenolic compounds that are naturally found in grape skins and seeds, as well as in other plants (e.g. tree wood). A formal definition of tannins is: 'Water-soluble phenolic compounds, having molecular weights lying between 500 and 3000, which have the ability to precipitate alkaloids, gelatin and other proteins' (Swain and Bate-Smith, 1962).

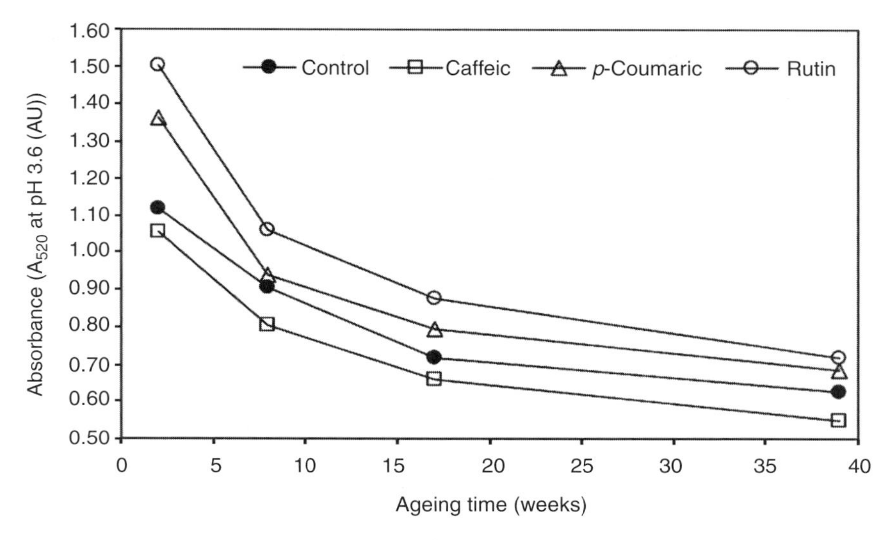

Fig. 3.9. The effect of adding various compounds to 'Cabernet Sauvignon' must prior to fermentation on the resulting wine's colour (reproduced with permission from Schwartz *et al.*, 2005).

Effectively, it is the ability of these compounds to precipitate proteins found in the saliva of our mouths that is important to the sensation of astringency. Increased precipitation of the saliva proteins translates into increased perceived astringency of the tannins (Gawel, 1998). In addition, tannins also may contribute to bitterness in grapes and, probably most importantly, juices and wines.

The basic building blocks (monomers) of grape tannins (also known as condensed tannins or proanthocyanidins) are the aforementioned catechin (C), epicatechin (E), epicatechin gallate (EG) and epigallocatechin. These monomers and their chains (polymers) are found in grape skins and seeds, as well as in green plant tissues such as the cluster stems (Sun *et al.*, 1999). The polymers can be two (dimer), three (trimer), four (quatromer) or more monomers in length; these short chains are sometimes referred to as oligomers. An example of a six-unit polymer might be represented by C-E-E-C-EG-C, which could also be referred to as a compound with a degree of polymerization (DP) of six.

Chains are formed in a polymerization process that occurs in the ripening grape as well as during the winemaking process and wine ageing (Haslam, 1980). In the grape at harvest, skin tannins tend to have a higher average degree of DP (more subunits), while those from seeds tend to have a lower DP (Cheynier *et al.*, 1998; Downey *et al.*, 2003).

For table grapes, excessive astringency through the presence of tannins can detract from perceived quality (chewing the skins or seeds of a grape will bring out the more astringent tannins). The same occurs for juice production, which may mean the product needs to be manipulated to reduce the amount of tannins (e.g. through fining or filtration). Greater maceration of the berry skins and seeds results in more phenolic extraction, so the method by which the grapes are harvested and the juice extracted has an effect on this.

In winemaking, grapes that are left on skins following crushing release the most phenolics. Red winemaking procedures take advantage of this, as flavan-3-ol monomers and polymers are extracted from the solid parts of the grapes first by water and later more and more by alcohol, assisted by heat, as fermentation proceeds. Examination of skins and seeds would show that phenolics are more easily extracted from the former as opposed to the latter (Sun *et al.*, 1999). Therefore, the composition of the grapes has a major effect on the wine product – most perceptibly in colour because anthocyanins are extracted, but also in flavours and mouthfeel effects from other compounds. In terms of sensory parameters, anthocyanins do not have mouthfeel characteristic (Singleton and Trousdale, 1992) and quercetin may have only a slight bitter or numbing attribute (Dadic and Belleau, 1973; Price *et al.*, 1995; Vaia and McDaniel, 1996), so the extraction of these has little impact on juice or wine taste or mouthfeel.

Adding complexity to this is that different types of tannins affect the product in different ways: flavan-3-ol monomers tend to contribute bitterness while the polymers tend to contribute more in terms of astringency (Gawel, 1998).

Grape aromas and flavours

With the exception of table grapes, where fruit texture is a concern, the aromas and flavours that are associated with grapes are their *raison d'être*. Grapes can be found to be bland, insipid, flavourful or overpowering to the taster, reinforcing the fact that there is a tremendously large range of characteristics and intensities in existence. From the neutral taste of 'Sultana' to the fruity muscat characteristic of 'Gewürztraminer' to the foxy flavour of 'Concord' to the acidic and bitter sensation of *V. riparia*, there are flavours to meet almost any need.

FOXY GRAPE How the term foxy came into use to describe the flavour of native American grapes is not clear. According to the Internet Wine Guide (http:// www.internetwineguide.com/structure/abwine/word.htm, accessed May 2008), it may have come about because the wild grapes that grew in North America were thought of as wild or feral by the European settlers, or that the word is a derivative of 'faux' or 'false', which was a term used to describe the grapes that came from seedlings of a desirable vine, in that they usually looked nothing like the parent. An excellent discussion on these and other possibilities for the origin of the term can be found in Pinney (1989). It seems, however, that there is no clear origin for its use.

It had been thought that the foxy character was due solely to the presence of methyl anthranilate in the grapes, but it has been shown that there is not a direct relationship between the amount of this compound and the intensity of the foxy aroma, rather, methyl anthranilate contributes more to the grapey aroma associated with them (Amerine *et al.*, 1959; Nelson *et al.*, 1977). In addition, methyl anthranilate has been found in *V. vinifera* wines (cv. 'Pinot noir'), albeit at low concentrations (Moio and Etievant, 1995).

However, being able to quantify and classify flavours and aromas is difficult, due to each person's own interpretation of how to describe them. At present, the analytical capability to separate and identify individual aroma compounds is expanding rapidly. As with other qualitative characteristics, we are moving from an era of using our senses to measure these aspects of grapes to a more quantitative one, such as through the use of gas chromatography (GC) and GC linked with mass spectrophotometry (GC-MS), the latter being a very powerful method of identifying individual molecules. The value of the human nose has not been lost to science, however, as the use of GC-olfactometry – where an instrument separates the compounds in a sample and a person measures them by rating each one's intensity and describing the quality of smell (via some descriptor, such as fruity, like a rose flower, etc.) as they emerge from the GC – is now common. This technique, in combination with parallel analysis of the sample by GC-MS, can lead to a much better understanding of how each compound matches with a smell and contributes to the overall aroma of a grape.

Aromas arise from many different chemicals, and in some cases the collective effects may be different to the individual ones. Additionally, many aroma compounds are present in grapes but are not odour-active, meaning that there is something that blocks our ability to detect them. In many cases this blocking is through glycosylation of the compound, and removal of the glucose from the molecule makes it readily detectable. The process of heating crushed grapes or grape juice enhances this process, as well as the presence of yeast during a fermentation, as they excrete enzymes that will cleave off the sugar (Delcroix *et al.*, 1994). Therefore, the spectrum of aromas and flavours in wines often greatly exceeds that found in the grapes themselves, or even the juice made from them.

In terms of primary grape aroma compounds that have been studied in detail, three stand out: terpenes, norisoprenoids and methoxypyrazines.

TERPENES Terpenes are a family of chemicals that impart fruity characters to grapes, mostly in the form of monoterpenes. Commonly found examples of these are linalool, geraniol, nerol and citronellol, which are found in cultivars such as 'Muscat', 'Sauvignon blanc', 'Gewürztraminer', 'Riesling', etc. In many cultivars, the monoterpenes dominate the ripe grape character of these cultivars. The smells they are associated with are tropical fruits, orange, rose, floral, etc. They are found in both the flesh and skin of grapes (Wilson *et al.*, 1986) and in two main forms: free (which includes some polymers called polyols (Williams *et al.*, 1980)) and glycosylated (described above). The former are sometimes referred to as Free Volatile Terpenes (FVT) and the latter Potential Volatile Terpenes (PVT).

The monoterpenes accumulate as berries mature, though the FVTs and PVTs can do so at different rates. Fruits that are not well exposed to the sun tend to have lower monoterpene levels, as do those grown in warmer areas (Ewart, 1987; Belancic *et al.*, 1997). This can result in grapes developing non-varietal characteristics, particularly when grown in excessively hot climates.

NORISOPRENOIDS Norisoprenoids are derived from carotenoids, which are generally orange-coloured pigments (e.g. as found in carrots) that assist in the harvest of light energy in photosynthesis. An example of this type of aroma compound is beta-damascenone, which has cooked apple- or raspberry-type characteristics.

Levels of norisoprenoids in grapes increase with light exposure and environmental temperature, and are found in such cultivars as 'Shiraz', 'Cabernet Sauvignon', 'Chardonnay', 'Semillon' and 'Muscat of Alexandria'. Some connection has been discovered between the levels of a carotenoid compound and those of 1,1,6-trimethyl-1,2-dihydronapthalene (TDN), which has been associated with an atypical ageing character in wines, particularly Rieslings (Marais *et al.*, 1992). As levels of the carotenoid compound go down, levels of the TDN released from the grapes increase, and high levels of TDN in wine relates to a kerosene-like character that appears in young affected wines, but is more associated with older, aged wines.

METHOXYPYRAZINES The previously discussed aroma compounds tend to be associated with the ripe end of the flavour spectrum, and their concentration tends to increase with time. The last group we will discuss, the methoxypyrazines, are associated with unripe grapes, and amounts tend to decrease with time.

Isobutylmethoxypyrazine and isopropylmethoxypyrazine are the most commonly found of the methoxypyrazines, and occur in cultivars like 'Sauvignon blanc', 'Cabernet Sauvignon' and 'Semillon'. They are an important component in grapes, as their high concentration in unripe grapes has been thought to be a deterrent to animals that might otherwise want to eat them (Guilford *et al.*, 1987), and their presence can go a long way to masking the riper aroma compounds that might be present. In winemaking it is usually undesirable to have much methoxypyrazine in the grapes as it is associated with aromas such as green capsicum or canned peas, which do not harmonize with well with the desired berry fruit aroma of ripe grapes.

This is very much a stylistic issue, as in some cases the green character has been found to be beneficial, as per the New Zealand 'Marlborough Sauvignon blanc' style. In this, the maritime climate has conspired to result in a mix of ripe aromas (e.g. from some thiols, which are sulphur-containing compounds some of which have aromas like passion fruit) and unripe (from methoxypyrazines) that has defined a new international style quite apart from the traditionally accepted style that was mostly fruit driven (Parr *et al.*, 2007).

One of the reasons why these compounds are of such importance is that the human sense of smell is very sensitive to them. Isobutylmethoxypyrazine can be detected at concentrations as low as 2 ng/l, or 2 parts per trillion (Buttery *et al.*, 1969). There has been much research into the factors that affect the rate at which it degrades as grapes ripen (Marais, 1994; Allen *et al.*, 1997; Hashizume and Samuta, 1999; De Boubée, 2004), but there is less research on what factors influence how much is created in the grapes in the first place. In the case of the former, both light and temperature seem to hasten the decline in methoxypyrazine concentration in the grapes, which is one of the reasons why grapes of relevant cultivars grown in warmer climates tend to have less of the unripe aromas. In the case of the latter, it has been suggested that methoxypyrazines are imported to the grape berries from the leaves (De Boubée, 2003), which, if true, would put another layer of complexity on its management.

In all cases, colour, flavour and aroma constitute some of the most important attributes that grapes and grape products can have. Compounds contributing to aroma (and flavour, but since the two are so closely linked, we refer to aroma only) were evolved in plants to attract pollinators or seed dispersers, and so are called secondary metabolites. These are produced only when the plant requires them, so we see them in flowers (to attract pollinators) or in fruit approaching ripeness (to attract animals that may move the seed around). In grape, flowers may or may not be scented since, though they do have nectaries, it appears that they are not always active (Pratt, 1971). Therefore, pollination by insects may or may not be of importance. However, of primary importance to consumers of grapes and grape products are those aroma compounds made in the ripening fruit.

ENVIRONMENTAL/CLIMATIC INFLUENCES

As with many organisms, the environment has a major impact on how the grapevine grows. Since the vine is a perennial plant, it exists under a set of varying conditions and, in a vineyard setting, the plants experience variation through both space and passage of time. Growth and development of the vine (and thus composition of the fruit) is modified by environment.

In order better to discuss how the environment impacts grapevines, it is useful to classify it. However, it is important to remember that the environment is an integrated system without specific boundaries and, just because we split it up into different categories, it doesn't exactly exist in that manner.

Climate has been split into three levels: macroclimate, mesoclimate and microclimate (Smart, 1982). Starting with the largest scale, macroclimate (also called regional climate) is defined as the general climate pattern as may be determined from a central recording station. A scale that could be associated with it is tens to hundreds of kilometres (Smart and Robinson, 1991).

The types of factors associated with macroclimate are (i) latitude (which influences sunlight hours, among other things); (ii) altitude; (iii) temperature and altitude versus latitude (e.g. for every increase in altitude of 180 m, there is a decrease in average temperature of 1°C); (iv) topography (large scale, such as mountains/landscapes); (v) length of growing season (time between the previous spring frost and the first frost of autumn); and (vi) quality of growing season (factors such as humidity, presence of relevant insects and diseases).

One step down from there is mesoclimate (also called topoclimate or site climate). Every factor that affects macroclimate also influences mesoclimate, but the scale being discussed is smaller, in the order of tens of metres to kilometres. Factors here include differences in elevation, slope, aspect or proximity to large bodies of water. These affect vineyard temperature, wind, rain, relative humidity (RH), frost-free period (down to 150 days, with 180 days preferred, Tukey and Clore, 1972), row orientation and slope (inclination and aspect), which can reduce frost risk, increase warmth and usually has faster-draining soils and increased air movement.

Soil type may also have an effect; for example, light-coloured soils contributing reflected light in an albedo effect.

The smallest scale of climate is microclimate, also known as canopy climate. Like mesoclimate, microclimate is a subset of the larger scale climates above it, so every factor that affects macro and mesoclimates also influences microclimate. More precisely, microclimate is the climate within and immediately surrounding a plant canopy, which means a scale of millimetres to metres.

Factors associated with microclimate include (i) bunch and leaf exposure and their temperatures; (ii) localized photosynthetic differences (e.g. sunflecks); (iii) irrigation dripper zones; (iv) fertilizer patches; and (v) plant water supply necessary for cell expansion/growth and stomatal function. Much of what the viticulturist does to vines on a seasonal basis relates best to the microclimate.

Trying to quantify climate

Once we have classified climate to help us understand it, we could also use some method of measuring the effect climate has on the growth of grapevines. A popular method of doing this is known as calculating the growing degree days (GDD) for a location.

Professor Bioletti of the University of California, USA was the first to use temperature and proximity to the ocean to divide California into six different growing regions for planting table and raisin grape vineyards in the early 1900s

(Winkler *et al.*, 1974). Later researchers found that this was not precise enough for the production of wine grapes, and set about improving the methodology used to create the regions. Using heat summation as a basis, Amerine and Winkler (1944) performed an extensive study using temperatures in areas already successfully growing wine grapes and quality parameters of the wines being produced. As a result of this, they defined five growing regions within California depending on the number of GDD accumulated. Each region produced grapes with distinct properties, even when comparing the same cultivar (Winkler *et al.*, 1974).

A GDD in this case is defined as the number of degrees the average temperature for a given day is above 10°C. This can be calculated on a daily basis and the results added to determine the number of GDD for a time period. In some cases, there may not be access to daily temperature data, so monthly average temperatures can be used, e.g. the monthly average temperature in °C minus 10°C, multiplied by the number of days in that month, with this calculation for each month summed over the growing season.

The threshold of 10°C is used for grapevines because there is little biological activity in plants below this temperature. An example of calculating the GDD for the month of February: if the average temperature for the month was 19.6°C, then GDD = (19.6−10)*28 = 269. Note that negative values of GDD are not included in the growing season summation. For a cool climate season the data look like that presented in Table 3.3. Data are usually presented in relation to the growing season, or generally April through to October in the northern hemisphere and October to April in the southern hemisphere, though in areas with a longer growing season (time between killing frosts), such as the eastern coast of New Zealand, it is longer.

Table 3.3. Growing degree day (GDD) calculation for the 2005−2006 growing season, Lincoln, Canterbury, New Zealand. GDD calculated on a monthly and daily basis are presented for comparison.

Month	Average temperature (°C)	Days/ month	GDD/ month	Cumulative GDD (monthly)	Cumulative GDD (daily)
September	9.5	30	0	0	29
October	11.1	31	34	34	81
November	12.7	30	81	115	174
December	17.9	31	245	360	379
January	16.5	31	202	562	580
February	16.4	28	179	741	760
March	12.6	31	81	821	844
April	13.6	30	108	929	953
May	9.2	31	0	929	966

Although access to monthly temperature averages is reasonably common, calculations done on a daily basis will better reflect what the vines experience and, similarly, those that integrate information on hourly, or even smaller, intervals are also available and are claimed to be the most accurate way of forecasting the suitability of a site for the production of a certain wine style (Smart, 2003). However, the availability of data with that sort of resolution is fairly limited, especially in a new region that is being evaluated for grape growing.

In comparing GDD calculations, we must know on what basis the data were calculated. For example, calculating the data on a daily basis gives a slightly different value for the same time period (see Table 3.3). In addition, because plants will begin to develop whenever temperatures are above the 10°C threshold, data from the months surrounding the traditionally defined growing season may be appropriate. In this case, for example, the cumulative GDD for the location given in Table 3.3 would be 29 plus 13 GDD higher, due to a few days in September and May (respectively) being warmer than 10°C. The late-season warm days can be important to those cultivars that are harvested at the very end of the season.

Gladstone (1992), in his book *Viticulture and Environment*, took a more detailed and physiological view of GDD and added several modifications to suit the grapevine, such as limiting the effect of very hot temperatures, as photosynthesis – and therefore vine productivity – stops as the environment becomes too extreme.

While 10°C has been settled on as a standard, it is not necessarily the most appropriate threshold temperature to use (as discussed earlier). Moncur *et al.* (1989) suggested that base temperatures for budbreak should be 4°C and for leaf appearance 7°C, based on studies with grapevine cuttings. However, for most uses, the base temperature of 10°C is adequate for the bulk of the season and, importantly, comparable with many of the data already available for many areas.

There are a number of other methods of estimating the suitability of an area for grape growing, including the mean temperature of the warmest month (MTWM) and the Latitude Temperature Index (LTI), developed by Jackson and Cherry (1988). Every method has its advantages and disadvantages, so it is important to know some of the limitations of whichever system is used.

Having an historical database of information related to the climate at any given site can be very useful when it comes to predicting when important phenological stages of grapevines (such as budbreak, flowering, bunch closure, etc.) will occur. Keeping records of these stages, along with seasonal weather information, can help to make better management decisions, saving money and time.

THE VINE AS A PERENNIAL PLANT (CARBOHYDRATE PARTITIONING)

Because the vine is a woody perennial, it is important to know something about how it survives from year to year, and possible interactions of this with management of the vines.

In the dormant season the vine survives by metabolizing stored energy. Starch is the primary storage form of carbohydrates in the permanent parts of the vine (Mullins *et al.*, 1992), and it is solubilized by conversion to glucose and fructose whereby it can be used for respiration and maintenance of the plant cells. Because there is no photosynthetically active tissue at this time, the vine is totally dependent on the stored reserves, not only to survive until the next growing season but also to invest in early shoot and leaf growth, which will then be able to produce photosynthates than can sustain further growth and, ultimately, contribute more stored carbohydrates.

As such, factors that affect the ability of the vine to make or store carbohydrates will potentially affect how the vine will grow in subsequent seasons. For example, if a vine is over-cropped, much of the carbohydrate the vine produces will go into ripening the fruit. Edson *et al.* (1993) demonstrated this with potted vines by adjusting crop load and looking at the effect of that on total vine dry weights. Figure 3.10 shows that the total amount of dry weight

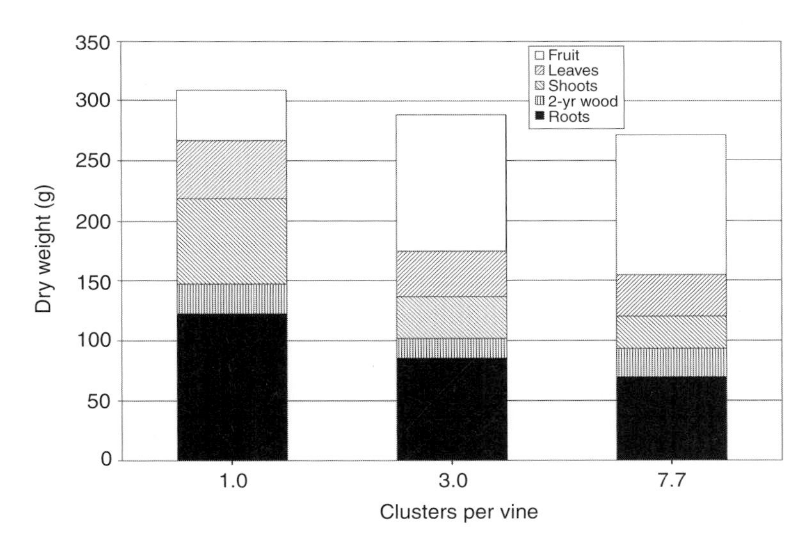

Fig. 3.10. Effect of different crop loads on total potted vine dry weight and its partitioning at the end of the experiment. Despite large differences in the amount of fruit between treatments, there is little difference in the total vine dry weights (data re-plotted and reproduced with permission from Edson *et al.*, 1993).

the vines produced (which is a measure of vine capacity) was similar (and not statistically different) across crop loads. Petrie *et al.* (2000b) demonstrated this as well, but with the added treatment of leaf removal. Results of that experiment showed that total vine dry weight was proportional to the leaf area present regardless of the presence or absence of crop, but that the presence of fruit drew away dry matter from the vegetative parts of the vine leading to much reduced leaf, root and shoot growth. Yet other researchers (Bennett *et al.*, 2005) have shown this in the field, where defoliated vines had much less soluble carbohydrate in the trunks and roots at the end of the season, which had follow-on effects in terms of the fruitfulness of the vines in the year following defoliation.

Balancing vegetative and reproductive growth

The discussion above highlights the importance of finding the right balance between the growth of vegetative and reproductive parts of the vine, given that the vine has a fixed capacity, or ability to make carbohydrates, in any given season. If there is too much crop, then the vine will decline in terms of shoot and leaf growth, and also have less energy to survive the dormant season and produce early-season shoots and roots in the following season. If there is too little crop, then vines will have excess growth that will have to be trimmed and the vineyard will not be sustainable economically.

The balance between the vegetative infrastructure and the commercial crop will vary depending on location, climate, type of grape being produced, management, etc. Therefore, there is no way accurately to predict what the numbers will be that make up the balance, and even from year to year that balance will vary, along with the weather and presence or absence of disease (and, to a lesser extent, the age of the vines).

In fact, the viticulturist's job is never finished: it is akin to walking a tightrope when there are unpredictable winds – constant, and intelligent, adjustment is necessary.

Climatic Requirements

In many areas, winter low temperatures define which grapes can be grown. The limits can be modified by lake influences, slope and many cultural practices. In other areas, the climate is too mild for easy commercial production of grapes, as their growth habit in these conditions encourages uneven cropping. Management decisions such as choice of specific cultivars, training system, ground cover, delayed pruning, burying vines in the soil, use of mulches and vine health factors such as crop load and insect and disease control (leaf health) are tools through which grapevines can be grown in less than ideal areas, potentially overcoming environmental limitations.

This chapter reviews some of the physiological requirements of grapevines with regard to their environment, with an eye to commercial production. Vines can survive almost everywhere, but managing them to the point where they produce a viable and economic crop in extreme climates can be the challenge.

COLD HARDINESS

Almost every climatic region grows grapes. Grapes are grown in extremely hot regions and very cold regions, but for a long time it has been known that some cultivars perform better in warm areas and others better in cooler areas. Cool versus cold climate locations produce different characters in the same type of grape grown, but cold climates create the additional challenge of grapevine survival through the winter period. Therefore, a realistic acknowledgment of the viticultural potential of a region is required: while it may be possible to grow grapes in a site, that does not mean that it is possible to make a living out of it.

Low temperature is one environmental factor that presents significant challenges, but also potential advantages. Cold climate viticulture is defined by the winter low temperatures in the region, a parameter that is a major concern of site selection. Data exist on the historical low temperatures in most regions likely to be planted to grapes. Figure 4.1 shows the 100-year minimum

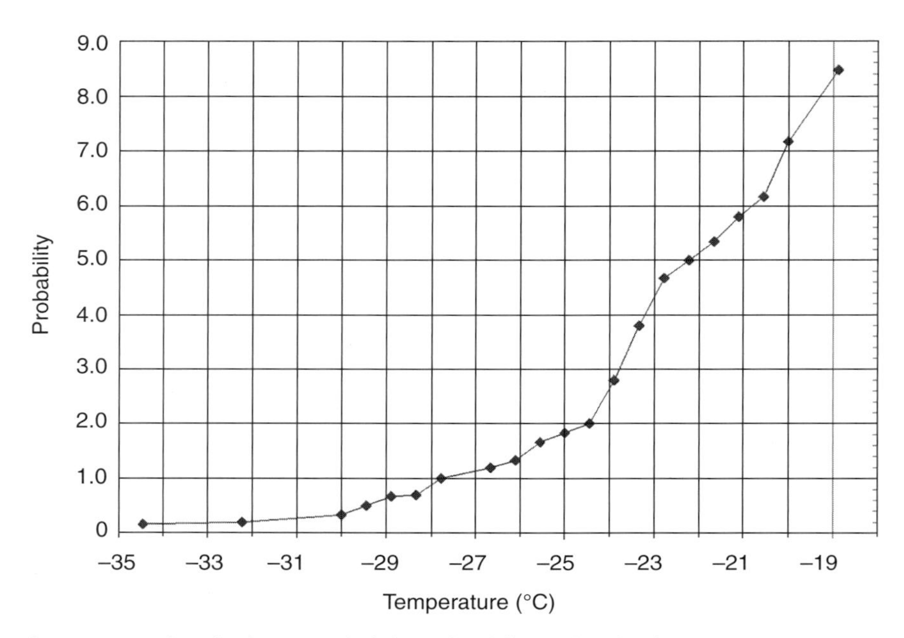

Fig. 4.1. One hundred-year probabilities for different levels of winter minimum temperatures for Geneva, New York.

temperature probability for Geneva, NY, USA. In this example, in half of the past 100 years the low temperature was −22°C or less, which is a pivotal temperature for native American grapevine survival. This information can be used to assess the degree of risk a prospective grower might be willing to take to grow a specific cultivar.

The effects of cold temperatures in the dormant season are not usually apparent until the following spring, when buds do not grow, or grow weakly and then stall. Shoots may fail to develop normally in the spring if one or more of the shoot primordia have been killed by the low temperatures (see Plate 16).

Plate 17 (left) shows a photograph of a cold-damaged vine well after budbreak. Its shoots developed normally for a short while and then faltered. Closer examination showed that the cambial layer in the cane had been killed during the winter (Plate 17, right). Since the xylem had not been affected, early shoot development was normal but, since there was no functional phloem in the cane, carbohydrates were not able to be moved to the developing shoots.

Grape species and cultivars differ in their adaptation to low winter temperatures (Pool *et al.*, 1990) and, as one might consider cool climate viticulture being defined by summer temperatures and frost during bloom, cold climate viticulture is directly associated with the minimum low temperature reached in the winter dormant period. Grape species have dormant survival ranges from a few degrees above freezing to less than 40°F below zero (−45°C) (Pierquet and Stushnoff, 1980). Grapevines enter winter by gradually developing hardiness (the ability to withstand temperatures that the growing plant could not survive). Grapes acclimate to winter temperatures at different rates: some harden off more rapidly than others and some lose cold hardiness in the spring more quickly than others (Wolf and Cook, 1992). Not all grape buds on a vine are killed at the same low temperature, so some proportion of the buds may survive to produce a crop or, failing that, allow the vine to survive and grow another year.

Grape buds that have had time to develop hardiness generally survive lower temperatures, so therefore the more basal buds (for all cultivars) and buds on canes that stop growth earlier (on early-maturing cultivars) tend to survive lower temperatures. Any environmental event that shortens the leaf activity season, such as an early frost, will delay the development of maximum hardiness (Howell, 2001). A gradual change in temperature, as usually occurs during the transition from summer to autumn to winter, is most desirable in maximizing cold hardiness, as periods of warmer temperatures can de-acclimate the buds and make them more susceptible to cold damage.

It is possible to measure the killing temperature of grape buds during development of hardiness and in the depth of winter through a freezing isotherm procedure (Wolf and Pool, 1987), which has been used to predict the potential damage caused by a freeze event. With this knowledge, viticulturists can then adjust their pruning levels to try to compensate for the damaged buds. Canes may also be collected and their buds examined microscopically for damage (Goffinet, 2004). Percentage bud survival can then be determined and pruning to a proportionally greater number of nodes delayed until the danger of cold temperature is reduced. As grapevines have compound buds at each node position (with primary, secondary and tertiary buds enclosed within one set of bud scales) and the degree of hardiness of the buds varies, the effects of some freeze events can be overcome at pruning.

The possibility of damage from low winter temperatures can also be managed. The first defence (aside from selecting a site in an area that does not become so cold) is in choosing grape cultivars known to tolerate the expected low temperatures of the region. The second defence is to protect the vines from damage by the low temperatures through avoidance. One way of doing this uses the same technique as used for spring frost mitigation (Chapter 6), as long as an inversion layer, where warm air is on top of the cold air near the ground, exists. Fans can mix the warmer upper air with the colder lower air and prevent the temperature from reaching as low as it might have otherwise done. However, if no inversion layer is present, there is no possibility of warming up the air near the vines.

Another method of managing winter cold damage is to leave multiple trunks, frequently three to five, each a different age so that if some are killed (a selection of trunk ages spreads the risk of damage) others can provide fruit during the next growing season. As the trunks of grapevines are just as vulnerable to winter cold injury as the canes and buds, if they are damaged they are susceptible to the development of crown gall (see Chapter 9), which can result in vine death during the subsequent growing season.

A more labour-intensive method is covering or burying the vines or parts of the vines. Canes trained to a low wire, or those trained to the fruiting wire, can be taken off, laid on the ground and covered with soil using a grape hoe or plough (see Plate 18). The soil provides insulation from the cold air temperatures, but is also kept warm by the large mass of soil beneath. Straw mulch applied over canes has also been used successfully for this purpose (Zabadal and Dittmer, 2003). This ensures at least a minimal number of bearing canes for the next season, although there can be a reduction in bud viability and cluster number as a result of this practice (Goffinet and Martinson, 2007).

Another approach does not protect the bearing canes but buries the vine crowns, assuring that new growth can arise from the crown. However, if there are killing temperatures, there will be a delay in getting a crop off the vines as new shoots will need to be trained up on to the trellis. Because burying protects the crown from cold damage, infection by crown gall-causing bacteria is also limited, compared with methods leaving the canes exposed.

In the methods involving burying of vines, if they are grafted it is possible that scion rooting may take place. This is where the scion develops roots from above the graft union and, as these tend to be quite vigorous, they can take over from the rootstock roots, resulting in an own-rooted vine. For this reason it is important to remove all soil from near the graft union following the winter period.

To summarize, grape production in cold climates is accomplished by (i) avoiding damage, usually by growing extremely hardy cultivars or by burying vines; and (ii) managing the damage by delayed pruning or leaving extra buds based on bud survival counting.

GROWING VINES IN TROPICAL AREAS

At the other end of the scale, vines are also grown in areas that do not experience temperatures even close to freezing. This creates some problems of its own, not because the grapevines have trouble surviving the hot temperatures, but rather because they do not produce grapes at a single time, which means batch processing of grapes is effectively impossible.

Grapevine growth is indeterminant, meaning that as long as conditions are favourable, the shoots will continue to develop (much like a tomato vine). If a grapevine is left in this state, shoots will elongate ad infinitum, with the basal part of the shoot turning woody and laterals developing at node positions. Fruit would still be produced by the vine, but largely from lateral shoots that eventually would be growing at a range of stages, with reproductive development spanning flowering to over-ripe fruit. This kind of situation is not compatible with traditional production practices, where grapes are harvested at one time and batch processed. In addition, since many disease management practices are designed to coincide with certain phenological stages, control of pathogens such as botrytis becomes problematic.

So much of the challenge with growing grapes in perpetually warm climates has to do with trying to synchronize vine growth and defend it from the wider range of pests and diseases that are present.

Several methods are used to coordinate vine growth. Drought stress, severe pruning and/or leaf removal of vines can provide the stimulus necessary to start growth of the compound buds over that of the lateral (Lavee, 1974), but these methods are labour-intensive and potentially harmful to long-term vine health.

Other methods are used, e.g. the application of chemicals such as hydrogen cyanamide (Shulman *et al.*, 1983), calcium cyanamide (Kubota and Miyamuki, 1992) or preparations such as garlic paste, which also works to enhance budbreak (Kubota *et al.*, 2000; Botelho *et al.*, 2007). Once bud development has been synchronized, commercial management of the vines becomes much more viable.

FROST TOLERANCE

Frost in this sense refers to below-freezing temperatures that occur near or after budbreak, as the vines emerge from dormancy. Green tissue, being much higher in water content, is more susceptible to freezing than woody tissues. This is illustrated by controlled-environment experiments where grapevine buds at different stages of emergence were tested for the minimum temperatures they could survive (Gardea, 1987). As shoots grow they have less and less tolerance to cold temperatures, so date of budbreak (which varies with site and cultivar) is an important factor in susceptibility to frost events.

The effects of freezing on green tissue is devastating, causing rapid cell death followed by visible translucency and then browning (see Plate 19). As with winter cold injury, not all shoots are equally affected, so damage can be patchy, in unpredictable patterns and with symptoms not appearing until well into the growing season.

Temperatures can also cause changes to shoot development without actually killing the tissue. Plate 20 shows changes to leaf colour that are caused by near-lethal freezing temperatures in the early part of the growing season, which are similar to symptoms of some types of herbicide damage.

Site selection is the best tool for managing frost risk, as it is costly to prevent and recover from. More information about frost management is given in Chapter 6.

HEAT AND LIGHT

In terms of a requirement for light and heat there is little needed for a grapevine to grow in the wild. In low light situations the vine grows upward and, once it reaches bright light it fruits, with the focus being the bringing of whatever seeds are in the fruit to maturation and then rendering the fruit enticing enough for something to take it away.

In the commercial sense, however, there is a need for heat in order that the vines have enough opportunity to produce carbohydrates that will go into both sustaining the vine from year to year and producing enough crop with adequate compositional parameters to keep the manager's business sustainable. Integral with this is the choice of cultivar, as some require more heat (and indirectly, light) to ripen to the point where they produce the flavour and aroma characteristics that are expected by the consumer. So, while there is a minimal requirement in terms of heat and light for the growth and survival of grapevines, if the goal is to produce a commercial crop the requirements become more complex.

The use of growing degree days (GDD, Chapter 3) or other heat accumulation tools is helpful in determining whether the selected cultivar and end use will match with a particular location. Ultimately, however, knowledge of what the end product is expected to be, how the vine interacts with the climate at the proposed location and to what extent management interventions are willing to be used are all necessary in making the most accurate assessment as to the success of the enterprise.

WATER USE

As for heat, the requirement of the vine for water depends on its situation. A vine can survive on very little water, but it will not be able to sustain a crop in doing so. Addition of fruit into the equation means a greater canopy area is needed to ripen it, which requires a supply of water to allow the stomata to be open and photosynthesis to occur.

Many grape-growing regions of the world rely on natural rainfall to supply water to their vines, but many are also reliant on water brought to the vines. If there is no mechanism to supply vines with water, the amount and quality of the crop is dependent on rainfall and soil water-holding capacity. If the viticulturist has a dry ripening season, then water application becomes a (potentially useful) management tool. A balance must be struck between applying enough water to maintain vine photosynthesis and fruit development without applying too much, resulting in overly vigorous vines that require excessive management.

Typical values of water use efficiency (WUE, the tonnes of dry matter produced per megalitre of water used by the plants) range from 10 to 30 for grapevines grown for wine production, though the value depends much on how the irrigation water is applied to the vines. Flood irrigation brings values down, while deficit irrigation results in higher values (see Chapter 6).

Too much water available to the vines is also detrimental to quality grape production and efficient management. Vines will grow to luxurious levels with ample water supply, to the point where fruit can be well shaded, which reduces quality (Smart *et al.*, 1988). To correct this requires many more management interventions, such as shoot thinning, hedging, leaf removal in the fruiting zone and there are carry-on effects in the dormant season, the vines requiring much more time to prune. Hence, managing the amount of water the vines have access to is important for commercial production.

SOILS

Grapevines can grow and produce crop hydroponically or in sand culture, provided that water and appropriate levels of nutrients are supplied with it. If soil serves its primary purposes – anchorage and reservoir for water and an appropriate range of nutrients (see Chapter 7), there is little direct effect of soil on grapevine growth and even fruit composition, though there is debate on this issue (Bodin and Morlat, 2006; Coipel *et al.*, 2006; Huggett, 2006; Van Leeuwen and Seguin, 2006).

However, the interaction of physical characteristics of soil and the origin of its parent material can have quite important effects on the magnitude of grapevine growth. For example, the chalk soils found in the Champagne region of France have a desirable combination of high water-holding capacity yet good drainage, porosity and permeability that makes them well suited to grapevine growth and grape production (Huggett, 2006).

The proportion of gravel in soil can have an effect on the root to shoot ratio, with high-percentage gravel soils reducing shoot growth relative to root growth (M.C.T. Trought, G.L. Creasy and G. Wells, 2008, unpublished results). This could be an advantage in maintaining fruit exposure with less canopy management.

TERROIR

A discussion of the environment and grapevine interaction must include aspects of terroir, which is a term coined by the French and for which there is no direct English translation. Perhaps as a result of this, there is much confusion as to exactly what it means, so it has ended up meaning different things to different people (Moran, 2006).

It has been attached to other food products (such as cheese), but is of much interest in the wine industry. Aspects of terroir include territorial, ecophysiological, oenological, identity, promotional and legal (Moran, 2006). Analysed in such detail it is an exceedingly complex topic and therefore one of great discussion.

At the crux of it all, however, is the idea that aspects of the environment (which can take into account the physical and social aspects) end up being associated with a unique style of a product. The style may be associated with flavour or aroma, marketing/branding or any number of other characteristics. In business, having a point of difference is key to marketing your product and retaining customers. Being able to say that your product is from a unique terroir is useful as a marketing tool as it provides a point of difference. The concept of terroir should be able to be applied to grapes in all forms, not just wine.

One difficulty in developing terroir arises from the necessity that products coming from an area considered to have a similar terroir should have some underlying common characteristic, usually some aspect of flavour, aroma, and (in the case of wine) mouthfeel. For grapes that are as heavily modified after harvest as those used in wine, the influence of the winemaker also comes into play in determining the characteristics of the product. Thus, defining a region as having a recognizable terroir is usually more than just agreeing that it comes from a specific area or land: it also depends on those making the wine having some common thread such that the wines that are made from that area do have a recognizable trait, which can form the basis of market differentiation. It is therefore easy to see that the concept of terroir can be the subject of many and lengthy debates!

CLIMATE CHANGE

Climate change is generally regarded as resulting in an increase in the global temperature and changes to the atmosphere's composition. The magnitude and rapidity with which these changes will take place is a matter of some debate but, nevertheless, it seems inevitable that, within the life of most vineyards, the conditions under which they were developed will not be the same for the life of that vineyard.

Two global changes that will occur are a rise in the average temperature and an increase in carbon dioxide (CO_2) concentration in the atmosphere. Both will have an impact on the basic process of photosynthesis. The rate of photosynthesis increases with an increase in temperature, and higher CO_2 levels mean that Rubisco, the enzyme catalysing the conversion of CO_2 to carbohydrates, will operate more efficiently. Thus, on the basis of this, we would expect to see that plants will be able to produce more biomass than they were able to under the earlier conditions. Note, however, that with an increase in temperature there is also an increase in the rate of respiration, which counteracts photosynthesis. However the increases due to Rubisco appear to outweigh the losses due to additional respiration (Chaves and Pereira, 1992).

However, it may be the changes to regional climate that could have the most practical effect on grape production. The limits of grape production are likely to move northward and southward in the northern and southern hemispheres, respectively, which seems to have been happening in Germany already (Schultz, 2000). However, some models predict an increase in the number of extreme weather events (Goodess *et al.*, 2003), which can have severe impacts on grape production: spring frosts that occur later into the season can wipe out grapevines and their crops; flooding can physically damage vines; wind can break shoots and longer times between rainfall events can cause non-irrigated vineyards to fall out of production or reduce the area that can be farmed, etc.

Importantly, there are impacts on the quality of grapes grown as well. With increasing temperature there are changes to fruit ripening, sugar and acid levels and, importantly, flavour and aroma compounds. It may be that in the future, areas that are considered as having a hot climate will no longer be able to grow grapes, and those that are considered cool will no longer be able to produce the cultivars they were well known for.

Another aspect of the environment that has changed is the amount of ultraviolet (UV) radiation that reaches the earth's surface. Normally, the ozone layer in the upper atmosphere filters out the short-wavelength UV radiation but, in areas affected by ozone layer depletion, more of this high-energy radiation affects humans and plants alike. Typical responses to more UV radiation are smaller leaves, shorter internodes and greater growth of lateral shoots (Barnes *et al.*, 1990; Keller and Torres-Martinez, 2004).

Plants do have mechanisms to cope with higher incident UV radiation, such as the formation of flavonols, alkaloids, waxes and free radical scavengers, or mechanisms to repair the damage caused to components of the photosynthetic pathway, DNA, membranes, etc. (Jansen *et al.*, 1998; Kolb *et al.*, 2003), but the impact of incident UV has already been shown to have effects on fruit amino acid content and aroma compounds (Schultz, 2000). If terrestrial UV is to increase in the future, viticulturists should be aware of the potential effects and management methods to avoid them. In some cases, research into how this is to be accomplished has yet to be done.

VINEYARD ESTABLISHMENT

Where to place your vineyard and how to bring it into production are critical questions in the process of developing a viable and quality grape production system. It is therefore important not to be hasty at this stage, as a mistake then will stay with the vineyard for the long term. Growing grapes is a high-cost venture so, to minimize risk, as much information as possible should be gathered at every opportunity. In this chapter, the process of putting in a vineyard is followed from first thoughts to making sure that the vineyard gets off to a healthy start, ready to produce as much high-quality crop as it can for the life of the vineyard.

SITE SELECTION

Probably the most important decision that is made for a vineyard is where it is situated. The site influences many factors that affect how the grapes will grow, how easy they are to manage and what kinds of grape products can be successfully produced from it. It is easy to see, then, why this decision should be thoroughly researched and well considered before choosing the spot: an inferior site results in an inferior vineyard producing inferior grapes.

In choosing the best site, then, it is important first to consider the market. Without knowing what the market wants to buy, there is no justification in putting vines in the ground. While this is not a book on market research, it must be emphasized that this information is vital to the longer-term success of any vineyard. It may be possible to get this information from a marketing body associated with the grape industry, or from a winery or packing house. From them it should be possible to get some indication of what the market is likely to want to buy by the time your vineyard is in production.

Once the market has been defined, then what needs to be produced and, therefore, where it needs to be produced can be determined. For example, if a market need for 'Sauvignon blanc' of a particular style has been identified, land in a region that can produce that style in an economic quantity should be sought. In doing this, the site with the lowest risk is a vineyard that is already known to consistently produce high-quality grapes. For example, if one were to want to produce 'Champagne', then land within the Champagne region in France must be used. However, these sites also cost the greatest amount of money, if they can be purchased at all! In some cases there will be reasons to look at planting in untested sites, where the risk is higher but the cost of the land is less.

The concept of matching the vineyard site to the grape product is an important one to work toward. This will also narrow down the potential areas that need to be investigated, and provide a basis through which to evaluate potential sites. Chapter 3 reviewed the environmental factors that are important to vine growth and productivity, and in site selection the pros and cons of a particular site must be evaluated with regard to the target outcome (i.e. grapes matching a particular set of quality criteria, refer to Table 2.2).

Now information about what will be required of a site in order to produce the target product can be gathered. For example, if the market demands 'Flame Seedless' table grapes, then a vineyard site that will be capable of growing, sustaining and ripening the 'Flame Seedless' variety is needed – there is no point in looking at a site that is too cold, does not have enough water or does not have access to the labour required to allow its production, etc.

To make the best decision about a potential site, the requirements of the variety must be determined and then compared with those that are available at the potential site. Common factors to be considered are heat accumulation, winter minimum temperatures, water availability and soil characteristics. There are modifiers that a site may also possess, such as a slope (related to air drainage and frost incidence), the slope's aspect (influencing sun exposure and temperature), proximity to a large body of water, elevation, etc.

It is difficult to say which of these many factors are the most important, as each may be essential in a particular aspect of grape production. A discussion on the most important factors follows.

Climate

The basic parameters the vines need, as laid out in Chapter 4, apply here. As temperatures are so critical to grapevine survival and production of the desired cultivar and composition of the grape, having access to long-term weather information at the site makes it much easier to make a confident decision about the site's suitability. Often, however, sufficient weather information is lacking, which increases the risk associated with developing that site. At a minimum one should be confident that enough heat will accumulate at the site to produce grapes of the desired composition, as well the knowledge that damaging winter, spring or autumn temperatures do not occur with a frequency that will mean economic ruin.

Large bodies of water can also assist in changing the mesoclimate. As water stores a lot of heat, in the winter it can increase air temperatures nearby and also act in causing air to circulate, which can prevent the accumulation of freezing air and grapevine damage (Newman, 1986; Shaw, 2001). In the spring, relatively cool air coming off the water can delay budbreak, meaning less risk of damage from early-season frosts.

The amount and timing of rainfall is relevant to the need for an irrigation system, or to aid vine spacing and rootstock decisions. Rainfall occurring close to and during harvest season can be particularly detrimental, as it can degrade fruit quality (e.g. through decreased fruit sugar and flavour concentration or increased disease incidence) and hinder human and machinery traffic in the vineyard.

Soil

Soil is an important consideration for site selection because, like climate, its basic properties cannot be altered much. The soil serves several purposes for the vine. At its most basic, it anchors the plant to the ground. However, it is also the medium through which the plant obtains the vast majority of its water and nutrients. Chapter 2 discussed what roots are, but here how roots interact with the soil is examined.

Grapevines can grow in many and varied soil types. There is no one soil type that is best for grapevines. Indeed, with the use of rootstocks a single cultivar, if grafted, can have roots that are completely different, and even then there can be particular rootstock/cultivar combinations that are best suited to certain sites. With this in mind, when evaluating soils it is best to confirm that the most deleterious characteristics are not present. These include flooding, impervious layers, excessively low or high nutrient levels or pH and insufficient organic matter. Soil testing (for nutrients and water-holding capacity at a minimum) is an integral part of evaluating a potential vineyard site, and should be performed in a comprehensive manner (Thomas and Schapel, 2003). To some extent, each of these soil-based problems can be mitigated with pre-planting management decisions.

Interacting with these are factors such as rainfall and/or water availability on the site. Irrigation water may be lacking or expensive. If the site has ample water supply for irrigation, then very sandy or gravelly soils can work well, whereas if there is not enough water for irrigation then they would not be. This is another demonstration of how complex and interrelated the components of the system can be.

So, from the overall point of view, it is very easy to identify potential locations for vineyards. What becomes more difficult is to decide which soils match which rootstocks and cultivars for which end uses. Regional soil maps, though widely available, are rarely detailed enough to provide enough information for rootstock/cultivar planting decisions. As soils can vary considerably over a short distance, having a soil expert walk the area and take soil cores as necessary is a valuable exercise, as is having a look at the soil profile in a soil pit (see Plate 21). Additional information can be gained by remote sensing techniques (Smith, 2002), such as aerial photography (both at visible and infrared wavelengths), satellite imagery or electromagnetic (e.g. EM-38) (Reedy and Scanlon, 2003) or ground-penetrating radar determination of soil water content (Huisman *et al.* 2003).

The water-holding capacity (WHC) of the soil is important in assessing a potential vineyard site. Together with the amount of annual rainfall, this provides the vines with their water supply for growth. The amount of water a soil can hold is the difference between the amount at field capacity, where any additional water drains out somewhere else and the permanent wilting point, where plants can no longer extract water from the soil. The values for these will vary depending on the soil type, with sandy soils not holding very much water but almost all of it being easily extracted by plants. High-clay soils will hold lots of water, but a larger proportion of it will be held too tightly for the roots to extract it. High WHC is not necessary for vine growth, but there must be an alternate supply of water for the vines if it is unlikely that natural rainfall will supply enough to satisfy vine needs.

Since plant rooting depth changes the volume of soil from which water can be taken, soil depth is also an important consideration. Increased depth means more potential for water and nutrient supply, which has effects on grape quality parameters (Coipel *et al.*, 2006). Note that, though vine roots can penetrate to significant depths and find water, there may be little in the way of nutrients available at those depths due to the lack of proper soil to carry them. Changes to soil depth or even texture across a site, as well as the presence of clay pans, are also causes of vineyard variability, which can complicate management of the vines due to differences in vine performance caused by the soil (Trought, 1997). The installation of tiles to drain excess water from the soil, as with perched water tables caused by impermeable layers, is important in soils that waterlog in the winter or spring as vines prefer well-drained soil (see Fig. 5.1).

Soil pH is relevant because of its influence on the availability of essential nutrients. At low pH, macroelements such as phosphorus and potassium become less available, while at high pH, microelements like iron and zinc become scarce (Marschner, 1986). *V. labrusca* vines are known to be more tolerant to acidic soils than is *V. vinifera* (Himelrick, 1991), probably because the former species evolved in an area with soils of lower pH. Soil pH can be altered through the addition of lime (increase) or gypsum (decrease), but these are temporary measures as the soil pH will tend to revert to its natural state over time (Himelrick, 1991). Nutrient status of the soil can also be changed, and soil organic matter is a large contributer to soil WHC as well as nutrient-carrying capacity. For more details on these aspects, see Chapter 7.

Fig. 5.1. This wet spot in a springtime vineyard is a problem for machinery and human traffic, as as well as keeping vine roots in anaerobic conditions.

If a vineyard is being replanted there can be issues with carry-over effects (Westphal *et al.*, 2002). These can range from depleted soil nutrients, nematodes or insect populations to pathogens left in the soil and remaining plant roots. Proper soil testing and preparation is essential to avoiding these problems.

Slope and aspect of land

In many areas, higher-quality grapes can be grown on sloping land (see Fig. 5.2). In cooler climates there is a benefit in terms of water and air drainage, as well as increased sun exposure if the slope faces the noon-time sun. However, sloping land also brings difficulties with machinery access and vineyard uniformity. There are many examples of very successful vineyards on both flat and sloping land but, in areas that are pushing the boundaries of economically viable sites, slopes are generally regarded in a positive way.

Other factors

Other factors to take into account include land cost/availability/zoning, likely pests and diseases, proximity to existing industry infrastructure, presence of skilled and unskilled labour, accessibility to markets or shipping ports, reputation of the region, pollution concerns from neighbouring activities or

Fig. 5.2. Sloping vineyard site in Oregon, USA. Excellent air drainage keeps cold air from accumulating near the vines.

previous cropping, water availability and quality, natural rainfall and its timing through the year. As some vineyard activities – such as bird-scaring, frost-fighting, early morning fungicide spraying, etc. – may not be expected by neighbours, it is best to resolve these issues before planting the vineyard to avoid conflict.

Knowledge of other plants and their phenological stages in relation to grapes can also help predict the performance of vines not yet planted. If there are reliable data for time of budbreak, flowering and ripening in grapevines and other crops, including ornamentals, in one region, the relationship between phenological dates of those other crops and grapevines can be used to predict when grapes will reach certain stages of development. For example, if it is known that in one region grapevine budbreak coincides with first bloom of 'Red Delicious' apple, then if that same apple is growing in another area that is being evaluated for grape production and the date that the apple trees bloom is also known, this leads to a good predictor of when budbreak will be for grape in that area. Advance knowledge of all the important phenological stages can be estimated using this technique, leading to a better risk assessment for the type of grapes needed to be produced. Unfortunately, the detailed information needed to be able to use this technique is rarely available.

In scouting new vine-growing areas, the use of aerial photographs to look for vegetation differences, Global Positioning Systems (GPS) for accurately marking out sites and elevation determination, and Geographical Information

Systems (GIS) to help organize the data into a useful form and plan out blocks, is increasingly common (Smith, 2002).

Being thorough and methodical in investigating a site is vitally important, but predicting its potential to produce quality grapes, even with the increased level of knowledge and experience available today, is still quite difficult. However, it is worth this effort to try to get it right since, once the vineyard is planted, alterations to many aspects of production become almost impossible. It is helpful to remember that management inputs can overcome many deficiencies a vineyard may have, but this may result in decreased fruit quality or increased site costs through the life of the vineyard.

At the end of all this, there are two questions to answer: (i) are there one or more factors associated with the site that will prevent it from growing the grapes the market needs? (ii) if so, then what grapes could be grown, how best should they be managed and will it be economic?

The alternative is to investigate another site. However, if the site is appropriate, then it is time to move to the next stage of development, determining how the site will be planted out.

SITE PLANNING

A vineyard site does not have to be rectangular, flat and with uniform soils. If a site is like this, then planning out the vineyard is a much simpler exercise! Most blocks require at least some compromises to be made in order to juggle all of the requirements of a vineyard. In addition to the grapevines themselves, provision may need to be made for headlands, access roads, buildings, washing pads, storage sheds, break rooms, offices, etc. All of these take up valuable space, and so their placement should be planned to take greatest advantage of the land.

To assist with planning this, it is useful to go back to the information gathered before and decide where the best locations will be for the grapevines. Land that is less suited to grapes is better used for other supporting structures such as buildings or access ways. It is common to develop some types of vineyards on slopes and, indeed, there are often a number of advantages to doing so. It is important to use the soil information to help with the placement of blocks, as erosion and potential vigour of the area are two important factors that impact on future management.

The block shape and size may be determined by a number of factors that may or may not be controllable. It can be defined by soil changes, access roads, source of irrigation water, row length, slope, etc. For many, minimizing the amount of variation within a management block is a desirable goal (Long, 1997). Thus, planning out blocks so that their management will be as uniform as possible helps in obtaining consistent results from season to season.

Row orientation for most vineyards will run somewhere near north to south to allow for even ripening and fruit exposure on both sides of the canopy.

However, east–west-running rows can be beneficial in raisin production, for example, for if the grapes are to be dried in between the rows, they obtain sun exposure for most of the day. In some cases, rows will run along the contour of a hill, to minimize erosion or facilitate use of machinery (see Fig. 5.3). Row orientation can also be changed to account for site-specific factors, such as in areas with more rainfall, running the rows in a direction that enhances morning sun exposure in the canopy to help dry out the fruit. Likewise, maximizing canopy exposure in the mornings rather than in the afternoons can be beneficial, as the latter tend to be hotter. This can be an advantage in a climate that is a bit too hot for the variety being produced, as it lowers fruit temperature and modifies the development of flavour and aroma compounds.

Preparing the site

Prior to planting, you may need to clear the land. If possible, do this in the season before you intend to plant and establish a cover crop in the meantime (see Fig. 5.4). This may help to reduce pest (e.g. nematode) and disease (e.g. root rot) populations and, in particular, will allow the control of problem weeds before the vines are established (e.g. cultivation and herbicide use in the season before planting will reduce weed seed populations (Brenchley and Warington, 1933; Schreiber, 1992)). If the land has been used for other crops, it is worth checking to see whether there could be carry-over effects from those to

Fig. 5.3. An example of a vineyard where the rows are contoured along the hillside instead of running straight.

Fig. 5.4. Mixed plants on a future vineyard site. Planting a cover crop prior to vineyard establishment can decrease weed problems in subsequent years, modify soil nutrients and structure and, in some cases, provide an interim cash crop.

grapevines. In some cases, microorganisms or their products build up in the soil and can hinder the establishment of the vines (Westphal *et al.*, 2002). This is also the stage to make any soil changes, such as ripping, terracing, adjusting nutrients or pH, etc. or to improve drainage through the laying of tiles.

If a source of water is needed at the site, then a well should be dug to ensure water supply before plans are finalized. If the yield of the well is below expectations, then a redesign of the vineyard may be necessary to accommodate it (e.g. incorporation of a storage pond, change to rootstock choice or a reduction in the number of vines grown). If water is necessary for grape production in an area under consideration, then purchase of the land should be conditional on the availability of water.

Shelter from wind

If the area where the vineyard is to be established has high winds, it is worth considering establishment of windbreaks (see Fig. 5.5). Compared with vines growing without protection from the wind, those sheltered by windbreaks exhibit (i) better budburst; (ii) larger leaves with more stomata; (iii) more clusters per shoot; (iv) longer internodes and longer shoots; (v) higher pruning weights; (vi) increased capacity to carry and ripen a crop; (vii) increasing

Fig. 5.5. A living windbreak in Canterbury, New Zealand. In this case, the trees in use are poplars. Another effect of windbreaks is evident here, too: the vines closest to the trees are noticeably smaller than the ones further in due to the trees competing for light, water and nutrients.

phenolic content in the fruit (Lomkatsi *et al.*, 1983; Dry *et al.*, 1989; Bettiga *et al.*, 1996; Smart, 1999; Neel, 2000).

Sheltered vines are also less susceptible to damage caused by whipping of the shoots and breakage (usually snapped off at the base early in the season). Sometimes loss of shoots can result in a potential 10–15% loss of productivity (P. Evans, New Zealand, 2000, personal communication). Studies that have compared the productivity of sheltered versus non-sheltered vines tend to show around a 15% increase in yields (Lomkatsi *et al.*, 1983; Bettiga *et al.*, 1996). One South Australian study (Dry *et al.*, 1989) suggested that yield gains are from increases in flower cluster number rather than from percentage fruit set or berry size, although gains in percentage fruit set could be envisaged in other situations.

Ideally, a windbreak is permeable to the wind so that the air moves both through and over the top of the obstacle (Cleugh, 1998). This slows the speed of the air in the vineyard and also provides protection from the fastest air, which is shunted over the shelter. The latter is provided for a distance of up to 20 times the height of the windbreak (Marshall, 1967), so windbreaks need to be high enough to justify the area of land that they take up. However, with their use comes competition with grapevines for water, nutrients (Cleugh, 1998) and even sunlight due to shading, especially at latitudes closer to the poles (Abel *et al.*, 1997).

Windbreaks can be made of artifical woven nets, but more often they are grown. Judicious use of windbreaks can help in areas that are frost prone by preventing cold air from moving into a vineyard area from above (deflecting it away), though this precludes the use of deciduous plants in their make-up. However, deciduous tree shelters better allow cold air movement through them in the dormant season, which can be helpful in frost-prone areas as well.

Windbreaks do require extra effort and cost to establish and maintain, take up land area that could be used to grow grapes and also can be a nesting site for pest bird species; however, in certain areas these drawbacks are countered by increased production and greater vine productivity.

Vine planting density

One of the early decisions that will need to be made in planting a vineyard is: at what density will the vines be planted? Across the world there is the whole gamut of spacings, from very high density, e.g. 1 m between vines and 1.1 m between rows (for example, Château Lafon-Rochet, Saint Estèphe, France has a planting density of 9000 vines/ha, which works out to this spacing) to low density 3 × 3 m systems (1100 vines/ha) or even greater spacings for some dry-land vineyards. How is the vine spacing determined?

In earlier times, vine spacing was determined by who, or what, was going to be working the soil and the vines. In days when people did all the work, it is said that vines were planted at densities that equated to tens of thousands of vines per hectare. Later, when animals began to be used for tilling the soil, between-row spacings were determined by how wide the animals were. More recently, especially in the New World areas, vine spacing has been determined by the width of machinery that was available to work the soil and plants (see Fig. 5.6), which were generally large-scale agronomic crops.

Another factor involved is the availability and value of the land. In Europe, land was in short supply given the number of people it needed to support, hence it was used intensively (a higher plant density means greater productivity per unit area). When people moved to the New World land was in abundant supply, so wider row spacings were seen as being convenient.

From the point of view of the vine, however, things are a bit different. In establishing the vineyard, row spacing should be determined by the predicted vigour of the site. In an area where there are no existing vines, this can be nothing more than an educated guess, which can sometimes lead to those guesses being wrong. The notion went that as the predicted vigour went up, so did the planting densities. The thought behind this was that, as the number of vines per unit area increase, the resources available to each vine (e.g. water, nutrients) decrease, thus leading to less vigorous vines.

This appeared to work in the Old World areas, such as practised in many Grand Cru vineyards, and so was emulated in many other, newer, grape-

Fig. 5.6. An example of a vineyard with wide row spacing, necessitated by the desire to use large machinery. Here, weeds are being controlled with an in-row mechanical hoe.

growing areas. However, the result in these new areas was sometimes very different, in that the vines remained very vigorous, and eventually declined into vinous messes producing poor-quality fruit. It is thought that this was caused by the different soils that are typical in many of the newer areas, in that they tend to be more nutrient rich than the farmed-for-centuries soils in the Old World and, in many cases, the soils are much deeper, leading to luxury levels of water and nutrients being available to the vines.

So the high vine density to combat vine vigour hypothesis proves not to be a rule, and brings us back to the question of how to determine planting density.

One of the current theories is that between-row spacing should be determined by the availability of equipment used to work the vineyard, and within-row vine spacing is the tool to really influence vine vigor. Closer vine rows result in more efficient use of the land area (meaning that if vines are cropped to the same level per unit of vine row, the total tonnage of fruit per unit of land for a narrow row-spacing vineyard is greater than that from a wide row-spacing vineyard), so the goal should be to have the closest row spacing that is still compatible with the equipment.

Within-row vine spacing will vary depending on the predicted vigour of the site (and rootstock/ cultivar combination) and how it's managed. Again, there are many different vine spacings used in practice, and there will be no one spacing that is ideal for any given situation.

For most vineyards using discreet rows of canopies, row spacing will approximate a 1:1 ratio with canopy height, which ensures that between-row shading is minimized. Figure 5.7 demonstrates one of the problems with close, but tall, rows. Therefore, if the vineyard has very close row spacing, the canopy height must be low to prevent between-row shading, and so the height of the fruiting wire must be low, too. Working with low fruiting wires has challenges in terms of equipment and labour, as the vines are so close to the ground working with them is very difficult. Row spacing is also influenced by trellis type used, with divided canopy systems requiring wider rows.

Fig. 5.7. Vineyard with close row spacing, but high fruiting wires and therefore tall canopies. Self-shading here is limiting the productivity of the vines as well as having a negative effect on fruit composition.

ROOTSTOCKS

The choice of rootstock is another pre-planting decision that lasts for the life of the vineyard, and so should not be taken without careful consideration. Rootstocks allow vines to perform better in soils of a particular type (e.g. high pH or very deep and fertile) or in those with an undesirable infestation (e.g. of phylloxera or nematodes). Because the producing part of the vine (the scion) is dependent on the rootstock for all things coming from the soil, there can be a considerable effect of rootstock on vine performance.

These effects can be categorized as being direct or indirect (Striegler and Howell, 1991). An example of a direct effect would be root production of plant growth regulators (e.g. cytokinins) that change shoot growth or vine fruiting. Other direct effects would be the ability of the roots to take up water and nutrients and to store carbohydrates and nutrients.

Indirect effects centre on follow-on changes caused by the direct effect: for example, increased uptake of water and nutrients will cause more luxurious vegetative growth and shading, which can lead to poor fruit quality. The fact that a cultivar on a particular rootstock has lower Brix, higher acid and less ripe flavours may not be caused by the rootstock itself, but the fact that the rootstock influenced crop load and canopy density, which led to the sub-optimal fruit composition.

The parentage of a rootstock has an influence on its characteristics. Historically, examples of commonly used species for the production of rootstocks included *V. rupestris, V. riparia, V. berlandieri, V. champini* and *V. vinifera*, while some were selections of a particular species. Most rootstock crosses were performed in the mid- to late 19th century in response to the introduction of phylloxera to Europe. Unlike the producing cultivars used, rootstocks are still commonly dioecious (having separate male and female plants), since the regular production of grapes is not a concern for them.

Selection criteria for rootstock breeding programmes were originally phylloxera tolerance and, importantly, ease of grafting to *V. vinifera* scions. Hence, *V. vinifera* was used in some crosses, though its susceptibility to phylloxera was the reason the other grapevine species were being used. AxR1 was one such rootstock developed (a cross of *V. vinifera* × *V. rupestris*), which performed well in Californian trials due to its high productivity (Lider, 1958). However, despite the fact that its resistance to phylloxera was already in question in Europe, as included in the report by Lider (1958), the grape and nursery industries used it widely as it did perform well in the field and grafting houses (Wolpert, 1992). However, over time and through the process of natural selection, a more aggressive strain of phylloxera was able to overcome the effects of the rootstock and cause the vines to decline. Eventually the vines had to be pulled out and replaced, at great cost in time, money and lost production (Himelrick, 2001).

The reason the rootstock failed was that its *V. vinifera* parentage tolerated a high population of phylloxera, which led to more opportunities for a natural mutation in phylloxera to occur. This eventually led to the appearance and identification of the new strain, or biotype (Granett *et al.*, 1985). Ironically, it was noted by Wolpert (1992) that Perold, in 1927 (Perold, 1927), described the similar downfall of AxR1 in South African vineyards due to phylloxera. In that 1927 publication he postulated that a 'new biological race of phylloxera has evolved' on the AxR1 roots, which predated the first confirmation of its occurrence by 58 years. There are now several different biotypes of phylloxera characterized, cautioning the grower again not to rely on a single rootstock, and to be informed about the potential problems that could appear during the life of the vineyard. Indeed, recent research suggests that native American rootstock species may also, one day, succumb to a new phylloxera biotype (Granett *et al.*, 2007).

Rootstock choice

Rootstocks that impart greater vigour to the scion are often used in situations where high crop loads are needed, and those considered to be devigorating used in areas with higher predicted potential vine vigour. However, there are no dwarfing rootstocks available for grapevines in the same way that there are for apple trees, for example. So, although there is a range of rootstock performance, giving recommendations for a given site is highly problematic, as each will perform differently according to their environment, the scion that is grafted on top, management and the end use of the grapes that are to be grown.

In the case of rootstock influence on the growth of the scion, a major management influence is through the use of irrigation. In a dry region, rootstocks with a steep root angle send roots further down into the soil profile and tend to find more water. Hence these rootstocks perform better in non-irrigated areas compared with shallower-rooting rootstocks. If irrigation is part of management, then both types of rootstock have the ability to perform well (McCarthy *et al.*, 1997). This highlights the importance of determining how a vineyard will be managed prior to planting as well as the challenges in interpreting the results of rootstock trials from around the world.

A look through the literature shows that the best performing rootstock changes according to these factors, so the final choice of rootstock should be determined with the best local knowledge available. In looking through much of the published results, some rootstock characteristics are reasonably common, however, and information for some rootstocks is presented in Table 5.1. It must be emphasized, however, that actual performance varies as described above.

It is worth noting that there is probably not a single, *right*, rootstock for any particular vineyard. It is particularly difficult to choose the best rootstock without having tested it at that site, and a rootstock may or may not perform well in producing the end product desired or under a certain management

Table 5.1. Selected rootstocks and related information.[a]

Rootstock	Parentage	Vigour	Drought resistance	Tolerance to lime	Tolerance to Phylloxera	Tolerance to nematodes	Tolerance to salinity
Freedom	1613C × *champini*	Moderate–high	Moderate–high	na	High	High	Moderate
Harmony	1613C × *champini*	Moderate–high	na	Moderate	Moderate–high	Moderate–high	Moderate
SO4	*berlandieri* × *riparia*	Moderate	Low	Moderate	Moderate–high	Moderate–high	Moderate–low
420A	*berlandieri* × *riparia*	Moderate–low	Low	Moderate	High	Moderate	Low
Ramsey	*champini*	High	Moderate–low	Moderate	Moderate–high	High	High
Riparia gloire	*riparia*	Low	Low	Low	High	Moderate–low	Moderate
44–53	*riparia* × (*cordifolia* × *rupestris*)	Moderate	High	Moderate	High	Moderate–high	na
101–14	*riparia* × *Rupestris*	Moderate–low	Moderate	Moderate	High	Moderate	High
Schwartzmann	*riparia* × *rupestris*	Moderate	Moderate–low	Moderate	High	Moderate	Moderate–high
3309	*riparia* × *rupestris*	Moderate–low	Moderate–low	Moderate	High	Moderate–low	Low
St George	*rupestris*	High	Poor	Moderate–high	Moderate–high	Moderate–low	Moderate
1616	*solonis* × *riparia*	Moderate	Moderate–low	Low	Moderate–high	High	na
AXR1	*vinifera* × *rupestris*	Moderate–high	Moderate	Moderate–high	Low	Low	na
99R	*berlandieri* × *rupestris*	High	Moderate–high	Moderate–high	High	Moderate–high	Moderate–low
5BB	*berlandieri* × *riparia*	Moderate	Moderate–low	High	High	Moderate–high	Moderate

[a]Sources of information: Anon (2007); Anon (2008); Arbabzadeh and Dutt (1987); Downton, 1985; Hardie and Cirami (1988); Hoskins *et al.* (2003); Lider (1958); McCarthy *et al.* (1997); Pouget and Delas (1989); Southy (1992); Stafne and Carroll (2006); Whiting and Buchanan (1992); Whiting *et al.* (1987).
na, not available.

regime. For these reasons, it is best to choose several rootstocks to plant out a vineyard to spread the risk. If the vineyard needs to be redeveloped in the future, the best performing rootstocks can then be used.

PROPAGATION

Grapevines can be propagated from seeds, cuttings, layers, grafts or by meristem culture. Sexual propagation, as for other plants, is through the formation of seeds. The genetic complexity of the genus *Vitis* assures that the vast majority of grape vines grown from seed will be new and unique plants. Therefore, all named varieties of grape are propagated asexually to ensure that the progeny have the desired properties of the original plants.

Growers can choose to buy their vines from a nursery, propagate their own vines asexually or even produce vines for sale. The choice depends on the availability of labour and plant material, what kind of vines are required (e.g. own-rooted or grafted) and the advisability of an additional enterprise. In most cases sufficient vines are required in order to make their purchase from an established producer of vines entirely justifiable.

Layering

Difficult-to-root cultivars can be propagated by layering (see Plate 22). Long canes still attached to the mother plant are partially buried in the row where a new vine is wanted, with the end of the cane protruding from the soil. The buried part of the cane roots readily and, when the new vine has developed a sufficiently large root system to be independent, the connection to the mother plant can be cut and the vine left to fill a gap in the vineyard row (or dug and planted elsewhere). Layering is a commonly used technique for filling empty locations in an own-rooted vineyard due to its ease and the fact that the young vines get assistance during establishment from the mother vine.

Cuttings

A cutting is a piece of a parent plant that will develop into a new plant when provided with an appropriate environment. Cuttings can be made from dormant canes or green shoots, although in practice most are from the former.

If own-rooted vines are needed, easy-to-root cultivars can be planted directly in the vineyard. Dormant unrooted cuttings can be placed into holes made with a steel rod. It is important to remember that the cuttings will need access to adequate water and freedom from weed competition in the early stages, as they have to develop an entire root system. Alternatively, the cuttings can be grown in a nursery for a season, dug out in the winter and planted in the vineyard as dormant rooted cuttings.

Grafting

The previously discussed methods produce own-rooted cultivars, normally only used in phylloxera-free regions or with phylloxera-resistant cultivars. Most vines planted in major grape-growing areas are grafted due to the presence of a soil pest or to take advantage of rootstock effects on scion growth.

The majority of commercially produced vines are propagated by bench grafting. A rootstock section of about 35 cm long is cut from dormant vines of the selected rootstock mother plant and all buds removed either manually or by a disbudding machine, which uses stiff-bristled brushes to remove the buds. The scion section is cut from healthy vines of the selected variety in lengths of approximately 35 cm for ease in handling but, before the actual grafting takes place, they will be cut into individual nodes.

If cuttings are not to be used immediately they should be buried in moist soil or treated with a botryticide and stored in high-humidity refrigerated storage near, but not below, 0°C, as the fungicide used may reach toxic levels if the water on the vines turns to ice (Nicholas *et al.*, 1992). These cold and moist conditions prevent desiccation of the cuttings, development of disease organisms and slow respiration in the tissues.

Before grafting the rootstock and scion, cuttings are soaked in water and the scion wood cut into one-node sections. These are matched in diameter with a rootstock cutting and the ends slotted with a special blade, giving a complementary 'V'-shaped or omega (Ω)-shaped cut. The two pieces when fitted together provide optimum contact between the cambia of the rootstock and scion, which will then 'knit' together the two pieces of wood. They may or may not be wrapped with special tape or dipped in wax to keep the union from drying out or coming apart too easily (see Plate 23).

The joined cuttings are placed in a moist medium (e.g. perlite, sphagnum moss, sand) and incubated at approximately 28°C for up to 4 weeks to encourage the development of undifferentiated cells (callus) at the graft union and roots at the cutting's base (see Plate 24). When the grafts have healed the cuttings are unboxed and can be planted outside, closely spaced. Maximum rooting will occur if the vines are well irrigated and the soil covered with plastic to prevent weed competition. Typically, the grafted vines are grown for a full season in the nursery, and in the winter dug out mechanically by undercutting, graded as to quality, roots trimmed and then bundled for delivery. Although they can be stored, the best results are from planting as soon after digging as possible.

Another grafting technique commonly used to change varieties in an established vineyard is field grafting. In this procedure, a small portion of a dormant cane (that includes a bud of the variety desired) is cut so that its cambium can be closely matched to the cambium on the trunk of the parent vine (see Fig. 5.8). Methods of doing this include T-budding and use of a cleft graft. This method has the advantage of enabling a vineyard to be converted to another variety quickly with very little loss in production, due to the existing and established vine root system backing up the newly grafted scion. However, it also has the disadvantage of its success rate being highly dependent on the weather and skill of the team doing the grafting (Nicholas *et al.*, 1992).

Other methods

Green, softwood cuttings can be rooted in mist beds in a greenhouse. An advantage of this is that many cuttings can be rooted in a small space, but it

Fig. 5.8. Top-grafted vine, showing the developing new shoot and the white tape used to keep the cambia of the cane and trunk closely associated as well as decreasing water loss.

does require special facilities and the softwood cuttings are more susceptible to disease than dormant cane cuttings (Dirr and Heuser, 1987). This method can also be used to produce grafted vines. The rootstock is rooted in a mist bed and then a section of the scion variety grafted to the rooted understock. These are grown in pots under controlled conditions until acclimated and then planted directly in the vineyard.

A new variety or clone that is originally selected from a single plant is propagated asexually to provide a small number of vines for further testing. The rapid production of large numbers of vines for commercial planting can then be accomplished by growing some of the vine tissue in sterile culture through micropropagation. Following the induction of many shoots in the culture tubes, each is harvested and transferred to sterile test tubes containing growth media. After rooting, the buds are acclimated to greenhouse conditions and then transplanted. These 'test tube plants' can be further acclimated to nursery conditions and then planted in the vineyard.

It is notable that vine health and, increasingly, pedigree, is an important consideration in the production or purchase of vines. Vine stock that has any of a multitude of diseases can result in low vigour and performance or, at worst, complete failure of vines (see Chapter 9), so using certified stock is a worthwhile investment.

PLANTING

Many of the usual vineyard structures, such as end posts, intermediate posts and wires, may or may not need to be put in prior to planting. The advantages of doing so are primarily those relating to cash flow but, in practical terms, it depends on how you plan on planting and training the young vines to the wire. Most mechanized methods of planting vines are incompatible with trellising structures, for example, so will need to be erected afterwards.

However, it is vitally important that all relevant site preparations have been completed prior to the vines going into the ground. If the site is not ready, it is best to delay planting until it is. Vines have small root systems and will not survive unless the soil has been prepared, weeds controlled and water supplied directly to them. An example of ensuring the latter would be matching irrigation emitter position with vine position. Many modern irrigation laterals have built-in emitters spaced at equal intervals along the tube. If these do not match up with the vine spacing, some vines can be left with the nearest emitter being too far away to supply water to the roots, leading to poor vine establishment or vine death. It is worth noting that irrigation tubing shrinks when cold water is passed through it, leading to a substantial change in overall length for long runs; this should be accounted for when laying out irrigation laterals.

Vines should be handled properly to ensure good establishment and a minimum of young vine deaths. Prior to planting the soil should be thoroughly watered, so that the soil is moist but not wet. Common ways to plant vines are with a mechanical furrow planter, an auger or waterjet (see Fig. 5.9), or by hand with a shovel. The use of tree planters is not recommended, as correct positioning of the roots is not possible: roots should not be bent upwards in the hole or furrow, as this discourages root growth and can result in root diseases, such as blackfoot (see Chapter 9) becoming established. Waterjets have the advantage of settling the soil closely to the roots and also providing water right where it is needed at planting; however, this method may not work in all types of soils (Hoag *et al.*, 2001).

In soils with a high clay content, the use of augers is not recommended due to possible glazing of the sides of the hole. This effectively creates a barrier to root growth beyond the hole created, leading to stunted vine growth. In other situations, however, this method is a good alternative to others (Fig. 5.10).

Once planted, the vines should be watered immediately, and often, depending on the condition of the soil beforehand. Record keeping is also

Fig. 5.9. Example of a recently planted vine with use of a waterjet. Note the abundance of water available, which also helps settle the soil around the roots.

Fig. 5.10. An augered hole for planting a vine. Note the friable sides of the hole, which will allow vine roots to grow unhindered.

important, as any last-minute changes to the original plans may be easily forgotten. Ensure that information as to variety, clone and location are all recorded in a safe place and perhaps even on the end posts, to avoid confusion by workers later on.

VINE ESTABLISHMENT

There has been considerable debate in viticultural circles about the best way to get the vines established and on to the trellising system. One of the basic decisions to be made is whether or not to train up the vines in the first year. The sprawling method (see Fig. 5.11) allows the vines to grow wherever they like on the ground, and little is done to them in the first season, aside from sprays to control disease. The advantages of this may be increased photosynthetic productivity of the vines compared with those trained up and with a single shoot, as the leaf area is greater in the former case. This is beneficial as the vine has more carbohydrates available, which means more shoot and root growth. Also, the trellising system does not have to be installed before planting, which creates less work in pre-planting preparations and increased flexibility in moving about the vineyard (e.g. for cross-cultivation). The downsides are that it is nearly impossible to control weeds when they are intermingled with the

Fig. 5.11. A recently established vineyard showing the vines without any supports, leading to a sprawling growth habit.

grapes, disease control can be problematic and the quality of canes produced may not be as great (e.g. possibly not as straight or as long as vines trained upward). To counter the weed problem some have used polythene, which controls the weeds and has the additional benefit of conserving moisture for use by the vine (see Plate 25). However, disposal of the plastic in later years can also be a problem.

The alternative to allowing the vines to sprawl is to train them up. This can be achieved through the use of individual stakes for each vine, or installation of end assemblies (or at least end posts), the fruiting wire and a fraction of the total number of intermediate posts, providing something to tie or secure other vine support to. Training the vines up is thought to be beneficial because the vine grows more quickly upward than downward (Kliewer *et al.*, 1989; Schubert *et al.*, 1999) and it allows the vine to be trained to the fruiting wire in the first season (if growth is sufficient). The downsides are that the vine may not have the same carbohydrate production as vines that are allowed to sprawl and, since they are up to the wire in the first year, there could be the temptation to leave a crop on the vines in the second year, which may have consequences to the health of the vine (see Chapter 3).

In addition, there are numerous ways to help the vines grow up to the wire and reduce some management requirements. Vines can be individually staked or string can be tied from the vine (some part of the vine that will not have any growth distal to it, to prevent girdling of the cane/shoot) to the fruiting wire. In either case, it will be necessary to fasten the vines to the stake or string to ensure the shoots grow upwards, which is a materials and labour requirement.

Vine shelters

The use of vine shelters has increased manyfold in recent years. These range from short plastic or cardboard sleeves (in the latter instance, empty milk cartons have been used), to narrow tubes of some kind, usually made of plastic, that are slipped over or wrapped around the vines (see Plate 26).

Simple plastic or waxed cardboard sleeves (e.g. milk cartons), perhaps 30 cm tall, are commonly used and are effective at protecting the vine from herbicide and perhaps some rodent damage. They are inexpensive, but do not provide all the advantages of using taller structures.

Taller plastic sleeves, ideally rigid, encourage shoots to grow up and straight to the wire, provide protection against herbicides and herbivory of the young plants, but also increase the rate of growth of vines so that they reach the trellis wire more quickly. The environment inside the structure is warmer and more humid, and also protected against the effects of wind (Wample *et al.*, 2000). Warmer temperatures increase the growth rate in the spring, when average air temperatures tend to be cooler. The greater humidity and reduced air movement places less water stress on the vines, allowing them to grow unhindered. The rigid shelters also guide shoot growth upward, meaning there is no training needed and reduced water use by the vines.

Disadvantages of shelters in general include their cost to purchase and apply, difficulty in controlling disease inside them and in accessing the vine (e.g. for shoot-thinning and watershoot removal), and possibly increased damage during frost events (although some have observed that the vines are protected against a frost). As well, the shoots that develop within the tube are often noted to be thinner, which some perceive as a weakness in the vine.

Disadvantages of using tall, rigid shelters include (i) their cost (although they can be used on more than one set of vines); (ii) the extra labour required to open and close them again when working on the trunks; (iii) the possibility of temperatures exceeding damage thresholds on hot and sunny days; (iv) the requirement of the presence of a fruiting wire or stake to steady the sleeve; and (v) perhaps more spindly growth inside the shelter (Due, 1990; Wample *et al.*, 2000; Olmstead and Tarara, 2001).

Regardless of the method of training used, it is important to train the vines to have a straight trunk. This improves the load-bearing capacity of the vine (thus reducing the load on the trellising system) and ensures that mechanization is more readily brought into the vineyard.

Young vine care

The goal for establishing young vines is to maximize vegetative growth, which includes the growth of the root system. To enhance this, the irrigation and weed management programmes should maximize water availability and minimize weed competition. Many newly planted vineyards will have bare ground to avoid competition from weeds, and soil moisture monitoring should be practised to keep water readily available to the vines in the volume of soil where their roots are positioned (i.e. shallow in the soil profile). Diseases should be monitored and controlled as needed.

Young vine trunks need protection from herbicide sprays because, not only will green tissue take them up, but the thin bark on the canes will also be able to absorb the chemicals. Hence the use of vine guards of some sort is desirable.

Any vines that do not survive should be identified quickly and replaced after a likely cause for their death has been determined. For example, if the irrigation emitter was faulty, it should be replaced before the new vine is planted. To attain maximum vineyard uniformity, it is best to replant failed vines as soon as possible in the same year, as long as the vines will be able to grow enough to survive the dormant season. Extra vines can be planted in between others at the ends of rows and used for replanting. This ensures the replants are as strong as the neighbouring vines when repositioned.

The strongest-looking shoot should be retained, eventually to form the trunk of the vine. This is done to focus the energy of the plant into a single shoot, ensuring that it grows as much as possible in both length and girth.

When the vines reach the fruiting wire there are the options of training the growing tip along the fruiting wire in one direction or cutting off the shoot tip above the wire, allowing lateral shoots to be trained along the wire in both directions. This sets the vine up for development of the head later on, during pruning.

During the first dormant season, vines should be pruned back to a point that is relative to their growth in the previous season. If the vines have grown a lot, several to many nodes may be left on the vine but, if it has only just grown past the fruiting wire, it may be pruned to just above the wire with the cane tied straight against the wire. It is important when tying vines to remember not to bind them in such a way that later growth will be girdled (see Fig. 5.12).

If vines have not grown much at all, it may be best to prune them back to the most distal and viable bud and tie them with string up to the fruiting wire or against a stake (see Fig. 5.13) or, if in a tall vine shelter, left on its own. This ensures that the next season's growth will start from as close to the wire as possible.

Vines that have been left to sprawl in the first season should have their strongest healthy shoot retained and tied up to the wire as already described.

If vines have grown well and already have been wrapped on the fruiting wire, it may be a good time to disbud the trunks. If this procedure is done well, it will

Fig. 5.12. It is important to tie canes distal to the last node, to prevent girdling in the subsequent growing season. Note that the end node has been cut through diagonally so that the bud has been removed, but there is still part of the node (and tendril in this case) there to prevent the tie from slipping off. The topmost bud that will grow in the following season is just below the fruiting wire.

only need to be done once – for the life of the vineyard. Figure 5.14 shows how this is done – all of the bud must be removed from the cane/trunk to prevent leaving any latent buds in the wood. Please note, however, that once a bud has been removed in this fashion, it will never form another shoot from there: the viticulturist must be certain that it will never be needed in the future. A general rule of thumb is to leave three to four buds below the fruiting wire so that, if the vine head needs to be re-established, there will be a source of shoots to do it. In climates with very cold winter temperatures, buds should be left at the base of the vine to be able to train new trunks, should there be frost damage.

Second season

In most cases, the management goal for the second season is the same as the first: to maximize vegetative growth, including growth of the root system. To

Fig. 5.13. Upward growth that young vines make during the season should be preserved by leaving the topmost viable bud so that the shoot will reach the fruiting wire as soon as possible. Here, the vine is tied to a stake, which is tied to the fruiting wire at the top.

ensure this, any fruit that the vine may set should be removed as soon as possible. The easiest time to do this is early in shoot development, when the flower clusters are easily seen and flicked off quickly by hand. In the case of tall trellising systems, a second year of growth may be needed to establish the vines on the trellis (see Fig. 5.15).

In certain environments, it may be possible for the vines to carry a small crop in their second year of growth. Usually, this would be in warm climates where the vines had made a lot of growth in the previous season and there was ample development of both shoots and roots. In areas where this is not possible, carrying a crop will be a drain on the vine, resulting in poor vine performance in future seasons.

In addition, it is important to replant any vines that haven't survived as soon as possible, again, to ensure that there is as little developmental difference between them and the more established vines.

Fig. 5.14. Disbudding young vine trunks. Secateurs are used to cut the entire bud out of the nodes along the lower part of the trunk. This prevents future, and unwanted, shoot development.

Fig. 5.15. Second-year 'Pinot gris' vines growing up to be established on a divided-canopy trellising system.

What looks like the beginnings of a proper canopy may appear in the second season, which ensures that the plants are investing lots of carbohydrate energy into the vegetative parts of the vine.

Irrigation and disease and weed control are still important in this year. Powdery mildew can easily appear on young vines as they grow rapidly, and maintaining adequate fungicide coverage can be difficult. Any disease on the vines will reduce the amount of energy they can invest in other parts of the vine structure and will have an impact on future vine performance, so therefore should be controlled.

By the time the second dormant season comes around, vine pruning will take a lot more time than in the first. Strong canes should be trained along the wire. The number of nodes left on the vines should depend on, and be proportional to, the amount of growth the vines have put on in the past season: larger vines should have more nodes left on them than smaller ones. Disbudding of the trunks should also proceed on vines as they grow large enough for removal of buds to be performed safely.

In instances of less vigorous growth, canes can be cut back to just above the wire and tied for a straight trunk and weaker vines similarly treated as for the first year. Spot fertilization or investigating why these vines have done poorly is worthwhile to ensure they catch up with their neighbours.

6

Seasonal Management

Once a vineyard has been established there is ongoing management to keep it in top producing form. In many ways this process starts at pruning and matching the vine to the trellising system, as this sets the vine up for producing fruit in the following year. There are then many other management interventions that can be used during the growing season, including those dealing with the canopy, any cover crop, water application and other methods that fine-tune the grape product. Finally, determining when and how to harvest the crop caps off a season of preparation.

PRUNING AND TRAINING

Pruning and training of the vine are two of the most important aspects for quality grape production, whether it be for table, wine or other uses. How the vine is pruned and trained interacts with where the vine is growing, the proposed end product and management style. The two practices are firmly intertwined.

In essence, through pruning and training, the vine is shaped and manipulated to fit within the volume of trellis (discussed later) that has been made available for it.

There are two main types of pruning that can occur: summer and winter. The purpose of summer pruning (removal of green shoot parts during the growing season) is mainly to open up the canopy, prevent self-shading or just to tidy up the appearance of the vines.

The real work in pruning happens in the dormant season, when a decision has to be made as to how much, and which parts, of the previous season's growth must be removed (see Fig. 6.1). The aims of pruning are to:

- Establish/maintain the vine in the desired form.
- Produce fruit of the target composition.
- Select nodes that will produce fruitful shoots.

Fig. 6.1. A partially pruned row of grapevines in Oregon, USA. Note how much material is removed in the process.

- Regulate shoot number/crop load.
- Regulate vegetative growth.

A key concept in thinking about pruning is balance. In the case of most grape growers, the balance is in preparing the vines to produce an economic crop, but also so that they will not spend so much of their energy ripening that crop at the expense of their ability to grow in the following season (a consideration in the cropping of any perennial plant). Refer to Chapter 3 for a discussion about the carbohydrate balance of the vine and how it is related to management.

As the pruner leaves more nodes on a vine, they leave not only more growing points (shoots), but also potential fruit (all the clusters on those shoots). Thus, pruning has a significant influence on the ability of a vine to grow and to produce a crop.

Let's look at this in more detail. If we approach a vine in the dormant season, we see canes that were the shoots from the previous season. On each of those canes are a number of nodes. At each of those nodes is a compound bud that contains three preformed shoots. Each of these shoots is capable of growing and each may, or may not, have flower cluster primordia on them. In most cases, only the largest of those preformed shoots will grow in the spring. So, if a vine was not pruned, it would seem like a very large bush, with shoots coming out at all angles and from almost everywhere on the old vine, each potentially carrying fruit (see Fig. 6.2).

Fig. 6.2. Unpruned vines in Napa Valley, California, USA, showing the many nodes left and the shoots growing at an early time of the season. These vines were due to be pulled out due to phylloxera infestation, and hence there was no need to prune them.

Now if we prune that vine back to just a few nodes, in the spring we will have a much more limited number of shoots coming out, as well as having a much more limited amount of crop – ideally this would result in the same number of shoots as nodes left, with each shoot having some crop on it. In this case, we have pruned off a lot of the potential vegetative and reproductive growth of the vine. In the spring, it will have available a much smaller leaf area with which to photosynthesize and make carbohydrates and, perhaps more importantly, it has a very limited fruit load. The result of this will be rampant growth of those shoots, the pushing of many latent buds (left over from previously removed canes or shoots) and also the growth of other preformed shoots in the compound bud (see Fig. 6.3).

Some of the most widely used guidelines relating to pruning decisions are published in the book *Sunlight into Wine* (Smart and Robinson, 1991), where vine balance can be estimated through examining the fruit weight to pruning weight ratio, the number of shoots per metre of canopy and the average shoot weight.

In the case of the fruit weight to pruning weight, a value of 5–10:1 is said to represent a vine that is in balance – that is, the amount of fruit it is producing is appropriate to the amount of canopy the vine has. This concept is not new, having probably been first proposed by Ravaz in 1930. This principle

Fig. 6.3. The head of a vine showing the growth of many (all but one of the visible shoots) unexpected (non-count) shoots. These lead to high canopy density, which delays fruit development and encourages disease.

is explained by experiments that looked at a range of crop loads and leaf areas on vines and came to the conclusion that approximately 5 – 14cm^2 of leaf area was needed to ripen each 1 g of fruit (see Fig. 6.4), depending on the trellising system and environment (Kliewer and Antcliff, 1970; Kliewer and Dokoozlian, 2005). As there is a relationship between shoot growth and leaf area, it is easy to make the jump from this to looking at pruning weight, which is the summation of shoot growth over the season.

Another way of looking at this is through the calculation of approximate dry matter production of a vine: (fruit weight × 0.25) + (pruning weight × 0.55), where the numbers that are used as multipliers are roughly equal to the proportionate dry weight of the fruit and dormant canes, respectively. This value can be thought of as a measure of the capacity of a vine (Winkler *et al.*, 1974), which is related to its ability to grow or produce grapes. So, in fact, the index is an estimate of the above-ground dry matter production by the vine for the season just past, which is a concept that we introduced with vine capacity in Chapter 3.

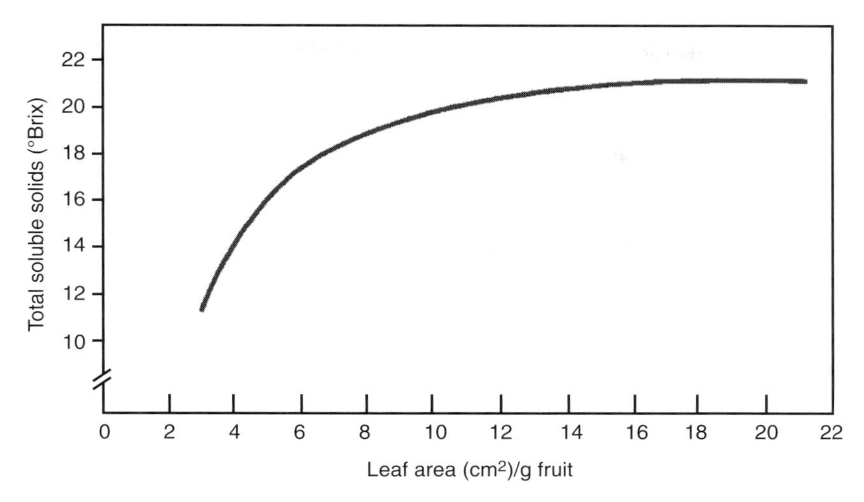

Fig. 6.4. Effect of a range of leaf areas/1 g fruit on fruit Brix after 4 years of treatment application to 'Thompson Seedless' vines. No appreciable increase of Brix is obtained at the very high values (from Fig. 1, Kliewer and Dokoozlian, 2005, with permission).

Some knowledge of vine balance can then be used to make actual pruning decisions. In pruning we will refer to count nodes and non-count nodes. Count nodes are those that are left on purpose during pruning *and* which we expect to produce one fruitful shoot (i.e. for the count shoot refer to Fig. 2.4). In practice, the first count node on a cane is not the first visible node, as the compressed nodes at the base of the cane do not normally push during budbreak. The shoot that arises from a count node is referred to as a count shoot and, in normal circumstances, this will be the primary shoot, which bears the most potential fruit.

A non-count node is any node that is not a count node. For example, the compressed nodes at the base of a cane are non-count nodes, as are those that are embedded in the trunk, head or cordon of a vine (latent buds). Similarly, shoots that arise from non-count nodes are termed non-count shoots (see Fig. 2.4). As well, secondary or tertiary shoots that may arise from a count node are also considered non-count shoots, as they are not anticipated to burst in the spring.

Types of pruning

The principal methods of pruning are cane and spur (also known as cordon pruning). With cane pruning, the fruiting shoots for the next season come from a length of the previous season's shoot, which is laid down and wrapped on the fruiting wire of the trellising system (see Fig. 6.5).

Fig. 6.5. Examples of cane-pruned vines.

Cane pruning typically uses up to four canes of 8 to 15 nodes in length on the vine, depending on trellis, vine spacing and cropping aims. In most cases, one or two replacement spurs are also left below the head, which is to facilitate cane choice the following year. Figure 6.5 shows the head in the centre of a bilaterally trained vine, but the head can be to one side on unilaterally trained vines (typically used for close vine plantings, where there is only one cane tied down). The goal is to choose canes that are (i) in a good location to retain the shape of the vine; (ii) well-exposed in the previous season, to give maximum fruitfulness; (iii) have good periderm formation; and (iv) not excessively thick or thin. The latter two parameters are associated with the cane's ability to withstand cold while dormant and increased fruitfulness in the following growing season.

Advantages of cane pruning are that vine fruitfulness is maximized and the number of potential non-count nodes on the vine is minimized, because there is less permanent wood on the vine and therefore less surface area from which latent buds can arise. As some cultivars are inherently less fruitful at the basal node positions on the shoot (Swartwout, 1925; Winkler, 1926; Buttrose, 1969), cane pruning minimizes the relative impact of this.

Disadvantages of cane pruning include (i) blind budding (places along the cane where no shoot develops), especially with long canes; (ii) limitations to between-vine spacing (in that the distance between vines is restricted to no more than twice the length of a typical shoot during the growing season); and (iii) uneven shoot growth along the cane. The latter phenomenon has been called the End Point Principle by David Jackson (1997), where shoots at the end of a cane tend to grow more vigorously than those in the mid-portion of the cane (see Fig. 6.6). Extreme examples of this arise when there is no shoot development in the middle parts of the canes, but there is at the ends and bases (see Fig. 6.7), which also demonstrates the Trunk Proximity Principle, where shoots tend to develop more vigorously near the base of the canes (Jackson, 1997). Cane pruning can also be more difficult when training workers to make

Fig. 6.6. Example of a cane showing signs of the End Point Principle, which states that growth of shoots tends to be more vigorous at the ends of canes compared with that at the mid-sections.

Fig. 6.7. This vine is showing an extreme case of both the End Point Principle and Trunk Proximity Principle: there is vigorous growth at the beginning and end of canes, but none in between.

the best decisions, resulting in loss of some of the advantages of cane pruning in the first place.

Spur pruning does away with continued use of canes by establishing a permanent arm, or cordon, from a cane (see Fig. 6.8). Note that, in establishing a cordon, shoots arising from the downward-facing bud on the original cane are removed, leaving every alternate and upper node positions present. Downward-pointing shoots do not grow so vigorously (Kliewer *et al.*, 1989) and, if shoots from both nodes were used, it would enlarge the fruiting zone, making some management practices more difficult. To counteract the loss of shoots from these positions, canes arising from the remaining nodes are cut back to (typically) two-node spurs. Note that, when training a cane that is destined to become a cordon, it should not be wrapped as tightly as if the vine is cane pruned every year. Too many revolutions around the wire can result in girdling of the cordon as it increases in girth (see Fig. 6.9).

Advantages of spur pruning are that it is easier to teach unskilled people how to do it and it is possible to partly mechanize the process using cutting bars (see Fig. 6.10), which means that the crew that goes in afterwards has much less brush-pulling to do. Spur pruning also results in the vine having more permanent vine material, which has been implicated in fruit quality (Howell, 2001). Disadvantages are (i) vine fruitfulness is limited (because the basal nodes on canes are typically not as fruitful); (ii) spurs can die out over time due to injury, disease or other problems; (iii) the fruiting zone can creep upwards through the seasons if there is a tendency for pruners to choose the distal nodes

Fig. 6.8. Example of a spur-trained vine, showing two node spurs.

Fig. 6.9. Cordon where the wire has become embedded into the plant. Eventually, the performance of the cordon will decline due to restriction of nutrient flow and, in extreme cases, the entire cordon may die.

on each of the spurs (this can result in very uneven or high fruiting zones); and (iv) the cordon needs to be renewed periodically due to damage, disease, etc.

Other factors to be aware of in spur pruning relate to the canes that were used to set up the cordons. Spur spacing depends on the internode length of the original cane, so initial cane choice is important. If internodes are too long, then spurs will be spaced too far apart and cropping will be limited. If they are too close together, there will be shoot shading, dense canopies and a host of other problems that are related (see the Goals and tools for canopy management section, on page 139). For most situations, keeping replacement spurs near the head of the vine is vital, because it is from those that renewal canes/cordons can be sourced. To avoid the upwardly creeping fruiting zone, pruners must be aware that shoots from the lower of the two nodes need to have preference over the upper. The low potential fruitfulness of spur-pruned vines can be overcome somewhat by altering the length of spurs retained, though this does have significant effects on the resulting canopy unless remedial measures (such as shoot thinning and leaf removal) are taken during the growing season.

No matter what the pruning system, there remain some tips to smooth the pruning process. If movable foliage wires are used in the trellising system, be sure to move these down before budburst as then they will catch the shoots as they are raised. If any trellis posts or wires have broken, your only chance to fix them is at pruning, before any tying down of canes has to take place.

Fig. 6.10. Example of a barrel pruner, which is used to remove unwanted cane material from spur-pruned VSP-type trellis systems quickly and easily. The two sets of rotating wheels cut the canes into small sections, which then fall from the trellis (image reproduced with permission from Hort Engineering Solutions, part of the Ormond Nurseries Group of Companies, New Zealand).

Pruning decisions

The basis for pruning grapevines is that each vine has a capacity to support growth (of shoots and roots) and ripen a crop. Each dormant season, an appropriate balance between vegetative production and yield must be found. Pruning is your first step in managing crop load and vine size, and so an important process that requires thought and reasoning. If too many nodes are retained on the vines, time-consuming canopy and crop management practices such as shoot thinning, cluster thinning and leaf removal may need to be done during the growing season. A vine in balance with site considerations, its crop load and management decisions will have moderate growth that fills the trellis and ripens a commercial crop.

Vines have evolved without something coming in and removing 90% of their season's growth each year. Vines produce far more nodes, each with a compound bud capable of producing three shoots, than they can support with the kind of growth needed to fill out typical trellis systems. However, if a vine is left with a large number of nodes, it will reduce the number of viable growing points, effectively pruning itself. Any growth that has not hardened off and formed an adequate periderm will not survive freezing temperatures and will be lost during the winter. In the spring, a smaller percentage of buds at those remaining nodes will produce shoots, i.e. percentage budbreak is reduced (see Table 6.1). This (along with other factors relating to the cluster) is why, if twice as many nodes are left at pruning, there won't necessarily be a doubling of crop (Christensen *et al.*, 1994). An increase of 100% in node number results in a relatively small 60% increase in shoot number and only a 22% increase in yield. Part of this is due to the decreased percentage budbreak, but also to factors related to cluster weight and berry number.

Note that budbreak on the 60-node vines is greater than 100%, because some nodes produce more than one shoot (two or three of the preformed shoots in the compound bud push instead of one) and because there will be a greater number of non-count shoots arising from the permanent parts of the vine (from latent buds). This is related back to vine capacity, where a vine has the capacity to push a certain number of buds each year and, if they are pruned to below that level, a greater number of shoots than nodes retained will be produced. If more buds than that level of capacity can support are left, fewer buds will break, percentage budbreak will decrease and growth of those shoots that do push will be weaker.

Matching pruning to vine capacity

Each vine in a vineyard may have a slightly different capacity due to the microclimate it experiences. Experienced pruners recognize this when they first look at a vine and judge its growth during the previous season. If there has

Table 6.1. Yield components for 'Sultana' raisin production given three different vine pruning severities (data adapted and reproduced with permission, Christensen *et al.*, 1994).

Yield component	60 nodes/vine	120 nodes/vine	Change (%)
Shoots/vine	67	108	+60
Budbreak (%)	113	90	−20
Yield (kg/vine)	5.31	6.49	+22

been a lot of vegetative growth, more nodes will be left than there were the previous season, which should result in smaller shoots with less vigorous growth and greater crop load per vine. If the vine has put on very limited shoot growth, fewer nodes will be retained to encourage more growth from those canes and reduce the crop load. Crop load in both cases will assist in balancing the vines: a large vine's increased crop load moderates vegetative vigour, and a small vine's reduced crop load puts more of the vine's resources toward vegetative growth (roots, shoots and trunk). However, for very weak vines, fruit should be removed altogether.

Experience has traditionally been the measure to balancing vine growth and pruning level. However, in the 1940s, scientists in New York quantified this by producing a formula that matches the number of nodes retained to vine capacity. This method is known as Balanced Pruning (Shaulis *et al.*, 1953), and can be a useful tool in learning the relationship between vine size and node numbers that should be left on the vine.

In Balanced Pruning, the number of nodes to be left on the vine is estimated by weighing most of the vine's canes produced in that year (more nodes than needed are left on the vine so that some can be trimmed off to reach the desired number after the measurements) and using that value in an equation. In this case, the capacity of the vine is estimated by just pruning weight. For 'Concord' in New York State, USA the formula is 30 + 10x, where x is the weight of prunings in pounds (1 pound is approximately 0.45 kg) (Jordan *et al.*, 1981). The first value is the number of nodes to be left for the first pound of prunings, and the second, the number of nodes to be left for each pound after that. For example, if the vine has been pre-pruned and there is 1.5 lb of canes, 30 + (10*0.5) or 35 nodes would be left on the vine. If the vine is larger and there were 3 lb of prunings, 30 + (10*2) or 50 nodes would be left on the vine. Thus vines that put on a lot of growth that season will have a large number of nodes left, vigour of the individual shoots will be less and the greater amount of crop on the vine will also slow growth, resulting in a smaller vine with reduced capacity the next time it comes to pruning. For a small vine, balanced pruning allows its capacity to grow from season to season. This formula was designed for use with mature vines (note that the minimum node number is 30): for very weak or diseased vines, the formula should not be used.

Both the location and variety were specified with the formula; vines growing in different areas and for different purposes may have a different formula. For instance, in New York where Balanced Pruning was developed for use with 'Concord' grapes, the formula was 30 + 10x (x in pounds). Under Australian conditions (and with metric units), the general recommendation of 30 to 40 nodes/kg of pruning weight has been suggested (Smart and Robinson, 1991). Since trellising (divided versus single) and vine and row spacing vary considerably from place to place, it is more informative to look at weight of prunings per 1 m of vine canopy rather than on a per vine basis, which allows for more direct comparison of values from place to place.

Regardless of where the vines are growing, the concept of the formula and tradition remains sound: larger vines should have more nodes and smaller vines should have fewer nodes left on them.

Other indicators of vine capacity are also used. One is to note how many 'effective' shoots grew during the growing season. This can be done at pruning time once you have learned to estimate the weight of individual canes. Smart and Robinson (1991) recommend that the ideal cane should have a weight of between 20 and 40 g during the dormant season. By weighing a variety of canes and getting an idea of what one in that range looks like, at pruning the number of canes that fall within this range can be counted. If a cane weighs less than 20 g it is not counted; if it is between 20 and 60 g it is counted as one. If it is greater than 60 g it is counted as two. The shoot counts are totalled up for the vine and the result is an estimate for the number of nodes to leave for the following season. If the number of nodes left the previous season was much higher than the number of 'effective' canes, then fewer nodes should be left on the vine as the vine did not have the capacity to push all of the nodes retained. If the number of nodes left the previous season was fewer than the number of 'effective' canes calculated, then many non-count or large shoots will have come out and the vine must have more capacity for growth available.

Performed over a series of years, larger vines will reduce their capacity and the smaller vines increase their capacity. Equilibrium is reached when you are leaving the same number of nodes every year. Shoot, root and trunk growth should be in balance with fruit production, which means a more uniform vineyard, facilitating vine management and resulting in an ideal balance between crop load and vegetative growth. Ultimately, this leads to higher quality of grapes and wine.

The basic procedure for cane pruning would look like this:

- Look at vine structure for any deficiencies that need to be corrected (e.g. poorly placed spurs, canes originating too close to the fruiting wire, poor cane growth near the head of the vine, etc.).
- Locate the position of those canes that are likely candidates for fruiting and those for spurs.
- Make main cuts, *but*:
- Leave extra canes to choose from in case the targeted ones break.
- Carefully remove the wood that was cut off without damaging the remaining buds.
- Make final cane and spur choices, cut to length and tie to wire.
- When cutting canes to length, cut through nodes, removing the bud and leaving part of the node swelling, which makes securing the cane to the wire easier.
- Remove any extra canes.
- Remove older wood that is no longer needed.
- Paint large pruning cuts with an approved fungicide or biocontrol agent to protect from *Eutypa* and other wood-invading pathogens.

Other considerations when pruning

1. Prune as late as possible: budbreak is delayed by late pruning (Antcliff *et al.*, 1957; Williams *et al.*, 1985; Friend and Trought, 2007), which can be an advantage if the vineyard or block is located in a frost-prone area. Yields have also been shown to increase with later pruning dates (Friend and Trought, 2007), which also may be an advantage. Delayed pruning and removal of pruned wood must be done carefully if the buds are swelling – they can be brushed off easily.
2. Choose canes that were in well-exposed positions: flower cluster initiation takes place in the buds at the base of each leaf near bloom, meaning that the following year's crop is beginning to be determined almost 18 months before it is actually harvested. When pruning and laying down canes, pick those that have been well exposed during the early part of the growing season.
3. Plan ahead for canopy density: overlapping canes will cause excessively dense canopies, which will lead to poor flower cluster initiation, spray penetration, air movement and fruit ripening. If more nodes need to be left than space available on the wire then, if possible, add another fruiting wire, arch the canes or perhaps the vines are too vigorous for the trellising system, in which case it should be changed (see Training and Trellising, below).
4. When is big too big? If single-canopy vines are producing near or over 1 kg of pruning per 1 m of row, it may be worthwhile considering changing to a divided canopy. Smart and Robinson (1991) have suggested that ideal values for pruning weights in these systems fall between 0.3 and 0.6 kg/m of row. If, for example, vertical shoot-positioned (VSP) trained vines are producing much over that, conversion to a divided canopy of some sort should result in more moderate growth over a larger area. This will return better-quality fruit (due to more fruit exposure to the sun, better airflow and better spray penetration), increased yields per acre and fewer management interventions, which should offset the cost of the conversion (Smart and Robinson, 1991). If vigour is still a problem (see Fig. 6.11) then other management techniques must be used to reduce it, for example, through the use of more aggressive cover crops, reduced irrigation, etc.

In general, if there is excessive shading in the canopy resulting in disease (because of poor spray penetration or increased humidity due to lack of air movement), poor fruit quality from lack of exposure and poor fruitfulness of the buds (due to a lack of light for flower cluster initiation), some management intervention is needed to correct it.

TRAINING AND TRELLISING

Trellising is the art of arranging the vine in space. The goals are to maximize light interception, minimize the number of additional inputs (e.g. labour,

Fig. 6.11. Very vigorous vines in an Oregon, USA Scott-Henry-trained vineyard. The upward-growing shoots are shading the adjacent rows as well as themselves; the downward-trained shoots have been cut off by the mower. These vines are in desperate need of some good viticultural management!

material costs, maintenance, etc.), provide physical support and be compatible with the management situation, which includes machinery (see Fig. 6.12). In choosing a trellising system, factors such as product end use, harvest method, mechanization, cost and labour requirements have to be considered, along with the predicted vigour/capacity of the vines at that site.

Self- or stake-supported

The simplest training systems are staked vines or bush vines. The vine is trained to a short trunk tied to a stake on which a head with several short canes is left. Once the trunk can support the weight of the canopy and fruit on its own, the stake can be removed (see Fig. 6.13). Alternately, the stake can

Fig. 6.12. An example of where there is incompatibility between mechanization and trellis design. An entire section of row has fallen over due to mechanical harvester damage. Trellis designs for mechanized systems should be robust enough to withstand the additional stress they cause.

Fig. 6.13. Bush vines in a Californian vineyard. Note the tyre tracks indicating cross-cultivation and the wide vine spacings.

remain and canes can be trained upward and tied to the stake (see Fig. 6.14). There are other forms of self-supporting systems, such as the basket vines, in which canes are woven together to form a low bowl shape, which may encourage the capture of dew, important in areas that cannot be irrigated (see Plate 27). Advantages of these types of systems are low capital cost, low vine densities suited to arid areas and the possibility of movement of machinery crossways through the vineyard. Disadvantages include low cropping capacity per hectare due to inefficient use of land area.

Single wire

Single wire trellising systems are used for a variety of purposes and are low cost relative to most other systems, due to only one wire being needed. Here, a single fruiting wire is used and shoots arising from the cane or spurs are free to grow in whichever direction they end up (see Fig. 6.15). In vine species with upright growth habits (most of the *V. vinifera* species), this creates a high hedge open at the bottom, whereas in those species with a trailing growth habit (e.g. *V. labrusca*) the hedge form is lower and less open at the bottom. Though this system is not compatible with close row spacings, the canopy is usually quite open, leading to good fruit exposure, spray penetration and relatively high crop loads.

Fig. 6.14. Individually staked vines. Note the clean cultivated soil in this young vineyard.

Fig. 6.15. These grapevines are trained to a single wire trellis. The ones on the left side are mechanically pruned and the ones on the right spur pruned.

The practice of minimal pruning can be combined nicely with a single-wire system. Vines are not pruned by hand, but rather trimmed by machine (see Fig. 6.16). The size of the cross-sectional box to which the vines are trimmed determines the approximate number of nodes left on the vine. Crop adjustment can be accomplished through trimming the vines after fruit set, when crop loads can be better estimated (usually, the bottom part of the vines is trimmed off in this operation), or later with a mechanical harvester (Pool, 1987; Fisher *et al.*, 1997).

Hedge-type

Vertical Shoot Positioning (VSP, Fig. 6.17) is a widely used trellising system for the production of wine grapes and results in the box-shaped hedge that is seen in so many vineyard calendars. A fruiting wire is relatively close to the ground (typically 90 cm at a row spacing of 2.5–3.0 m), dormant canes wrapped to it (the horizontal cane version of this is called a Guyot system) and shoots are trained upwards from it during the growing season. This results in a fruiting zone above the cane, which localizes the crop relative to most of the canopy. Fruiting wires can number one or two, with one to four fruiting canes being typical.

Additional wires are needed to maintain the shoot growth in its vertical position, adding to the system's cost. Labour is also required to position the

Fig. 6.16. A comparison of minimally pruned vines with hedge-type vines in the early part of the season. The minimally pruned vines have many more, but shorter, shoots than the hedge vines, which can lead to greater productivity due to greater capacity for photosynthesis.

Fig. 6.17. Wine grapes growing on a Vertical Shoot Positioned (VSP) trellis. Note the catchwires, made visible by the leaf removal in the fruiting zone, that are used to contain the shoots within the hedge-like shape.

shoots within the wire and keep them there, as vine growth habit and wind tend to make them fall out. Cultivars or clones with an upright growth habit are much better suited to VSP, as it is much easier to keep them within the wires. Typically, trimming of the sides and tops of the canopy is also necessary, as often VSP is used in areas where vigour is greater than the system can contain. To allow for more nodes to be laid down between vines, the canes can be arched rather than tied to the horizontal fruiting wire (see Plate 28), which also assists in levelling out shoot growth along the cane. Names of these variations are Pendelbogen, Umbrella Kniffen or simply Arched Cane.

Training to a Sylvoz system is a way of increasing the number of nodes left per metre of row without having to divide the canopy by a more expensive trellising option. Vines are cane or spur pruned to a mid-height fruiting wire, but additional canes are left trained downward and tied to a lower wire (see Plate 29). This spreads out the shoots and fruiting area – the former being beneficial for canopy density, but the latter making it more difficult to target fruit sprays and perform effective leaf removal operations.

Divided canopies

Divided canopies were developed to address the concerns of maximizing crop and accommodating high vigour. The first significant breakthrough in this area was made by Nelson Shaulis and his research crew of the Geneva Experiment Station in New York, USA (Shaulis *et al.*, 1967). The Geneva Double Curtain (GDC, Fig. 6.18) addressed the issues of increasing productivity (through better light interception per unit land area) and mechanization of 'Concord' grapes, which were primarily grown for juice production. The system is effectively two single wire systems set side by side. Due to the trailing growth habit of *V. labrusca* vines, two distinct curtains of shoots develop (though usually with the help of some shoot positioning). GDC has been adapted for use with other grapevine species, though its best application remains with trailing growth-habit cultivars.

Other horizontally divided canopy systems have been developed, such as the Lyre system, which is similar to the GDC in that it simulates two close-spaced rows of vines while only being a single row. The Lyre is so called due to the instrument-like arrangement of the two VSP-like trellising systems (see Fig. 6.19) when viewed end-on, which was chosen to optimize light interception and therefore productivity (Carbonneau and Casteran, 1987). However, the capital costs of establishing this trellising system are high in comparison with that of most others, so the perceived benefits must be worth the additional expense.

A popular divided canopy system is the Scott-Henry (see Fig. 6.20) and its derivatives (e.g. Smart-Dyson) (Smart and Robinson, 1991). Rather than splitting canopies horizontally, this system splits them vertically, with one set of

Fig. 6.18. The view down the centre of a Geneva Double Curtain (GDC) trellis, showing cane-pruned vines and the distinct distance between the two fruiting wires.

Fig. 6.19. Looking down the inside of a row of vines trained to a Lyre trellis. It is an advantage if a person is able to walk down between the canopies and work on the shoots. In any case, there should not be any shoots bridging the interior of the two sides.

Fig. 6.20. Scott-Henry-trained vines in Mudgee, New South Wales, Australia. Of note is the distinct band of light running through the middle of the canopy, which demonstrates good practice with this system, as air should be able to move between the upper- and lower-trained shoots.

shoots trained upward and one downward. An advantage of this system is that it is easy and relatively inexpensive to retrofit to a VSP system, which is useful if the eventual vigour of the vines has been underestimated at planting.

Some generalizations on divided canopies would be that light penetration is increased through their use, which in part accounts for their beneficial effects on improving crop load and fruit composition. They also allow for more flexible matching of vines to potentially vigorous sites and stretch the yield:quality envelope. In many cases the cost of establishing or converting to a divided canopy system are justified given the benefits received over the life of the vineyard.

Other trellising systems

Overhead systems (e.g. pergola or parronal) are the most efficient in terms of land use in that almost 100% of the incident light is captured and used by the vines. This leads to high yields per unit area of land, but there is a cost in terms of trellis construction and difficulties in working with the system. A series of wires is strung between post supports and the vines trained along the top (see Plate 30), usually with a low number of vines per hectare compared with the previously mentioned trellising systems.

Overhead trellising is used in fresh, wine and raisin grape production, to varying extents. Adoption of this system in raisin production is growing as a result of techniques allowing raisins to be dried on the vine (DOV) rather than in between rows (Christensen, 2000).

Munson 'T'-type trellises are used extensively for fresh grape production as well as for wine grapes and raisins. As its name suggests, a single post supports a horizontal brace on which three wires are strung up and down the rows. Shoots are trained on the top of the trellis wires and the fruit hangs below (see Fig. 6.21). Variations of this have the top brace at an angle, where shoots are trained up and along it, allowing the fruit to hang underneath, but with better access for management and harvesting. In the case of the Swing-arm Trellis (Clingeleffer and May, 1981) the 'T' portion of the trellis also pivots, to improve vine mechanization and productivity.

There are many variations on these themes for providing a structure on which the grapevine can grow, each adapted to a particular situation or need. For a review of these see Freeman *et al.* (1988) and Smart and Robinson (1991).

For those vineyards that are intended to be managed by hand, research such as that by Kato and Fathallah (2002) will be of increasing interest, as it has been demonstrated that certain types of trellising systems are more user-friendly than others, in terms of risk of musculo-skeletal disorders such as those associated with repetitive strain injuries.

Fig. 6.21. A 'T'-type trellis, which encourages the fruit to hang freely under the canopy. This allows good access for sprays as well as for handling.

End assemblies

Trellises support much of the weight of grapevines and their fruit, as well as resisting movement caused by wind and vineyard machinery. The structures at the ends of the rows must resist the along-the-row tension caused by the weight on the trellis, as the intermediate posts carry the strain of the vines, fruit and even the accumulated weight of rainwater on them (Mollah, 1997).

The end assemblies must transfer the tension in the wire (which is, in large part, determined by the spacing between intermediate posts (Freeman *et al.*, 1992)) to the ground and, as such, need to be more robust than the intermediate posts. The main load-bearing unit of an end assembly is called the strainer, and its ability to transfer the load to the soil depends on the trellis height, strainer diameter, leverage affected by its depth in the ground, and the soil itself (Smart and Robinson, 1991; Freeman *et al.*, 1992; Mollah, 1997). Three main types of end assemblies will be discussed here that should be useful for most vineyard situations.

The tieback end assembly (see Fig. 6.22) is a system where the load is transferred though the end post to an anchor in the ground. The wire used to tie

Fig. 6.22. Tieback end assemblies. Note that two loops of wire are used to link the strainer to the end post and that the loops have a wire tensioner to adjust the strain on the posts.

the anchor to the end post should be equivalent in strength to the sum of all wires used in the row, where foliage wires count as half because they are not under as much tension (e.g. if there is one fruiting wire and four foliage wires, then the wire to the anchor should be equivalent to three wires) (Smart and Robinson, 1991). The tieback is cost efficient and allows for vines to grow unhindered right up to the last post of the row, but does require some additional headland for equipment turning compared with other systems.

A diagonal stay system (see Fig. 6.23) transfers the tension into the row using an angled brace, which can be made of wood or metal. The foot at the end of the diagonal component acts as a pivot for the strainer post, so that the tension attempts to twist the strainer up and over the diagonal stay. This is an effective system that is inherently stronger than a tieback one (Freeman *et al.*, 1992, Mollah, 1997), and also requires less space outside the last vine in the row for headland. However, due to additional materials used, it does cost more, and there is some interference for vine training caused by the diagonal stay.

Fig. 6.23. Diagonal stay assemblies in a VSP vineyard.

The box end, or horizontal stay, assembly (see Fig. 6.24) is regarded as the strongest, as the design keeps the force on the strainer vertical rather than horizontal, distributing it over a larger area (Smart and Robinson, 1991). Therefore, this post needs to be larger that the others. Though strong, the design is expensive to build and vines growing within the box are much harder to manage.

The failure of end assemblies is usually caused by either (i) insufficient strainer diameter or strength, leading to breakage (see Fig. 6.25); (ii) not

Fig. 6.24. Many rows of box end assemblies in a New Zealand vineyard. Note how constrained the growth of vines is inside the box compared with those in the row.

Fig. 6.25. A free-standing end assembly that has failed due to inadequate strength. The adjacent row has already been retro-fitted with a diagonal stay assembly in recognition of the original oversight.

putting the strainers into the ground far enough, which leads to them being pulled out; or (iii) soil not being strong enough to bear the load, which leads to strainers tilting or moving through the soil (see Fig. 6.26). A thorough analysis of predicted trellis loading and soil strength, before determining the type of end assembly to use, will prevent costly repairs and lost crop in the future.

VINEYARD FLOOR MANAGEMENT

There are many options available for managing the soil surface in vineyards. For many years the soil was regularly cultivated to discourage weed growth – a bare earth policy (see Fig. 6.27). More recently there has been a shift from this to establishing certain types of plant growth between the vine rows. This could be due to either increasing numbers of options for controlling non-grape plant species, a change in attitude about managing soil health or increasing awareness of the sustainability of management practices.

In terms of impact on the vines, a bare earth policy would seem to be the best option. Other plants growing around the vines compete for water and nutrients, and can also hinder air movement through the vineyard. Bare earth is a simple policy for determining success rates and results in the lowest spring frost risk (Dethier and Shaulis, 1964) in areas so affected. However, continued tillage of the soil is bad for both soil structure (increasing the chance of erosion)

Fig. 6.26. An end assembly post moving in the soil. This could be due to the weakness of recently disturbed soil in this newly developed vineyard.

Fig. 6.27. Cultivated vineyard rows, which leads to less competition for water and more nutrients for the vines.

and microorganism populations, and herbicide use is expensive and can have adverse impacts on the environment (Doran, 1980; Pankhurst *et al.*, 1995; Roper and Gupta, 1995).

Therefore, some form of planting in the vineyard system can have beneficial effects, such as reducing erosion, dust and undesirable weed development and increasing water infiltration rates, soil organic matter, quality of soil structure, populations of beneficial organisms and vineyard accessibility (for, e.g. workers, tractors and harvesters), especially in wet conditions (Hartwig and Ammon, 2002). Cover crops can also be useful as a tool to manage soil water availability and alter grapevine shoot growth. Disadvantages of cover crops include the need to prepare soil, buy seed, and plant and establish the cover crop. In dry areas, its use of soil moisture can be a drawback, as well as introducing competition for nutrients. A vegetative cover on the soil also lessens heat accumulation during the day and release during the night, which can increase the risk of spring frost damage (Dethier and Shaulis, 1964). Vine root development is altered and some plants can harbour pests such as rodents, diseases, insects and nematodes.

Careful selection of an appropriate cover crop is then warranted; there are many options available. Annual plants are usually shallower rooting and therefore less intrusive on the vines and, if left to seed, do not have to be re-sown. Perennials have better staying power, but may be more expensive to establish.

A popular option for cover crops are grasses, such as brome, rye, blue and fescue or even a grain crop (see Fig. 6.28). The ability of a cover crop to die back in the summer is usually an advantage as, when conditions are at their driest, it is not competing with the vines for water and nutrients. Cover crops must also be able to survive in between the vine rows, where there is usually no supplemental water. In areas of low nitrogen, legumes may be used as a cover crop to help manage soil nitrogen availability. More recently, flowering plants have been used in vineyards as they attract beneficial insects, resulting in fewer pests (Landis *et al.*, 2000; Nicholls *et al.*, 2000).

In areas with excess available water in the soil, deep-rooted plants, such as chicory, can be planted to help draw water away from the vines (Caspari *et al.*, 1997).

In some cases, a volunteer crop is allowed to become established as a cover crop, which has the benefit of being low cost and will be made up of those plants that are hardy enough to survive in between the rows. However, this volunteer crop may have undesirable characteristics, such as adding nitrogen to the soil, not dying out during the summer months or not achieving sufficient ground coverage to allow foot and vehicle traffic during the wet season.

Fig. 6.28. Oats sowed in alternate rows of a mature vineyard. These are grown to be incorporated later into the soil. In a young vineyard they can be grown to provide shelter from wind damage.

Additional costs are associated with cover crops, such as the need to mow (particularly in the spring in frost-prone areas, as a short grass cover crop has minimal impact on frost incidence) and to keep the cover crop from encroaching on the vines or growing at the wrong time (usually resulting in the need to use herbicides).

During the vine establishment phase it is not a good idea to keep cover crops in the vineyard, unless water and nutrients are not limiting. They should also not be used in areas of severe frost risk or where soil water is at a premium. However, once the vines are established, cover crops can be planted if appropriate (see Fig. 6.29).

Non-living cover crops also exist in the form of mulches, which can be organic or inorganic, and can result in improved weed control, water conservation, soil structure, water infiltration rates and root branching. However, they can also increase vine vegetative vigour, shelter rodents and increase costs (Penfold, 2004). The use of mulches is often worthwhile during the establishment phase of vineyard development, as it can aid vine growth (see Table 6.2, Fig. 6.30) or in areas of low water availability.

Mulches are taking on other roles as well, in altering the light environment of the vines and fruit. Spreading of mussel shells under the vine rows (see Plate 31) was found to improve 'Pinot noir' wine quality, with minimal impact on other performance factors (Creasy *et al.*, 2006). These preliminary results are very promising, and have the bonus of providing an end use for a waste product.

Fig. 6.29. Recently seeded grass in a 2-year-old vineyard.

Table 6.2. Effect of using different types of mulches on the second-year pruning weights of 'Pinot gris' vines growing in New Zealand. Cocomulch is a combination of brown plastic and coconut fibre (from G.L. Creasy and G. Wells, 2003, unpublished data).

Pruning weight (kg)	Control	Black plastic	Cocomulch	Compost
Mean	0.071	0.116	0.307	0.281
Standard error	0.020	0.019	0.048	0.039

Fig. 6.30. Three-year old vines with a difference. Those on the left were grown for the previous 2 years with black plastic mulch while those on the right were the control, with obvious effects in this non-irrigated vineyard.

FROST MANAGEMENT

For some growing areas, the occurrence of early-season freezing temperatures is a real risk and, in some cases, late-season frosts as well. The latter tend not to be as common as the former but, if they do occur prior to harvest, the premature death of the vine leaves means that there is no carbohydrate source for ripening the fruit and providing the vine with stored energy to overwinter and begin growth in the following season. Early-season frosts of importance are those that occur after budbreak, when developing and green tissues are much more susceptible to freezing damage.

Types of frost event

To ascertain ways of managing frost, a review of the types of frost events is required. The type with the fewest options to manage is the movement of below-freezing air into an area, or advection frost (Cornford, 1938). In areas with surrounding mountains or hills covered with snow, the freezing air moves down from the force of gravity (cold air is heavier than warm air) in a katabatic flow, enveloping the lower areas.

A second type of frost is a radiation frost, which occurs on clear and windless nights as heat from the surface of objects is lost to the atmosphere (see Plate 32). If there is little or no mixing of air, the heat lost from the surface collects such that a layer of warmer air (the inversion layer) develops above. Since, in this case, there is warmer air above the freezing air, a strategy for reducing the freezing temperatures near ground level is to mix the two layers.

Passive control strategies

The best way to deal with frosts is to avoid them through site selection, which has been dealt with in an earlier chapter. Cultivar choice can also have an effect. Late budbursting cultivars such as 'Riesling', 'Muscadelle' and 'Dolcetto' are a better choice in a frost-prone area than early-budding cultivars like 'Chardonnay' and 'Pinot noir'.

Given that the coldest air is closest to the ground, a simple solution is to increase the height of the buds from the ground. Data from New Zealand suggests that a cane/cordon height of 2 m (for example, as may be found in a high GDC system) could provide an extra 4°C of warmth than the fruiting buds on a vine trained to typical VSP, at 90 cm from the ground (Trought *et al.*, 1999).

Active control strategies

The type of freeze event determines which control strategies are most useful. Radiation frosts have more options when it comes to preventing damage. Many take advantage of the inversion layer that develops over the vineyard by mixing the layers, for example by the use of helicopters. These have the advantage of not requiring a huge capital investment, but the disadvantage of being relatively expensive to run when they are needed. Therefore, these are best used in areas where frost is not a regular concern.

Air can also be mixed with engine-powered fans mounted high on a support pole (see Fig. 6.31). These are fixed high enough to draw down the warmer air from a typical inversion layer and mix it with the colder air.

Other innovations for mixing air include (i) portable, ground-based fans that blow cold air at ground level up into the inversion layer, resulting in

Fig. 6.31. Two frost fans are visible in this picture of a pre-budbreak vineyard. The snowy mountains in the distance are a source of cold air and potentially frosty conditions.

mixing (see Fig. 6.32) (Augsburger, 2000; Guarga *et al.*, 2003); (ii) tractor-mounted fans (heated or unheated); and (iii) forcing air through the drip irrigation system, which causes it to heat as well as cause air to move through the vineyard (Annabell, 2001).

Other methods are available that add heat to the vineyard environment which are useful for both radiation and advective frosts. Burning straw bales is one such technique, or the use of fuel burners, also known as smudge pots, in vineyards (see Plate 33). However, for a number of reasons these are no longer favoured. The method by which they work is sound, however. Aside from providing radiant heat, they also create a mixing of the inversion layer as the hot air rises, meets the upper layer, cools and then settles back to the earth, effectively circulating the air. Point sources of heat, if too strong, can cause such an upward rush of hot air that it travels right through the inversion layer (Trought *et al.*, 1999), causing complete loss of the warmer air, with potentially catastrophic results back on the ground.

It may not seem intuitive, but the addition of water to the vineyard system can also add heat in a useful way. The most effective systems apply water over the vines, adding heat (the heat of fusion) as the water freezes. As long as there is liquid water freezing around the plant tissues, their temperature will not fall below −0.6°C (Evans, 2000), which even green grapevine tissues can tolerate. The result of this can be spectacular (see Plate 34). Application of water should continue until the ice is melting because, if it is not, there is no heat of fusion, and the temperature of the ice and vine tissues can drop precipitously (Evans, 2000).

Fig. 6.32. A forced cold air system called Selective Inverted Sumps, developed in Uruguay. The fans pictured are portable, but others can be built in vineyard low spots to access the coldest air.

Water can be applied over the whole vineyard area, or be targeted towards the vines. Micro-sprinklers (see Plate 35) limit the amount of water used compared with that of impact sprinklers; alternatively, in-row sprinklers can be used, which saves further on water consumption (Trought *et al.*, 1999).

Use of the methods mentioned so far is fairly simple in concept, but complex in execution. Those interested in learning more about these frost protection systems are referred to Rieger (1989), Snyder *et al.* (1992), Trought *et al.* (1999), Evans (2000) and Poling (2007).

Other methods

There are a number of other methods of note for decreasing the occurrence of freezing temperatures. Already mentioned in this book are techniques such as keeping cover crop height to a minimum and practising delayed pruning. Another makes best use of the heat that the soil absorbs during the day by maximizing the amount that is released at night. To do this the soil must be free of weeds or other plants and also be moist, as this darkens the soil and also increases its heat-carrying capacity. The soil surface was traditionally packed, but at least one study has found that cultivated soil releases slightly more heat at night (see Fig. 6.33; Creasy, 2004).

Fig. 6.33. Examples of cultivated (upper) and packed (lower) soils used to assess their effect on fruiting cane temperatures during frost events (from Creasy, 2004). The influence of frost cloth tents over the canes was also evaluated, but was found to make little difference. Note how dark the moist soil is, contributing to the amount of energy that can be absorbed during the daytime.

CANOPY MANAGEMENT

Canopy management, in simple terms, is the manipulation of shoots, leaves and clusters to optimize conditions for producing grapes for a specific purpose.

Goals and tools for canopy management

The goal for canopy management is to optimize light interception by both leaves and fruit. To what level this is done depends on the purpose for which the grapes are being grown. In terms of achieving the optimum exposure, site selection is still the first step. Selecting an appropriate site has a life-of-the-vineyard impact, so choosing one that is compatible with the production goals is important. For example, planting a vineyard on a particularly vigorous site will result in much additional time and labour spent correcting the excess vigour, which shades the fruit, through shoot trimming, shoot thinning, leaf removal, etc.

The next important opportunity to affect your canopy management is through scion and rootstock choice: both should be matched to the site (or vice versa) and its potential for vine growth, e.g. choosing a shallow-rooting rootstock for a deep fertile soil to minimize the vigour effect.

Vine planting density is also an important pre-planting decision that affects canopy management. Spacing that is too wide can result in vines that cannot sustain enough growth to fill the trellis. Spacing that is too close can result in vigorous growth and crowded canopies. If there is potential for very vigorous growth, a divided canopy should be considered right from the start.

Row orientation is also important. North–south-oriented rows have east- and west-facing sides and therefore the sun will spend half the day shining directly on one side and half the day shining on the other. East–west-oriented rows have a north and a south side to the canopy, but only the north side receives direct sunlight. Fruit from the continually shaded south side of canopies ripens more slowly and may be more susceptible to disease than the other side, because of the lack of sunlight (Naylor *et al.*, 2003). Therefore, fruit from E–W-oriented rows tends to be less uniform.

Cover crop, irrigation and fertilization can be management tools that influence the vegetative growth of the vine. Vines will grow best in a vineyard with luxurious levels of water and fertilizer, and with bare soils. Thus the canopy management goal is to manage canopies to fit within the trellis, using any management tool available.

Leaf petiole analyses can be used to monitor the nutritional health of the vines, and soil nutrient tests can supplement this information. This information should be used to apply to the vines only those nutrients that are needed.

Limiting the amount of water accessible to the vines is an effective way of reducing vegetative growth. This can be either directly, by managing irrigation inputs, or indirectly, by managing between-row cover crops. Monitoring and maintaining soil moisture levels to appropriate levels for desired vine growth is essential in minimizing the number of canopy management interventions that are needed.

Shoot thinning

In terms of active, in-season, canopy management, shoot thinning is an effective tool for reducing canopy density. If a vine has been pruned to two-node spurs, non-count shoots may still arise even if the vine is pruned to approximately the right number of nodes to achieve balance. These non-count shoots generally are not fruitful, increase canopy density and should be thinned out.

Shoot thinning is especially important in the head area of the vine, as there are many latent buds embedded in the wood that can develop into non-count shoots, crowding the renewal area.

The best time to perform shoot thinning is early in the season, when the shoots are about 15 cm long – they are easy to see and come off easily at this stage. Later in the season the base of the shoot becomes more firmly attached to the cane or spur and the shoot must be cut off. More non-count shoots can grow as the season progresses, but management of cover crops, irrigation or other techniques could be used to decrease vine vigour and prevent this. The effect of shoot thinning alone as a canopy management device can be significant (see Fig. 6.34).

Fig. 6.34. Shoot thinning can have a positive impact on vine canopy density. An unthinned 'Pinot meunier' canopy (upper) shows little light coming through from the other side. The shoot-thinned canopy (lower) demonstrates a much more open canopy.

Shoot thinning, in the ideal scenario, is a temporary fix to problems with the canopy. If shoot thinning is necessary every year the vines may be out of balance, and making a more permanent change to bring the vines into balance would be prudent (e.g. a trellising change, removing or planting more vines, changing the cover crop, etc.).

Shoot positioning

Another vital aspect of canopy management for most trellising systems is shoot positioning, which must be practised diligently to get the best out of them. Often, what is desired is a range of shoots positioned perpendicularly to the fruiting wire, either straight up (e.g. for VSP) or straight down (e.g for half the canopy of a Scott-Henry trellis or for Geneva Double Curtain). Shoots growing out into the row, down and below the fruiting wire (for upright trellises like VSP) and also, very importantly, shoots growing along the row, should be repositioned as early as possible since these contribute to shading and also tend to develop strong lateral growth (see Fig. 6.35). The out-of-line shoots have implications for the effectiveness of shoot topping, as described below.

Fig. 6.35. Shoots in a VSP that are not vertical. These horizontal shoots are the source of many leaves in the fruiting zone, which leads to high canopy density, fruit shading, poor spray penetration and thus increased risk of disease.

Foliage wires are an important part of many trellising systems. If these wires are lowered at pruning time, in the spring, when the shoots reach 25 cm or so in height, the wires can be lifted into place to guide and support the developing shoots. Additional wires can be lifted up to support further growth, and wires can also be shifted up the trellis as needed. This lessens the amount of hand-tucking that is required. Note that, in areas with high winds, catch wires also help prevent shoot breakage.

Shoot topping and hedging

Shoot topping (summer pruning or hedging) is another temporary-fix management tool. As long as growing conditions persist, the vine will continue to grow: under ideal conditions a shoot can grow to many metres long in a single season. This would be enough to grow up and out of the top of a typical trellis and back down to trail on the ground. Clearly, such long shoots would limit access through the vineyard as well as go against the goal of optimizing exposure of the maximum percentage of leaves and fruit.

With a well-balanced vine, shoot growth will slow as the vines approach fruit set and stop by the time the fruit is pea-sized. Shoot length at this point wouldn't be far outside a typical trellis and shading would not occur. However, in most vineyards vines are able to grow much more than this and, to prevent shading between and within the rows, any parts of the shoots growing outside the trellis volume should be cut off (see Fig. 6.36).

It is important not to cut the shoots too short, however, as typically there must be ten to 14 leaves on each shoot to support the development of a typical crop load and the rest of the vine (Howell, 2001). Most trellis designs should accommodate a sufficient amount of shoot.

There is the possibility of lateral growth after initial topping. Lateral shoots will tend to develop from the shoots after topping, so their development should be monitored to determine when to top again. Regrowth should not extend outside the trellis volume to the point where it begins to shade other parts of the vine. In some vineyards shoot topping may need to be performed many times in a season – an expensive proposition.

Leaf removal

There are a number of reasons for leaf removal on vines. One follows along from the primary goal of canopy management: to optimize the exposure of fruit to the sun. In general, this hastens fruit maturity, increasing soluble solids and decreasing titratable acidity (TA), and results in increased varietal aroma, flavour and colour (Smart and Robinson, 1991).

Fig. 6.36. Trimming shoot tops and sides with a mechanical hedger, which maintains the canopy in the shape necessary to prevent between- and within-row shading.

The amount of leaves to be removed varies with the desired end use of the grapes. For wine grapes 60% or more exposure of the clusters to the sun is regarded as the most beneficial (Smart and Robinson, 1991). In some vineyards, 100% leaf removal in the fruiting zone is practised (see Plate 36), though this can result in sun/exposure damage to some of the fruit (Kliewer and Lider, 1968). Optimum leaf removal opens up the canopy around the fruit sufficiently to obtain the benefits described, but not so much that direct sun can harm the fruit. For some production regions or end uses, less direct exposure to the sun may be warranted as grape flavours may be changed too much.

One major outcome of leaf removal is a reduction in the incidence of botrytis infections (Gubler *et al.*, 1987). For this, it is often necessary only to remove leaves from just the east side of a N–S-oriented row. This exposes the fruit to the morning sun, which will help to dry out the area sooner, reducing the number of infection periods. On an E–W-oriented row, it is usually best to remove leaves on the side of the canopy facing away from the sun to improve light exposure. If the canopy is wide or thick, leaves from both sides may need to be removed to obtain a sufficient improvement in airflow and exposure.

The best time to remove leaves depends on the situation, but it should not be performed prior to fruit set, or crop load and flower cluster initiation may be affected (Poni *et al.*, 2005). Earlier leaf removal ensures that the berries develop in an open environment, which is beneficial as if it is left too late (e.g. until after véraison), berries may be burned through this sudden exposure to the sun (see Plate 37).

Other methods

Most active management of vines concerns manipulation of the fruit, shoots and leaves. However, aside from the use of cover crops and control over irrigation, the roots can also be managed through pruning. Much as the pruning of canes limits vine capacity, because there is a balance between root, shoot and fruit growth, if the roots are also pruned there can be a decrease in shoot growth. In practice this is accomplished by running a strong tine through the soil down the vineyard row, breaking vine roots that are growing out towards adjacent rows (see Fig. 6.37). Ripping can be carried out intensively (two sides of each row), or less so by ripping every other row.

Fig. 6.37. Root ripping, or the breaking of roots that are growing out between the rows. This practice prunes growth below ground in much the same way that canes are pruned above ground, and also limits the capacity of the vines.

Things to avoid

1. Wide canopies: a thick canopy presents difficulties in two ways. One, it is difficult to manage a thick canopy. Exercises such as leaf removal are difficult to perform as interior leaves need to be targeted and it may require more time to remove enough leaves if the canopy is wide. Two, it presents problems in disease control as a thick canopy decreases light reaching inside, reduces air movement, increases relative humidity within the canopy and decreases spray penetration. All of these can contribute to the increased risk of disease (Thomas *et al.*, 1988). Additionally, more fruit will be within the canopy where it won't be well exposed to light, and it won't mature as quickly or accumulate as much colour.
2. Dense canopies: for reasons of exposure and disease incidence already mentioned, a dense canopy works against the goal of a high-quality crop (see Plate 38). Dense and shaded canopies also reduce bud fruitfulness the following year, resulting in smaller potential crops and a tendency for the vine to move to a vegetative cycle (Smart and Robinson, 1991). Use shoot thinning and leaf removal as temporary measures, and investigate ways of reducing vegetative growth in a more permanent way.
3. High shoot densities: a contributor to dense canopies, this commonly occurs due to over-pruning (and thus the development of many non-count shoots) or leaving too many canes on fruiting wires. This may be an indication that dividing the canopy may be a better management option. Alternatively, you may be able to change your irrigation scheduling or use of cover crops to reduce the amount of water available to the vine.
4. Large gaps in the canopy: areas of the trellis not filled with canopy are effectively wasted. Gaps can be caused through missing vines or inadequate growth. Replant any missing vines, especially early in vineyard development. If the vines cannot maintain adequate growth to the fill the canopy (both along the row and up to the top of the trellis), then ensure that the vines are healthy, receiving enough water and nutrients, etc. If nothing is amiss, then the vine spacing may have been too wide, and adjustments to other management practices – such as removing/changing cover crop, irrigation and fertilization plans, etc. – should be made to increase vine capacity.

Quantifying change in the canopy

There are many reasons why the ability to acquire hard data on canopy characteristics, such as density and fruit exposure, would be useful: (i) in testing the effectiveness of shoot thinning on reducing canopy density; (ii) in monitoring changes in irrigation management on shoot growth and fruit exposure; and (iii) in performing quality control on contracted vineyard labourers, etc. In each case having some figures to back up personal observations is helpful.

The Point Quadrat method of evaluating vertical canopies is one such tool (Smart, 1988). In this a thin rod, simulating a ray of light, is pushed through the canopy at fruiting zone height. Each intersection with a leaf or cluster is recorded, and the process repeated at intervals of 20 cm down the row. If the rod makes no contact with anything, then the record for that point is considered a canopy gap. This simple, if somewhat tedious, operation can yield useful data about canopy characteristics, such as (i) leaf layer number (LLN, the total number of interceptions with leaves over the total number of insertions); (ii) interior leaves (the total number of leaves surrounded by either a leaf or a cluster over the total number of leaves); (iii) interior clusters (ditto, but for clusters); and (iv) percentage canopy gaps (the total number of gaps over the total number of insertions).

The Point Quadrat data align nicely with other means of measuring canopy density, including the use of light meters (Smart, 1988). This system's advantages are (i) only basic equipment is needed; (ii) it is objective; (iii) it is not difficult to train people to collect the data; and (iv) it gives useful information. The major disadvantage is that is requires quite a bit of time to collect the data. However, in those instances where differences in canopies need to be quantified, the Point Quadrat method is worth serious consideration.

More recent methods of evaluating canopy density include (i) the optical recognition of canopy gaps – usually using a digital camera in the visible or infrared range to detect areas of the canopy through which light can pass (known as the gap fraction, Dobrowski *et al*., 2002); and (ii) the amount of green within the canopy zone. It is to be expected that there will be advances in this area, which will allow the routine collection of data on canopy status in a vineyard, even as the season progresses.

The results of good canopy management are;

1. An open canopy.
2. A uniform canopy, which is an easy-to-manage canopy and one that can produce a higher quality of fruit.
3. Less disease: powdery mildew development is inhibited by sunlight (Willocquet *et al*., 1996), and botrytis and other diseases need moisture and high humidity for optimal development (Thomas *et al*., 1988), and thus opening up the canopy decreases the length of time when conditions are favourable for disease organisms as well as improving spray penetration.
4. Higher yield: better shoot exposure means less loss to disease, greater fruitfulness and a greater ability to ripen a crop.
5. Higher-quality fruit: not only can there be more fruit, but it can all be of better quality: with good canopy management, there will be an increase in percentage soluble solids, colour, fruit aromas and flavours and decreases in TA and herbaceous aromas and flavours (Smart *et al*., 1988).
6. A higher-quality end product.

IRRIGATION

The management of water application to vineyards is a valuable grape production tool in many regions around the world. In some areas, natural rainfall is sufficient for commercial production of grapes, and in other areas there is sufficient rainfall even for the establishment of young vines. In those places where there is a seasonal net water deficit (rainfall and soil-available water is not enough to support vine growth through part of or a whole season), application of supplementary water is necessary. If water supply is inadequate, both the vine and the crop can suffer (see Fig. 6.38).

In some instances, water may be applied by hand, such as for vine establishment. Other methods of delivering water to vineyards are through flood irrigation and some type of tubular reticulated water system. Though common for some crops, in general overhead sprinkler irrigation is not recommended for grape production due to the adverse effect it has on disease incidence (though this type of system can be used for frost management or to delay budbreak (Lipe *et al.*, 1992)). Under-vine micro-sprinklers, however, can be used in vineyard systems as the canopy and fruit does not become wet.

Flood irrigation (see Fig. 6.39) has the advantage of not requiring great capital expense, but its limitations are that is it practical only in level areas, requires flattening of the vineyard floor and subsequent maintenance, can result in uneven water application and is inefficient compared with targeted irrigation methods (Peacock *et al.*, 1977b). A modification of flood irrigation, twin-furrow irrigation (see Fig. 6.40), uses less water than flood irrigation but has similar drawbacks (Christen *et al.*, 1995; Neeson *et al.*, 1995).

Fig. 6.38. Pre-véraison berries showing signs of severe water stress, which will be affecting cell division and enlargement.

24

26

25

Plate 24. Grafted vines in the nursery, after incubation at 28°C. The buds start to push and shoots form during this time, adding a bit of colour to the nursery.
Plate 25. Use of plastic mulch during vineyard establishment, which conserves moisture and reduces weed competition. Disposal of the plastic once it starts to degrade is a problem, however.
Plate 26. Montage of different types of vine shelters.

Plate section supported by John Coleman of Plasma Physics Corporation.

27

29

28

Plate 27. A trellis-less vine in Greece. Vines are woven into a basket shape to encourage collection of early morning dew, and spaced far apart to so the amount of soil water available to the vine is maximized (Photograph reproduced with permission, L. Brenner).

Plate 28. Vines trained to an Umbrella Kniffen-style system in Margaret River area of Western Australia. The canes are arched over wires placed at the desired height over the vine heads.

Plate 29. Sylvoz trained vines, where their short canes are tied to a lower wire.

30

31

32

Plate 30. An overhead vineyard trellis system growing Sultana grapes for fresh consumption in Chile. Note the fresh shoots on the ground, which are the result of shoot thinning to open up the canopy.
Plate 31. Mussel shells as part of a research trial at Neudorf Vineyards in Nelson, New Zealand.
Plate 32. Shoots that have been damaged in a radiation frost. Only the tips have been affected as heat loss from these was greater than the leaves below, which were protected from direct radiation loss by the shoot tips.

33

34

Plate 33. A row of fuel burners in a Californian vineyard. Though common in years past, they are becoming less used due to cost, convenience and environmental concerns.

Plate 34. The aftermath of frost protection using sprinklers. Inside the ice the vine tissues are still unfrozen (Photographs courtesy, D. Darlow).

35

36

37

Plate 35. Micro-sprinklers mounted atop each vineyard post deliver water evenly over the vineyard.
Plate 36. Leaf and shoot removal in a VSP vineyard. Management of these vines is to remove all of the leaves in the fruiting zone.
Plate 37. Berries that have been damaged from late season leaf removal. Direct exposure to the sun and heat has caused blackening of the berries.

Plate 38. The interior of a dense and wide canopy. Note that the shoot carrying the lower cluster has lost all its leaves due to the lack of light.

Plate 39. An example of severe chlorine toxicity in an Australian vineyard. The marginal chlorosis is not diagnostic of this however, as these symptoms can appear through a number of different ways.

Plate 40. A broken end assembly post caused by passing machinery, because the post was slightly out of line with the row.

41

42

43

Plate 41. Botrytis infection contributing to 'noble rot' of some Riesling grapes. Note the shrivelled but whole berries.

Plate 42. Botrytis has infected frost damaged shoot tissue, creating a source of inoculum for later in the season.

Plate 43. Point infections of Botrytis on Sauvignon blanc berries. It is likely that the entry point was a defect or a stomatal pore in the berry skin.

Plate 44. Powdery Mildew on left, grapevine leaves, showing leaf curling; right, shoots, and lower, overwintering canes (dark areas).

Fig. 6.39. Flood irrigation in a New South Wales, Australia 'Semillon' vineyard. The land must be levelled quite precisely for flood irrigation to be efficient.

Fig. 6.40. A modification of flood irrigation: twin-furrow. This system uses less water than the basic one, but has similar drawbacks.

Targeted irrigation, usually using a series of pipes and tubes to deliver water to the vines, is now common in many viticultural regions. Establishment costs can be high, but there are savings in the amount of water used and advantages with the precision to which water can be applied to vines.

Typically, a small tube is laid down under each row and drippers (also called emitters) – either built in or inserted at appropriate intervals – deliver water at a known rate. Tubes can be tied to a wire or simply rest on the ground (see Fig. 6.41). Older-style emitters often required the former method, as soil particles could be sucked back up into the tube as it drained, but the modern irrigation laterals are quite robust and simply need to be kept away from machinery paths.

In areas with high temperatures and low water availability, irrigation tubes can be buried underground, delivering the water directly to the root zone and further increasing the efficiency of application (Camp, 1998).

Fig. 6.41. Irrigation lateral tube with built-in emitters. The short distance between emitters ensures an evenly irrigated line down each row, which is especially advantageous with newly planted vines.

Methods of monitoring soil moisture

In grape-growing areas that need supplementation of water supply, investment in some means of measuring soil water will benefit management of the vineyard. Observing the trends of how the soil profile dries out allows advance planning as to when to schedule irrigation, which can save on water and pumping costs and maintain vine growth at a preferred rate. Addition of too much water simply multiplies the work necessary to keep the canopies in good shape: more shoot trimming and leaf removal at the very least.

At its most simple, the amount of moisture in the soil can be followed through simple calculations of data available in many weather reports. If the soil starts the season at field capacity (where no more water can be held by the soil), evaporation pan data (a standardized method of measuring water loss from liquid form into the atmosphere) can be converted to evapotranspiration that occurs from plants. The formula for this is

$$\text{evapotranspiration (ET)} = \text{evaporation pan figure } (E_{pan}) \times \text{crop factor } (k_c)$$

The crop factor adjustment accounts for the difference between water loss from an open pan of water and water loss from plant stomata, with values typically ranging from 0.2 to 0.5 (Nicholas, 2004). The crop factor will vary between plant types, and also with the stage of development of the plants, as small grapevine shoots early in the season transpire less water than a full canopy later on. However, if a cover crop is present, the overall crop factor may not change much over a growing season (Yunusa *et al.*, 1997).

The water lost to evapotranspiration can be calculated on a daily or monthly basis (at whatever frequency that E_{pan} data are available) and subtracted from the field capacity value, which has added to it any rainfall and/or irrigation applied. Figure 6.42 shows an example of a season's soil moisture as estimated through these types of calculations. In this example, irrigation needed to be applied to keep the plant-available water from dropping below zero. Even though there is quite a bit of rainfall in the winter and spring, none of this is usable in the soil profile because it is already at field capacity. Note that there is still a deficit in soil moisture at the end of the season that may not be replenished by winter rainfall: if this was the case, additional irrigation would be required in the following season.

As an indirect method of keeping track of soil moisture this has its advantages, as no equipment is needed; however, your particular site may not have the same weather conditions as the site from which the data come, which can lead to significant errors in determining the amount of water left for the vines. Unless this method is combined with periodic verification with some form of direct soil water measurement, it can only be used as an indication of irrigation needs. More precise management techniques will require other methods.

The amount of water in the soil can be measured, or the water status of the plant can be estimated. Quantifying the former can be done through measuring soil moisture tension, which is the amount of force required to extract water

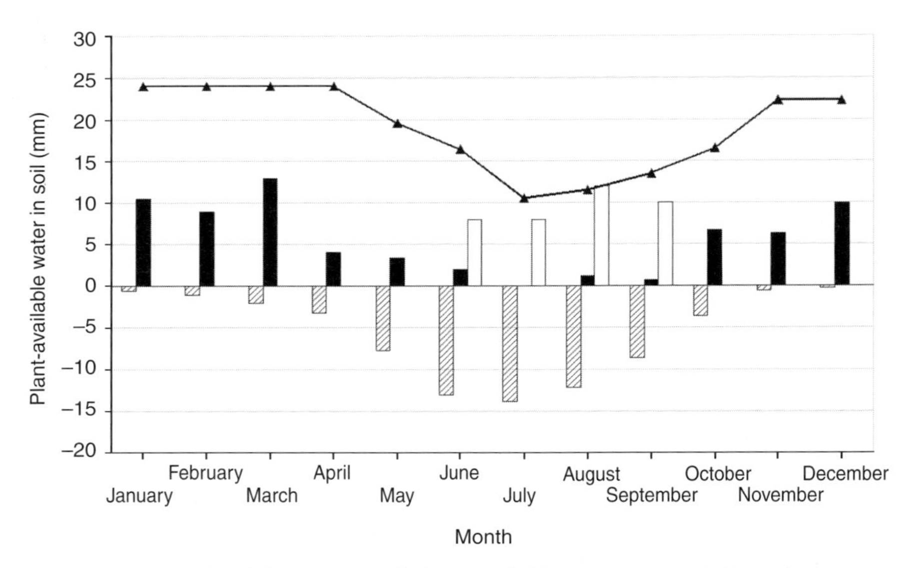

Fig. 6.42. Example of the amount of plant-available water in a soil through a growing season (line) and how evapotranspiration (hatched bars), rainfall (black bars) and irrigation (white bars) affect it. At field capacity the soil has 24 mm of water available in the profile, but this dips to as low as 10 mm in July.

from the soil particles (this is the same force that a plant root has to contend with). A tensiometer (a partly filled water tube with a porous cap on the bottom and a vacuum gauge on the top) measures this directly when it is buried to the appropriate depth in the soil. As the soil dries out, water moves from the porous cap into the soil, creating a vacuum in the tube, which registers on the gauge. These are fairly simple devices that work well at higher soil moisture contents (Tollner and Moss, 1988). However, because grapevines are fairly efficient at removing water from the soil, tensiometers are not ideal for use in vineyards.

Another commonly used instrument relies on the electrical conductivity of a substance changing as the water content within it changes. Gypsum is often used as the conducting material, and a block of it is buried in the soil. As the soil dries so does the block, and the change in its electrical conductance can be recorded. These sensors are more accurate in drier soils than tensiometers (Zazueta and Xin, 1994), and have common application in vineyard situations. As an electrical output is being measured, the data can be logged and transmitted to remote stations, automating some of the monitoring process. However, the blocks have a finite life, which is shortened when used in acidic soils, and it can be difficult to convert the conductance output into actual required amounts of water to apply (Johnston, 2000).

Other methods measure the dielectric constant of a soil (which varies with moisture content), through capacitance or time-domain reflectometry (TDR).

Capacitance is measured with a probe lowered into a tube that has been carefully installed into the soil, and TDR with stainless steel rods inserted precisely into the soil. Both of these methods measure only a small volume of soil and require precise placement of the tube/rods, but they can determine volumetric soil water content (making it easier to calculate the amount of water needed) over a wide range of values (Whalley *et al.*, 2004; Plauborg *et al.*, 2005).

Perhaps the most accurate method of determining soil moisture is with a neutron probe, which measures the number of neutron particles that are reflected back to a source from the soil, with a wetter soil reflecting back more neutrons. As such, this method also gives an indication of volumetric soil water content and has the added advantage of being able to measure a much larger volume of soil than capacitance or TDR methods (Zazueta and Xin, 1994). Because of this, the tubes into which the probe is inserted do not have to be so precisely installed, leading to more consistent measurements from tube to tube. However, the cost of these machines is high, taking readings is a slower process (up to 32 s each), the unit must be operated manually and, as a radiation source is used, the operator often must be licensed. As a result of the convenience factor, there is a tendency for growers to use other methods (Nicholas, 2004).

Methods of monitoring plant water status

Other than measuring what's in the soil, the water status of the vine can also be measured which, since it directly measures the plant rather than inferred plant status from the soil status, may be more useful. For many years a pressure chamber has been used to do this in research (Scholander *et al.*, 1965; Waring and Cleary, 1967; Ritchie and Hinckley, 1975), and now the technology is well within the grasp of many growers (Choné *et al.*, 2001; Williams and Araujo, 2002). With this method, a leaf is cut off the vine and the blade placed into a chamber that can be pressurized (see Fig. 6.43). The petiole of the leaf is left protruding from the chamber and the pressure inside the chamber increased (usually from a compressed gas cylinder). Since the xylem of plants is under tension the fluid inside retreats back toward the leaf blade when it is cut from the plant. As pressure is applied to the leaf blade, the fluid in the xylem is forced back up the xylem to the cut surface. When this happens, the pressure inside the chamber is equivalent to the tension that was inside the xylem before the leaf was removed. The greater the tension in the plant, the less available water and the greater the water stress.

Because there is a large variation in the water potential of plant tissues over the course of a day (Hardie and Considine, 1976), the time at which the measurement is taken must be standardized. Traditionally, pre-dawn measurements were used as this was the time at which the plants had the most available water (i.e. were under the least amount of water stress). Some

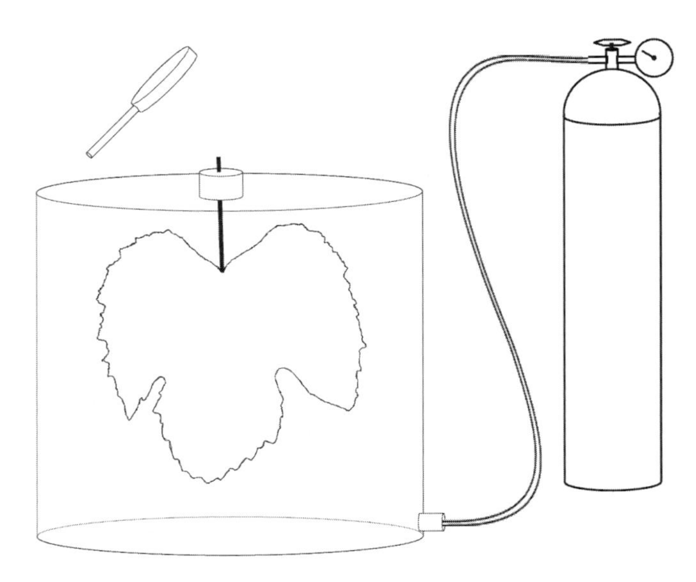

Fig. 6.43. Diagram of a pressure chamber as used to evaluate plant tissue water potential.

research suggests that midday measurements are more accurate (Williams and Trout, 2005), but others propose that any measurement times are indicative (Williams and Araujo, 2002) or that there are varietal differences in the response to water stress, which limit its broad application (Schultz, 2003).

As the level of tension in the xylem vessels increases (i.e. the water stress in the vine is increasing) water will be drawn out of the other plant tissues, causing a decrease in stem diameter (Klepper *et al.*, 1971). This can be measured through the use of accurate transducers and related to vine water status (Goldhamer and Fereres, 2001; Intrigliolo and Castel, 2007).

Water movement through the plant can also be measured through water loss via leaves (stomatal conductance, Escalona *et al.*, 2002) or from water movement up the stem (using, for example, heat-pulse methodology (see Fig. 6.44; Green and Clothier, 1988). The former technique is difficult to automate (Intrigliolo and Castel, 2007), but the latter has potential for use in commercial situations (Patakas *et al.*, 2005).

Scheduling water application

Once the level of vine water stress can be measured, a system needs to be devised through which the optimum time for application of water can be derived. The simplest strategies involve addition of water when a threshold has

Fig. 6.44. A sap flow meter attached to a vine trunk. The device applies a pulse of heat to the trunk and measures the amount of time it takes for the heat to travel up the trunk to the detector. The time is related to how fast the sap is flowing and, therefore, to the rate of water used by the canopy.

been reached. For example, if neutron probe results show that the soil moisture is reaching the point where vines will be unable to draw water from it (or some other threshold set by the vineyard manager), then water is applied. With this system the vines are always in a situation where there is no water shortage – often an advantage where maximization of quantity is desired. However, in those situations where the amount of yield is not the main focus, or where the vegetative growth needs to be managed, a strategy that leaves the vines under some water stress is often employed.

Two examples of this are the use of Regulated Deficit Irrigation (RDI) and Partial Rootzone Drying (PRD). RDI was first used on fruit trees (Chalmers *et al.*, 1981) and involves giving the plants less water than they use in ET. This reduces shoot growth because vegetative growth is more sensitive to water stress than is reproductive growth (Coombe and McCarthy, 2000). A basic RDI programme would be to maintain water availability until following fruit set, then in the period of cell division in the fruit limiting the amount of water, followed by provision of adequate water supply during ripening (Nicholas, 2004; Fereres and Soriano, 2007). This results in efficient use of water while not affecting yields in comparison with a full-water control.

PRD is a type of RDI, but using spatial separation of water application rather than just lowering the amount (Dry, 1997). Water is applied to one side of a vine and the other side is left to dry out, whereupon the watered side is reversed and the process is repeated through the season. This results in a reduction of shoot growth, attributed to the production and transport of abscisic acid (ABA) from the roots to the shoots (Stoll *et al.*, 2000). ABA has been found to decrease the stomatal conductance of the leaves, leading to reduced photosynthesis and shoot growth. There has been extensive research into the potential of PRD to maintain grape quality while not affecting yields or vine health; however, on a use of water basis, there appears to be little advantage in using PRD over RDI (Gu *et al.*, 2004), especially considering the additional infrastructure that is needed to make PRD work.

Effective measurement of vine water status is essential for strategies that employ near to detrimental soil water deficits, as slight under-application of water can result in severe vine stress and loss of productivity, or even death (see Fig. 6.45).

OTHER MANAGEMENT PRACTICES

Girdling

Girdling is an old practice, as Coombe (1959) credits a report by Lambry in 1817 as the first one describing the technique. Girdling is the removal of a ring of bark 3–6 mm wide from the trunk, arm or cane basal to the cluster (see Fig. 6.46). The result is that the carbohydrate materials produced in the leaves accumulate in those parts distal to the wound, including the clusters of flowers or fruit. Girdled bands usually heal over within a few weeks (Jensen *et al.*, 1981), but that is enough to increase fruit set if performed at bloom; or, if performed at the stage of rapid fruit growth (immediately after fruit set), to increase berry size by 10–90% (Coombe, 1959; Jensen *et al.*, 1975). However, the magnitude of the response depends on cultivar. Additionally, girdling can also hasten maturity by improving colour and sugar accumulation if carried out later in the season (Peacock *et al.*, 1977a).

Fruit thinning

Thinning of all types can be used for crop reduction (see Fig. 6.47), but frequently other goals are desired, such as removing diseased or fruit exhibiting delayed

Fig. 6.45. 'Flame Seedless' vines that had been watered at 25% E_{pan} for the previous two seasons. Note the lack of fruit and the very limited vegetative growth. The vines will survive, but only just.

Fig. 6.46. Example of a freshly girdled trunk (left) and one that was so treated in the previous year, but that has since healed (right). Photographs were taken at véraison in a table grape vineyard.

Fig. 6.47. An example of an overcropped young vine. Excessive fruit for the size of the vine leads to poor fruit quality and vine carbohydrate status and reduced growth in the following years.

ripening. The main types of hand thinning are of flower clusters, clusters and berries. Hand thinning is selective, and the worker can pick the best size and shape of clusters for the market, which saves time later on during harvest.

Flower cluster thinning

This can be carried out early in shoot development because the clusters are easily visible before many leaves have expanded. The early removal of flower clusters can increase slightly both harvest berry size and Brix (Dokoozlian and Hirschfelt, 1995). Some cultivars produce three to four clusters per shoot, and usually the basal cluster is the largest. Removal of the distal clusters is beneficial for basal cluster development and for crop control of the whole vine. Although this stage is the optimum time for hand thinning because the clusters are not obscured by leaves, it is not the optimum time to remove clusters in varieties that tend to form overly compact clusters, as this will only accentuate the problem.

Cluster thinning

This involves the removal of clusters after fruit set, so the berry number of the remaining clusters is not affected. The major purpose is to remove excessive crop load and balance the leaf area with the crop load. It is used in table, raisin and wine grape vineyards. Small, misshapen and excessively large clusters can also be removed, which is beneficial in table grape production; in wine grape production dropping any crop that is lagging in maturity can also help final harvested fruit composition.

Berry thinning

Removal of parts of the cluster after berry set is primarily used in the production of table grapes. Quality improvement comes from controlling cluster size, looseness and shape. Because the rachis contains less fruit, clusters are usually less compact and the reduction in flower number (with flower thinning) results in larger berries that mature earlier (Ahmad and Zargar, 2005). In some cultivars individual branches within the cluster are removed to improve cluster appearance and reduce compactness. Some varieties benefit from cluster shortening to produce round clusters that pack easily.

Application of plant growth regulators

Gibberellic acid (GA) sprays have several uses in table grape production. If applied during mid-bloom, flowers are thinned and result in less compact clusters (see Fig. 6.48) and larger berries, but a second GA application at fruit set (at the time of girdling) can also increase berry size (Weaver and McCune, 1959; Kasimatis *et al.*, 1971; Harrell and Williams, 1987). However, not all varieties respond in the same way to girdling or GA applications, so each variety must be studied for optimum treatment, timing and suitable response.

HARVEST

As a viticultural tool, the timing of harvest has perhaps the most impact on grape composition. Therefore, knowing what the grapes are going to be used for is vital in determining when and how to harvest. For example, early harvest of wine grapes means that flavour, sugar and acid profiles are substantially

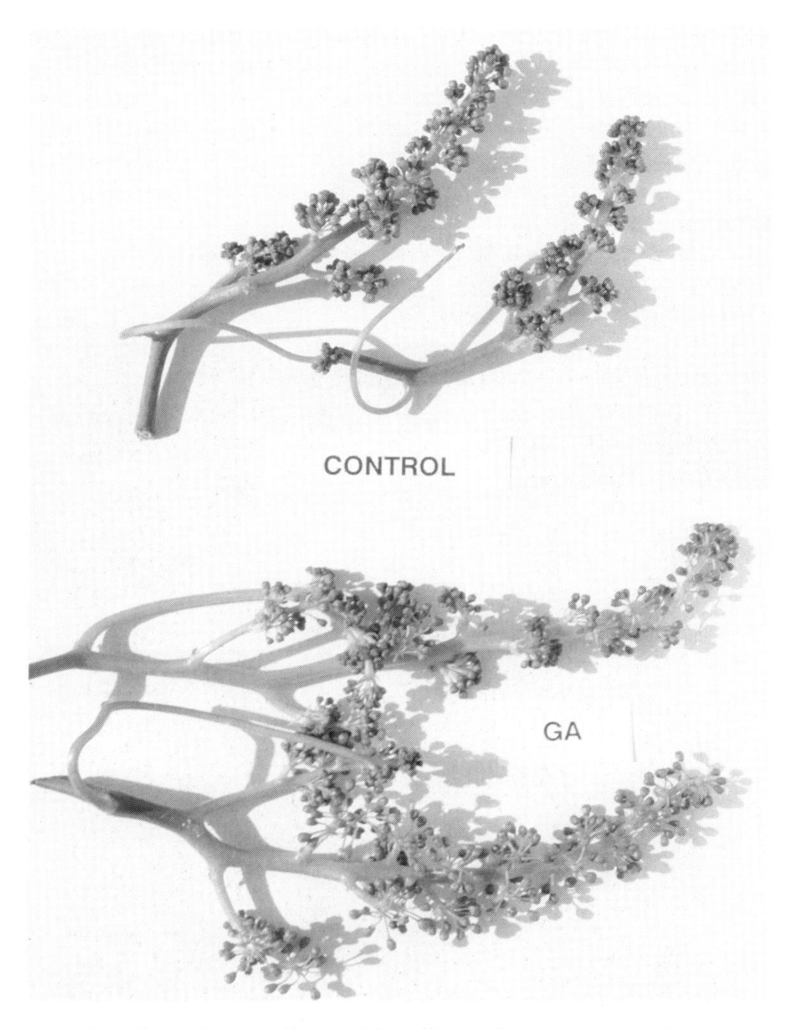

Fig. 6.48. The effect of a pre-bloom gibberellic acid (GA) spray on 'Pinot noir' clusters. Compared with the control, the treated clusters are longer and each flower has a longer pedicel, leading to a less compact cluster.

different, and suited to styles such as sparkling wines or verjus (unfermented juice of unripe grapes, used in cooking). Late harvesting of grapes can result in raisins or those suited to producing dessert wines, which have high residual sugar. The products can be very distinct, but in some cases come from the same cultivar.

Harvesting method is also of importance to product quality. For fresh consumption, delicate handling is required for premium fruit, as the bloom (waxy covering) of the grape should be preserved. Fruit are hand harvested and placed in protective containers for shipping to market.

For some wine styles, hand harvesting is a must (see Fig. 6.49); for example, for sparkling wine production the fruit needs to be intact before processing starts, or harsh phenolics may be extracted from the rachis and skins (Jackson, 2000). For many table wines, however, machine harvesting (see Fig. 6.50) is acceptable as the breaking of the berry skins and their mingling with the juices helps to extract desired compounds.

Fig. 6.49. Hand harvesting of grapes is a labour-intensive process. In this vineyard, there is the added complexity of unfastening the bird netting in order to access the grapes.

Fig. 6.50. Mechanical harvesting of wine grapes. These harvesters can be self-propelled (as pictured here) or towed by a tractor. The grapes are conveyed to a bin travelling in an adjacent row.

Nutrition

Grapevines have the same basic nutritional requirements as other higher plants, but what is a nutrient? In this chapter we will be talking mainly about mineral substances that the vines need to grow and produce fruit, but oxygen, hydrogen and carbon are also vital and are derived from water and carbon dioxide (CO_2) in the atmosphere.

Luckily, in most cases mineral nutrients are provided by the environment without our interference. However, best practice dictates that problems should be anticipated and corrected before they occur. The area of vine nutrition is an excellent example of why this is so, as deficiency in boron or zinc, which may not be apparent visually, can result in low percentage fruit set, leading to poor vine productivity. In contrast, vines that are deficient in nitrogen have low amounts of growth and usually have light green leaves, symptoms that are readily apparent. Vines that have been growing well for some years can, relatively quickly, develop a problem that can be traced back to nutrient imbalances that have come about from that nutrient having been mined from the soil by the roots, but not replaced by the manager.

Because the soil is a complex medium and the uptake of nutrients by plants is, in some cases, unregulated and in others carefully metered, determining when plants are deficient in, or suffer from excess amounts of, nutrients is not terribly straightforward. Compared with a simpler nutrient delivery system, hydroponics (where plant roots are grown in liquid where its composition can be easily altered), there is less certainty in knowing whether a nutrient that is applied to the medium will become immediately available to the plant. In dealing with soil nutrients it is important to remember that whatever is done to amend a nutrient may take a long time to show an effect and, if something has been done incorrectly, recovery time may be very long. Therefore, the process used to determine how much and of which nutrient is to be applied is very important.

NUTRIENT ANALYSIS AND CORRECTION STRATEGIES

In most cases the soil can provide sufficient amounts of nutrients for plant and fruit growth, at least in the shorter term. Since soils differ in their ability to supply needed nutrients (this can be thought of as a reservoir of nutrients), the soil type is important in site selection. Under the same cropping scheme, some soils will be exhausted of their nutrients before others. The question becomes: how can the soil be tested to determine when a nutrient threshold is approaching?

This, and the amount of nutrients needed by vines, is determined by soil analysis, knowing the capacity of the soil to supply the nutrients and by plant tissue analysis. Generally, plant nutrient deficiency symptoms occur after the most beneficial time for application of nutrients, e.g. after fruit set or near the end of the growing season. Thus, preventative monitoring and proactive application of nutrients is essential for continued vine health and productivity. The long-term goal is to maintain the nutrients at adequate levels despite changes in seasonal environment, crop loads or catastrophic weather events (such as heat, frost or hail).

What constitutes 'adequate' for producing grapevines is not necessarily the same in different vineyards. As soil characteristics, weather patterns, management systems and cultivars change so much from place to place, the values that are acceptable for production may change (Wolf *et al.*, 1983; Christensen, 1984; Schreiner, 2002). Referral to a local source of knowledge is helpful in interpreting soil and plant tissue test results.

Grape mineral nutrition management should be implemented in a long-term programme: one that reflects the ability of the soil to regularly provide optimum nutritional elements and considers the amount of nutrients that are removed by the crop. The latter can be significant, even when assuming that all vine prunings are left in the vineyard: up to 2.4 kg nitrogen (N), 3.2 kg potassium (K) and 0.4 kg phosphorus (P) are contained in each tonne of grapes. If prunings are removed (for example, due to concerns about them acting as a disease inoculum source), the amounts of these nutrients removed per tonne of prunings are 2.68, 2.76 and 0.46, respectively (Winkler *et al.*, 1974). New vine growth in the spring can acquire nutrients from either stored reserves or the soil. Up to 50% of the N and P, 15% of the K and 10% of the calcium (Ca) and magnesium (Mg) used may be from stored reserves (Schreiner *et al.*, 2006).

Testing and nutrient addition

Before vines are planted, there is little option but to test the soil as an indicator of what the nutrient status will be. It is still a useful exercise, however, as gross deficiencies such as pH or individual nutrients are more easily corrected prior to the vineyard being established. Soil collection for testing should always be done as specified by the laboratory doing the analysis, for results will vary according to the depth of soil taken (Jackson and Caldwell, 1993), as well as to the time of year. Consistent collection methods ensure that the recommendations made from the results will be as useful as possible. Despite all care, however, because vines have such extensive and deep rooting systems, getting an accurate representation of the soil and its nutrients that are actually available to a vine is practically impossible (Robinson, 1992). Nevertheless, it remains as good a tool as we have for evaluating potential vineyard soils, and should not be overlooked.

Similar to the situation for measuring plant water status versus water in the soil, measuring the nutrients in the plant tissues is a better method of knowing what the vine situation is. However, there are potentially many different parts of the vine to sample, and each can contain a different level of nutrients (Christensen, 1984; Schreiner, 2002). Additionally, the time of the season also causes variation (Smith *et al.*, 1972; Christensen, 1984; Schreiner, 2002). Because of this, there has been some standardization in the procedure.

Most commonly sampled are leaf petioles; since they are basically a conduit, they are thought to represent a snapshot of plant nutrient status at the time of sampling. Leaf blades are also used as they represent the photosynthetic workhorse of the plant. Timing of sampling is also important, typically at either flowering or véraison. Robinson *et al.* (1978) carried out a comparison of these methods and timings and found that petiole sampling at flowering generally gave the best indication of vine nutrient status, at least for the irrigated vineyards in Australia that they were surveying. It should be noted, however, that vine N status is not reliably indicated through tissue analysis (Amiri and Fallahi, 2007).

Results from frequent plant analysis (yearly is ideal, though in established vineyards with a management history it can be stretched out to once every 3 years) can be used to predict trends in plant nutrient status and then to make adjustments to the fertilizer programme before any symptoms appear. This, again, is the overall goal of plant nutrition management – to keep the vines from entering into a limiting situation.

Because of the complexity of the plant–soil system, there is not a blanket recommendation for application of fertilizers in vineyards. Some soils, such as those with a high sand content, are commonly N deficient, and therefore the application of N is essential for continued production of high-quality grapes (Spayd *et al.*, 1993). However, application of N to vines growing in heavier and younger soils is often not beneficial as it raises its level even higher, resulting in more vigorous growth and decreases in fruit quality. Fertilization programmes should be developed on a site-by-site (or even block-by-block) basis, using information about soil and plant nutrient content.

THE NUTRIENTS

Important plant nutrients are often categorized into macronutrients and micronutrients, so called because of the rough amounts of them that are required for healthy growth. The macronutrients are carbon (C), hydrogen (H), oxygen (O), N, P, sulphur (S), K, Ca and Mg. The first three come from air and water and will not be discussed further. Micronutrients (just as important, but not used in high concentrations) are iron (Fe), manganese (Mn), copper (Cu), zinc (Zn), molybdenum (Mo), boron (B) and chlorine (Cl). A brief review of each of these follows.

With reference to pictorial symptoms of deficiency or excess, readers are referred to the excellent resources of Pearson and Goheen (1988) and Nicholas (2004). Many times the term chlorosis will occur in describing symptoms: care should be taken in interpreting chlorosis, as it can be caused by shade, ozone damage, sun scald, drought, wet soil, *Eutypa*, winter injury and some pesticides, as well as by nutrient deficiencies (Jordan *et al.*, 1981). Always take into account as much information as you can obtain about a problem before diagnosis.

Macronutrients

Nitrogen

This element is a key part of many plant cell functions because it is used in making the amino acids used to build proteins, which are used to make the enzymes responsible for much of the work done in plants as well as making structural components, etc. (e.g. RuBiSCO and chlorophyll in photosynthesis). If it is lacking, vine leaves will turn completely yellow (a form of chlorosis), lower their rates of photosynthesis and cease to grow (Ryle and Hesketh, 1969; Keller *et al.*, 1998; Chen and Cheng, 2003). The root to shoot ratio will also change, with lower N levels resulting in more roots produced for the same amount of shoot (Marschner, 1986). Excess N leads to overly vigorous growth (significant lateral development, long internodes and thick canes) and large and dark green leaf blades. This can have both indirect and negative effects on vine fruitfulness (Winkler *et al.*, 1974).

Nitrogen is a mobile nutrient in the plant, so can be scavenged from one organ and transported to another that needs it more. For example, labelled N fed to vines at the beginning of growth was found in varying amounts between the various vine organs as the season progressed (see Fig. 7.1) (Conradie, 1990).

To correct deficiency, N should be applied before budbreak and possibly a second time at bloom, as these are the times when grapevines take up most N (Conradie, 1980; Schreiner *et al.*, 2006). However, the forms of N that are available to plant roots (nitrate, NO_3^- and ammonium, NH_3^+) are mobile in the soil, so if excessive water falls or is applied it can be washed (leached) from the soil profile (Bergström and Brink, 1986). Therefore, in areas with springtime rainfall, N is best applied in multiple smaller amounts of fertilizer rather than as one large dose, and starting in mid- to late season rather than in early season, to maximize vine uptake (Peacock *et al.*, 1989).

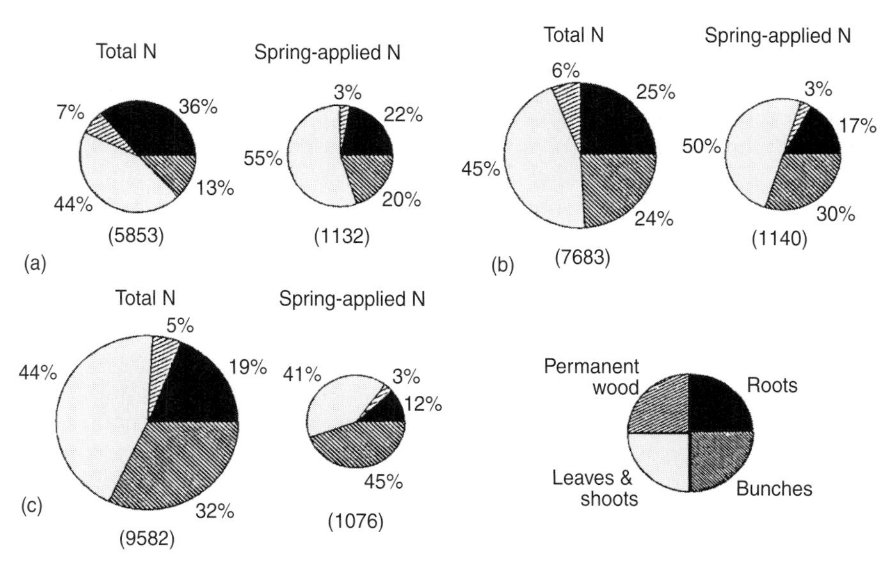

Fig. 7.1. Remobilization of spring-applied nitrogen in sand culture-grown grapevines. The charts indicate the relative proportioning of nitrogen between permanent wood, roots, shoots and leaves and fruit (when present) at harvest (A), start of leaf-fall (B) and end of leaf-fall (C). The numbers in parentheses indicate the total amount of nitrogen/vine (mg) (adapted from and reproduced with permission, Conradie, 1990).

Phosphorus

This major plant nutrient is a constituent of phospholipids and nucleic acids used in cell building and is therefore abundant in meristematic regions such as actively growing shoot and root tips. It is also involved in sugar metabolism, respiration and photosynthesis. Vines are rarely deficient in this nutrient, possibly due to efficient capture from the soil and its remobilization within the plant (Jackson, 2000). Deficiency symptoms appear as yellowing of interveinal areas of older leaves, which can turn red if the deficiency is severe (Nicholas, 2004). Unlike N, P is not mobile in the soil, so there must be exploration of new soil to secure it. Fortunately soil is generally able to supply large quantities of P, which roots take up in the form of orthophosphate (PO_4^{-3}). In those soils lower in P there are often mycorrhizal associations with the roots that assist in its uptake (Possingham and Groot Obbink, 1971). Phosphorus is frequently applied to maintain vineyard cover crops, but application of P fertilizer can result in lower K levels in the vine (Conradie and Saayman, 1989) and so should not be applied without consideration of the available K supply (Haeseler *et al.*, 1980).

Sulphur

Sulphur is taken up by roots in the form of sulphate (SO_4^-), though it can also be taken up by leaves when it is in its gaseous form (De Kok, 1990), which is the active form of S that prevents powdery mildew infection. Many plant proteins contain S: for example, cystine is a major S-containing amino acid. Sulphur-containing compounds called thiols have been found to be important determinants of white wine aroma, lending box tree, citrus or cooked leek characteristics to the mix (Tominaga *et al.*, 2000). However, S is rarely deficient in grapevines as it is frequently applied in a spray for control of powdery mildew.

Potassium

Potassium is a very important element in protein and fat synthesis, enzyme activation and as an osmotic charge balancer. Its uptake form is the cation, K^+, and is used in large quantities by fruit crops, especially. The quantity removed with a moderate crop of grapes is frequently greater than that of N (Winkler *et al.*, 1974). Xylem exudate, sap that bleeds when vines are cut close to budbreak, is high in K (Glad *et al.*, 1992), which is evidence that the roots are actively absorbing soil nutrients.

Vines with K deficiency usually develop interveinal or marginal chlorosis, or both. At more severe deficiencies, marginal necrosis and leaf cupping develops (Nicholas, 2004). False K deficiency (also known as spring fever) is a temporary appearance of symptoms on basal leaves prior to bloom and is associated with cool and cloudy weather conditions, which cannot be corrected through the application of K fertilizers (Christensen *et al.*, 1990). It is thought that N reduction is lowered under these conditions and an accumulation of ammonium ions results, which interferes with potassium's role in protein synthesis (Maynard *et al.*, 1968). In sand culture, excess of K results in interveinal chlorosis of basal leaves, slightly reduced shoot growth but unaffected root development (Li, 2004). There are also problems with high juice pH associated with excess K in grapevines (see Fig. 7.2; Mattick *et al.*, 1972), so K should be kept within moderate levels in the soil.

Potassium deficiency symptoms are known to be corrected by potash (which is any of a number of K salts) application (Shaulis, 1961), and any deficiency should be corrected as soon as symptoms become visible. Potassium chloride (muriate of potash) is usually the most economical source of K, but in growing areas with high salt concentrations it can result in chloride toxicity, so potassium sulphate is the recommended alternative in this case. Deficiencies of K in low-pH soils may be caused by excess application of Mg (Bates *et al.*, 2002), as these two cations compete for the same uptake channels in root tissues.

Calcium

Known as the cement of plant structure, Ca is a component of the middle lamella that holds plant cells together and maintains cell integrity (Epstein,

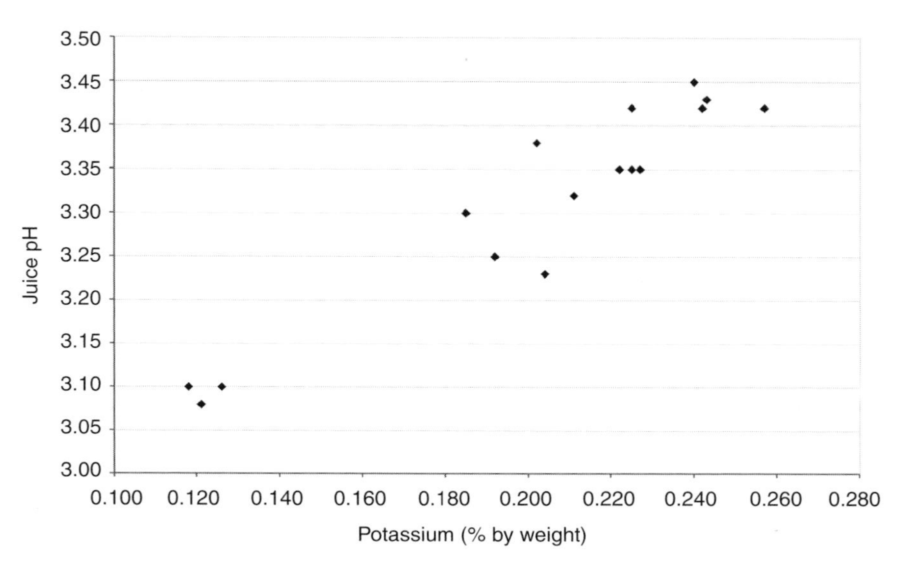

Fig. 7.2. Relationship between grape potassium and juice pH for 'Concord' grapes taken from vines of differing potassium status. As potassium concentration in the fruit rose, so did pH, in a linear fashion (adapted and used with permission, Mattick *et al.*, 1972).

1961). It also participates in both carbohydrate translocation and N utilization (Gossett *et al.*, 1977). Calcium is infrequently deficient in vineyards (Nicholas, 2004), but applications of calcium sulphate (gypsum) are used to increase soil pH and to treat high-sodium soils (Winkler *et al.*, 1974).

Magnesium

The central mineral element in the chlorophyll pigment is Mg, which assures the importance of this nutrient in plant growth. Magnesium also functions in chlorophyll synthesis, enzyme activation and can play a similar role to Ca in membrane stability. It also is important in the partitioning of carbohydrates from the leaves to the roots (Cakmak *et al.*, 1994). Deficiency is most severe on basal parts of the shoots due to its remobilization to newer parts of the vine. Symptoms of moderate deficiency are chlorosis between the large veins, which remain green, as does the tissue at the leaf margins (Pearson and Goheen, 1988; Nicholas, 2004). Severe deficiency results in chlorotic leaf margins and red coloration in red grape varieties (Nicholas, 2004). Magnesium deficiency is most common on very low-pH soils and is increased by excessive K fertilization (Bates *et al.*, 2002). Regardless of pH, however, high soil K interferes with the uptake of Mg (Wolf *et al.*, 1983) and should be avoided. To correct deficiencies, soil pH can be raised (particularly with dolomite, also known as calcium magnesium carbonate – $CaMg(CO_3)_2$) or magnesium sulphate ripped into the soil.

Micronutrients

Iron

Although not present in the chlorophyll molecule, Fe is required for chlorophyll synthesis and is also involved with electron transport in respiration. Iron deficiency is observed first in early-season chlorosis in the youngest leaves, which causes reduced photosynthesis (Smith and Cheng, 2007). Iron shows very low mobility within the plant, so symptoms appear in developing plant tissues. Iron chlorosis is often ephemeral, appearing when growth is rapid and soils are cold and more likely to be wet, which limits the ability of the vine to take it up (Tagliavini and Rombolà, 2001). As the soils dry out and warm up, supply from the soil improves and chlorosis no longer occurs in the newly forming leaves. Chlorosis can occur even when the soils have adequate Fe, for different reasons, however (Mengel *et al.*, 1984; Davenport and Stevens, 2006). Chlorotic symptoms are greatly increased in high pH (alkaline)-soils, so plant nutrient analysis under these conditions will detect low concentrations of Fe. Iron chlorosis-resistant rootstocks are available for grapevines, but often result in excessive growth and reduced grape yield (Tagliavini and Rombolà, 2001). Therefore making sure soil pH allows for Fe uptake by the rootstock desired is important. Vesicular-arbuscular mycorrhizae can ameliorate the soil immediately surrounding the roots (the rhizosphere), altering its pH and increasing nutrient availability (Bates *et al.*, 2002), and thus providing another reason to encourage their presence. In most cases Fe deficiencies are more easily corrected by iron salt or iron chelate sprays rather than by soil amendments.

Manganese

Manganese is essential in the synthesis of chlorophyll and plays a role in enzymatic reduction–oxidation reactions that power most functions in the plant tissues. Manganese (and Fe) deficiencies are usually found on high pH-soils, with symptoms of Mn deficiency identified by interveinal chlorosis on shaded and mature leaves, as this mineral can be remobilized within the plant. Manganese sulphate applied to the soil under the vine will correct Mn deficiency, although plant analyses should be conducted in following seasons to track changes in plant concentration. The use of mancozeb for disease control can maintain Mn levels in plants (Deckers *et al.*, 1997), as mancozeb is a coordinated product of Zn (1.9%), Mn (15%) and ethylene bisdithiocarbamate ion.

Copper

Copper is a non-protein component of several oxidizing enzymes such as ascorbate oxidase and tyrosinase, and forms an important part of the electron-transport chain in photosynthesis. Deficiency symptoms for Cu in grapevines are similar to those of Fe-induced chlorosis and, like Fe, it is not moved from one place to another within the plant. Copper is generally very toxic to many organisms, and one reason that it is not frequently found to be deficient in vineyards is that Bordeaux mixture (copper sulphate and hydrated lime) is very often applied to vines in the control of downy mildew. It is possible for high Cu levels in the soil to reduce root growth and development (Wainwright and Woolhouse, 1977), but analysis of above-ground parts of plants does not seem to indicate high soil and root Cu status (Brun *et al.*, 2001). Therefore, if high Cu levels in soils are suspected, the nutritional status of the root tissues should be determined.

Zinc

Zinc is important in the synthesis of the growth regulator indole acetic acid, is a component of many different enzymes and plays an important role in pollen and fruit formation (Marschner, 1986). Zinc is not very mobile within the plant, so if the nutrient becomes deficient part-way through a season, younger rather than older plant tissues will show symptoms. Zinc deficiency is characterized by abnormal development of internodes ('zig-zag' growth pattern of shoots), interveinal chlorosis in early summer, small leaves (also known as little leaf) and a widened petiolar sinus (see Fig. 7.3). Straggly clusters with undeveloped shot berries and generally poor fruit set are also common, and can occur without the appearance of leaf or shoot deficiency

Fig. 7.3. Shoot symptoms of zinc deficiency. Note the small, misshapen leaves that have an open sinus, short internodes and jagged leaf margins.

symptoms (Nicholas, 2004). Application of zinc sulphate or zinc chelate sprays can alleviate symptoms.

Molybdenum

Molybdenum is a component of the enzyme systems that carry out reduction–oxidation reactions. Enzymes that require Mo include nitrate reductase – which is essential for the incorporation of root-supplied N into cell components – and other enzymes involved in cell function (Kaiser *et al.*, 2005). It also appears to be important in pollen formation and fruit set (Williams *et al.*, 2004). Molybdenum is a mobile nutrient within plant tissues and required in only very small amounts, even in comparison with other micronutrients (Marschner, 1986). In grapes the major economic symptoms of deficiency are shot berries (millerandage or hen and chicken) and leaf chlorosis, caused by a lack of usable N from nitrate reductase. Molybdenum deficiency has been identified as the cause of the 'Merlot problem' in South Australia, where foliar sprays of the nutrient applied at bloom greatly improved vine yields (Williams *et al.*, 2004).

Boron

In grapes, boron is important in cell division and in the normal development of pollen and pollen growth, so it follows that boron-deficient vines have poor fruit set. It is an important component of cell walls (Thellier *et al.*, 1979) and is not mobile within the plant. Boron is limiting for growth in many grape-growing regions of the world (Cook *et al.*, 1960), and symptoms of deficiency include zig-zag development of shoots (similar to those for Zn, Fig. 7.3), internode swellings (Cook *et al.*, 1960) and leaf chlorosis, later turning into necrosis, though this tends to occur on older leaves as compared with Zn deficiency (Nicholas, 2004). Boron deficiency is also associated with dry soils, but is easily corrected by soil and/or foliar applications of soluble borates. Boron toxicity symptoms are misshapen and small leaves on new growth (see Fig. 7.4), brown spots on the margins of older leaves and sometimes death of the shoot tips as a result of movement of B from the roots upwards in the transpiration stream (Marschner, 1986). Boron toxicity may occur through the use of high-boron-containing Potash as fertilizer or in high-boron irrigation water. Because B toxicity is as much of a concern as deficiency, plant analysis should be used to maintain sufficient concentrations (Shorrocks, 1997).

Chlorine

Chlorine is abundant in nature (for example, between 4 and 8 kg of chloride/ha can be supplied through rainfall (Marschner, 1986)) and is important in photosynthetic reactions as an activator of enzyme reactions and as a charge-balancing ion (Marschner, 1986). Very little is required by plants, but it is usually found at much higher levels than this. Chlorine is very mobile within the

Fig. 7.4. An example of altered leaf development on shoots of 'Italia Pirovana' in Chile caused by boron toxicity.

plant. Very few incidences of Cl deficiency are reported – the usual concern is with toxicity (see Plate 39). Increasing levels of Cl supplied to vines result in reduced shoot growth (Groot Obbink and Alexander, 1973), and the critical toxic concentration for grapevines has been determined to be 400 mg Cl/kg soil (Xu *et al.*, 1999), though this can vary depending on variety and environmental situation.

Mechanization

The cultivation of grapes is a highly labour-intensive process. Unlike agronomic crops, the mechanization of horticultural crop management has been slower due to the more complex nature of the plants and production of their fruit. The proportion of harvest costs within the total cost of production for crops was higher for grapes in the 1960s (Shepardson *et al.*, 1970), and already by that time it was recognized that finding sufficient labour to harvest the fruit was problematic.

Practical research on mechanization of harvest started in the 1950s, on the west coast of the USA, with researchers in California developing a device that would cut clusters hanging from an inverted-'L' trellis. An innovative thought, but of limited practical use because the clusters did not always hang in the right place to be cut off efficiently (Winkler *et al.*, 1957). In the late 1950s, researchers at Cornell University, USA also started developing a mechanical harvester that used mechanical vertical shaking to dislodge the fruit from the vine (Shepardson *et al.*, 1962, 1970). This developed into a commercial device in 1967 that was particularly aimed at harvesting 'Concord'-type grapes grown on the GDC trellising system (see Chapter 6; Shepardson *et al.*, 1970). The advent of the modern horizontal-type shaking system had its beginnings in the mid-1960s through the efforts of two 'Concord' grape growers. Their innovation was to use plywood boards to knock a vertical trellis, dislodging the berries, which were then collected by a conveyor belt (Morris, 2000).

From these beginnings have developed a wide range of machinery that help with not only harvest, but also shoot positioning and thinning, canopy trimming, leaf removal, crop reduction and pruning. Application of pesticides is also highly mechanized, but will be discussed in more detail in Chapter 9. The ultimate goal of an all-mechanized sustainable vineyard is currently not realized, but work is continuing in this area (Morris, 2006) and will surely succeed.

A key to mechanization is vineyard uniformity, as often machines are not very forgiving of driver error or trellising structures that get in the way (see Plate 40). It is the earlier effort that goes into developing as uniform a vineyard as possible that pays off in terms of greater ability to reach management goals, such as precise leaf removal in the fruiting zone. Lasers are now commonly used to make sure that rows are put in straight, with tolerances measured in millimetres so that machinery works more efficiently.

Currently, there are limitations to mechanization in terms of physically being able to traverse or access some vineyards. Most machinery developed for vineyards is not very amenable to steep slopes, though shallower ones can be adjusted for with the use of hydraulics to keep the machine level while the wheels are at different heights on the incline.

WEED MANAGEMENT

Between-row management of cover crop or weed growth in modern times is through the use of tractor-mounted mowers. These have the advantage of being much quicker than traditional methods (use of animals as grazers or for powering cutting tools such as sickle-bars), which is especially helpful in areas where frost is an early-season occurrence as the vegetation needs to be kept as low as possible. However, machinery does also have disadvantages, such as fuel use and contribution to soil compaction, which can be mitigated by using smaller machines (such as four-wheel bikes) or reducing the number of passes through the vineyard by performing more than one operation at once. There are mowers that deposit the clippings under the vine rows rather than leaving them between the rows, which can assist with weed control under the vines as well as increasing soil organic matter over the longer term.

Hydraulically powered mulchers (see Fig. 8.1) are also in use, which can eliminate the need for pruners to remove brush from the vineyard manually or by tractor. These break the prunings up into small pieces that have less potential to release disease spores (Seyb, 2004), are more quickly degraded and therefore do not carry disease inoculum sources in the following season. Dealing with the prunings in the vineyard also means that they do not have to be disposed of through burning or burial, and the nutrients contained are returned to the soil ecosystem.

In management systems where synthetic herbicides are not able to be used, weeds have been controlled through the use of hand labour. However, in larger vineyard developments it is impractical to do this, so several tools have been developed to help mechanize the process of cultivation under the rows, which disrupts weed development. Examples of these include tools such as the mechanical tiller, a hydraulically driven rotary head mounted on an articulated arm (see Fig. 8.2). The rotor retracts automatically or semi-automatically using trunk sensors and fingertip controls that move the head away to prevent damage to the vines.

These devices can provide good control of weeds providing they are used often enough. They are much faster than hand weeding, but do require tractor movement through the vineyard, and repeated cultivation causes a decrease in

Fig. 8.1. A narrow, hydraulically driven mulcher suited to moving between vineyard rows.

Fig. 8.2. A mechanical weed management tool that uses a pressure-sensitive controlled rotary head and tines to allow under-row cultivation, mowing, sweeping and many other in-row jobs (image of the Weed Badger® courtesy ©Weed Badger Div., Town and Country Research & Development, Inc.).

the number of roots in the surface soil layer if the tillage head is allowed to run too deeply (Saayman and Van Huyssteen, 1983). Properly used, however, they can be a viable alternative to other methods, particularly in vineyards where the use of herbicides is to be avoided.

Similarly, devices that hill-up and take away soil from under the vines are used to bury and then unearth winter freeze-sensitive canes in those areas where vines are too sensitive to cold temperatures to survive in the open. This procedure can be quite effective when grapes are being grown in a region outside of the optimal climatic requirements.

Other options for weed control include (i) repeated mowing under the vines (see Fig. 8.3); and (ii) the use of directed heat in the form of flame (Ascard, 1998), steam (Melander *et al.*, 2002), hot foam (Collins *et al.*, 2002) and even microwaves (Sartorato *et al.*, 2006). These too can be effective in certain situations, but require frequent repeated use and are fairly resource intensive (Pinel *et al.*, 1999).

CANOPY MANAGEMENT

The management of grapevine canopies can now be completely mechanized, but this is very dependent on the trellising/training system in use. In the

Fig. 8.3. An under-vine mower where the mowing heads are pushed aside by vine trunks as they move down the row. This sort of system works only with vines that have significant trunk development.

minimal pruning system there is removal of only the bottom part of the canopy, which mainly serves the purpose of crop control. With GDC, devices have been developed that both thin shoots and brush the shoots into the two distinct canopies the system requires (Pool *et al.*, 1990; Morris, 2006). Brush-type devices can also be used to rub off non-count shoots on vine trunks, which is otherwise a laborious undertaking.

For hedge-type systems tractor-mounted cutter bars are useful for trimming shoot tops and canopy sides (see Fig. 6.36). For leaf removal, a movable head that either sucks in leaves toward a cutter bar or blows high-pressure air out to rip leaf blades from the petioles can be used to obtain more fruit exposure and a more open canopy (see Fig. 8.4).

The systems that use pulsed air to remove leaf blades do have one disadvantage in that the cut plant material often stays in the canopy or can fall onto the fruit, creating potential hot spots for disease development later in the season.

CROP MANAGEMENT AND HARVESTING

In addition to taking ripe fruit off of vines, mechanical harvesters can also take it off earlier in the season, as a form of crop reduction (Pool, 1987; Clingeleffer, 1993a; Pool *et al.*, 1993). Originally, the harvesters were tested in the same configuration as harvest, which resulted in fruit being removed, but also fruit being damaged that, subsequently, did not ripen properly (Pool *et al.*, 1993; Petrie and Clingeleffer, 2006). More recent suggestions have been to remove

Fig. 8.4. Mechanical leaf removal equipment. Left, a system that uses a vacuum to suck leaves into the reciprocating blades. Right, a system that used pulsed, high-velocity air to remove the leaf blades. Compared with well-practised hand leaf removal, these systems are more likely to leave petioles and portions of leaf blade in the canopy.

the crop from only a section of the canopy of minimally pruned vines, accomplished by removal of some of, or repositioning of, the harvester's beater bars, which has given good results compared with hand thinning (Petrie and Clingeleffer, 2006). An added benefit of this is that it is possible to weigh the fruit coming off, which may assist with calculating potential vine yield.

Raisins have been mechanically harvested for many years (May *et al.*, 1974), and special training systems have been developed to increase efficiency (Christensen, 2000). More recent research has developed a system for producing dried-on-vine raisins using conventional trellises and mechanical wine grape harvesters (Peacock and Swanson, 2005), which makes the practice much more grower-acceptable as less capital expenditure is needed for specialized equipment and trellising.

Mechanical harvesting of wine grapes (see Fig. 6.50) has been commercially practised for some time, and is now an accepted way of obtaining quality grapes at low cost and at any time of day or night. Most reports agree that a well-set-up mechanical harvester will provide fruit quality that is close to that of hand-harvested fruit (Johnson and Grgich, 1975; Clary *et al.*, 1990). Modern equipment with further refinements (see Fig. 8.5) are now approaching the ability to harvest berries with little or no skin breakage, which may lead to further changes in how the industry perceives the practice.

However, this labour-saving device is not without its challenges. Because the machines are not discriminating when it comes to harvest, a second set crop (fruit arising from lateral growth, which is far behind in terms of development) and anything that is dislodged from the canopy can be caught and end up in with the grapes. Therefore, for mechanical harvesting to be effective, viticultural management up to this point must ensure that fruit that is not ripe is not in the canopy at all, as well as any other items that may be collected during the process. Though material other than grapes (MOG) is found in hand-harvested fruit as well, the size of these items tends to be smaller in mechanically harvested fruit (Petrucci and Siegfried, 1976), which makes it harder to remove and can lead to greater extraction of unwanted substances during winemaking. Nobel *et al.* (1975) reported that the inclusion of leaves as part of MOG did not significantly affect the acceptability of wines, but others (Wildenradt *et al.*, 1975; Huang *et al.*, 1988) have reported deleterious effects. It is likely that wine style and winemaking practice both have a significant effect on this, which is why varying results have been reported.

One advantage of mechanized harvest is that cluster size and position within the canopy is no longer as great a concern, which can lead to greater freedom in developing trellising/training systems. Hand harvesting is most efficient when clusters are large and within a distinct zone in the canopy, but with machine harvesting this does not matter (Petrie and Clingeleffer, 2006). This lends itself to the adoption of minimal pruning (Clingeleffer, 1993b), where mechanization is a great cost and time-saver, and allows for more flexibility in getting the fruit in when it is ready.

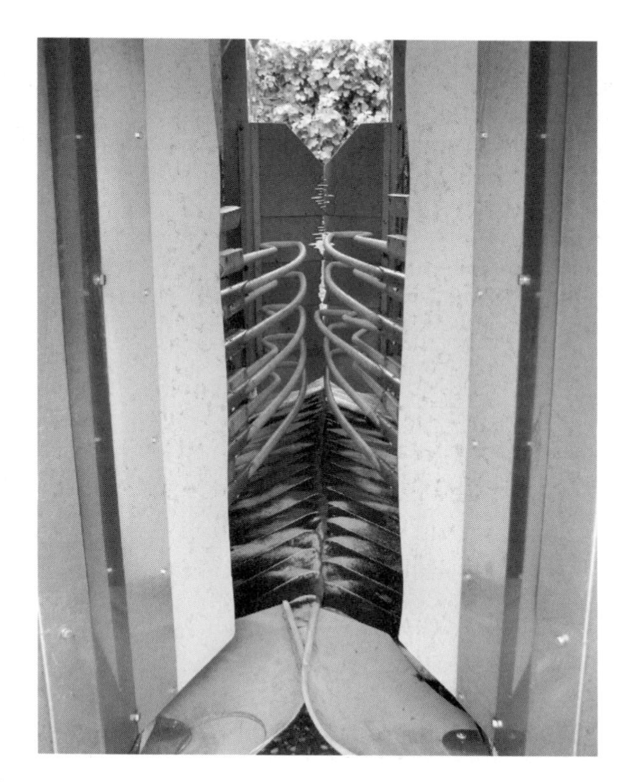

Fig. 8.5. The interior of a modern mechanical harvester. Note that the bars are removable, curved and flexible, which allows for adjusting to fruiting wire height and lessens the physical impact on the vines.

PRUNING

Once mechanical harvesting had become common, the most time-consuming aspect of grape production became pruning (Morris *et al.*, 1975; Pollock *et al.*, 1977). Mechanizing this was accomplished through hedging the growth (Shaulis *et al.*, 1973; Pollock *et al.*, 1977) or cutting hanging canes back (Morris *et al.*, 1975). The latter mimics the result of hand pruning (and may require hand follow-up to adjust the node numbers left on the vine), while the former generally results in greater numbers of nodes per metre of row.

Using machines to assist hand-pruning techniques, such as described above, through the use of a barrel pruner (see Fig. 6.10) – or more recent innovations such as a simple, tractor-mounted bar that lifts the foliage wires up and out of the canes – or a set of rotating rubber tyres that pull cut canes up and out of the trellis, means that pruners no longer have to extract canes from the foliage wires, which can be a cost- as well as a labour-saving measure: recall that initially it was a shortage of labour that precipitated the move into mechanized vineyard practices in the first place (Shepardson *et al.*, 1970; Patrick, 1983). However, attempts at adapting machines to existing trellis and training systems have met with limited success. For a completely mechanized system the problem needs to be turned around: how should a trellis/training system be set up to best be mechanized?

The practice of hedging, which is relatively easy to mechanize and practise on simple trellising systems, can result in good fruit quality and reduced production costs. However, there can be minuses due to increased levels of botrytis in cooler growing areas caused by more shoot growth and fruit shading (Clingeleffer, 1993a, 2000) if vine vigour is not managed correctly. In a non-irrigated vineyard in Spain, an 11-year study using simulated hedge pruning found little effect of the practice on fruit quality, but a significant increase in yield as well as in vine capacity (de Toda and Sancha, 1999). For 'Concord' grapes destined for juice the practice of mechanical pruning was found to be a commercially sustainable practice after a 6-year study in Michigan, USA (Zabadal *et al.*, 2002), where there are short summers and cold winters, though additional hand work was necessary to maintain good canopy structure. Reynolds (1988) reported similar results in Canada using simulated mechanical pruning. In many areas, hedging, as a way of mechanizing pruning, has worked very successfully and has been adopted widely, particularly in Australia (Clingeleffer, 1999).

The promise of viable robotic pruning (and its potential to cope with current and varied grape production systems) has been around for some time (Ochs *et al.*, 1992; Gunkle and Throop, 1993; Monta *et al.*, 1995). Challenges are many, as the natural variation found between vines is high, vines change through the growing season and the environment itself is not necessarily very robot-friendly (Kondo and Ting, 1998). However, research into this area continues and, as advances in sensing and control improve, we will no doubt see this emerge as a commercial option, though probably only in specific situations.

ENVIRONMENTAL MONITORING

Mechanization carries through to other tasks in the vineyard as well. Remote and real-time monitoring of many environmental or even vine characteristics is now possible thanks to the use of wireless technology. Weather stations, monitoring combinations of air and soil temperatures, relative humidity, solar radiation, wind speed and direction, rainfall, etc. (see Fig. 8.6) have been available and affordable for some time. The information has been used for calculating disease infection periods, projected growth of vines and time applications of weed sprays. Similarly, monitoring of plant and soil water status is possible, and can be used in the determination of irrigation frequency.

Fig. 8.6. An automated weather station, which can send data in real time over a cellular phone network. Not shown with this unit, which has a solar panel for power, temperature, humidity, wind speed and direction sensors, are the soil temperature sensors and a pole with the rain gauge and solar radiation sensor.

Advances in miniature electronics will surely result in a greater ability to monitor climate variables, both in terms of the number of characteristics and also the intensity or frequency with which they can be determined in a given area. Developing methods to be able to use all of this information to its best advantage is the next big challenge, which to date has taken the form of using geographical information systems.

FULLY MECHANIZED SYSTEMS

Further developments in trellis design are leading to systems that lend themselves to greater degrees of mechanization, and therefore to labour and cost reduction. The Movable Free Cordon system is an adaptation of the single-wire trellising system that uses bowed vine trunks and a cordon wire that is free to move up and down through slotted guides installed at every post. Harvest is facilitated by the ability to shake the vines up and down with minimal impact on the trellis system or direct contact with the grapes, which detach about 1 m from the impact head (Intrieri and Filippetti, 2000). Mechanized pruning is kept simple because there are no foliage wires. Curiously, this system is quite similar to the first mechanized harvesters developed in the late 1950s, which also used an up-and-down motion to pick the fruit (Shepardson *et al.*, 1962, 1970).

The Morris–Oldridge system of mechanized vine management incorporates all vine manipulations needed for production on 12 major trellising systems, with the promise of balanced cropping and improved economic viability (Morris, 2006). Such systems are still in their evaluation phase, but results appear to be promising. However, true innovation with reference to trellis design and mechanization opportunities has yet to be produced.

Higher levels of integration of all of the options for automation and reduced human labour input should be a goal for the future. With increasing competition in the marketplace, decreasing economic margins and more costly or inadequate numbers of personnel to work in the vineyard, increased use of mechanization will be essential, but this must not come at the cost of quality. Therein lies the future challenge.

Grapevine Pests, Diseases and Disorders

There are a wide range of situations where grapevines and their fruit can be put in jeopardy. This is important not only because of vine health reasons, but also for the economic health of the business. Therefore, minimizing vine exposure to the adverse effects of biotic and abiotic origin is essential.

Pests cause or have a significant potential to cause loss. There are many, many different organisms in the environment, but none are considered pests unless they cause a problem. Despite this, grapevines have many pests and thus problems. A large proportion of ongoing vineyard costs is devoted to managing diseases and pests, and with good reason. Many pests are difficult to suppress and sometimes it is best to adopt a tolerance approach and accept some damage. The point where the cost in money or time is greater than the economic return from the practice is called the economic threshold. The economic threshold, however, varies with pest, location and end use of the grapes.

Damage to grape vines can occur in the roots and in the above-ground parts, the trunk, arms, cordons, canes, shoots, leaves, rachis and the berries. Pests of grapes include diseases (bacterial, fungal and viral), insects, arthropods, pollution, mammals (including humans) and birds. Disease organisms have ecological niches: some flourish under wet, damp conditions (e.g. downy mildew), some under dry conditions (e.g. powdery mildew), some in heavy soil (e.g. phylloxera), and some in sandy soil (e.g. nematodes). Grapes are grown in most of these environmental conditions. Pests can cause acute damage to vines or cause symptoms that are frequently described as 'decline' where vines become progressively weaker and ultimately stop producing. Examples of classic decline pests are phylloxera, nematodes and eutypa.

Some grape pests are problems everywhere grapes are grown, while the significance of others, although profound in their localized areas, is not nearly as high on a global scale. Many of the regionally limited pests are contained because of climatic limitations on the disease, pest or vector of the disease. For example, Pierce's disease (of bacterial origin), does not proliferate within the vine in cool conditions (Feil and Purcell, 2001), and the geographical location of the vector (such as the glassy-winged sharpshooter) is also limited by temperature (Hoddle, 2004). This chapter will concentrate on the more ubiquitous, global, pests: those causing problems in many grape-growing areas. Be aware, however, that localized pests can be the major problems in any given region.

Many diseases have been introduced from distant grape-growing areas; for example, major diseases of grapes were introduced into Europe from North America where the wild grapes were tolerant; examples of this follow. The potential exists for more pests to follow grapes to new growing areas, so the lessons of history should be heeded.

Powdery mildew is a fungal disease introduced to Britain with plants probably collected from the tropics, and first reported by a gardener named Tucker, after whom it was originally named (*Oidium tuckerii*). From there it made its way to France in about 1847, causing widespread damage: production losses of up to 80% due to the berry splitting and the degradation of grape flavours it caused. In 1854 the eventual control method for powdery mildew was promoted, being based on observations that Mr. Tucker made about its similarity to a peach tree disease that was being controlled with an aqueous mixture of sulphur and lime sprayed on the trees (Pearson and Goheen, 1988). Sulphur is still the main fungicide used for powdery mildew control today.

Phylloxera, the root louse native to eastern North America, was first introduced into Europe around 1863, and gradually ate away at grape production. The French government offered a reward of 300,000 francs in 1873 to anyone who could find a way to stop phylloxera. Prize-induced treatments ranged from the ridiculous to the universally lethal, but that prize was never claimed (Campbell, 2004). The failure to come up with a workable solution led to another startling decrease in wine production in France: from 83 million hectolitres (hl) in 1875 to 23 million hl in 1889. The best solution so far has been the grafting of European *V. vinifera* grapes to American rootstocks (initially) or to hybrid rootstocks. Phylloxera has spread to almost all grape-growing areas of the world, causing the replacement of millions of vines as the own-rooted plants succumbed to the pest.

With the introduction of phylloxera-resistant rootstocks, other North American native pests were introduced as well, leading to a third wave of disaster for the French vignerons. Downy mildew, imported with the American grapevine material that was used to combat phylloxera, was first reported in France in 1878 and caused another downturn in French grape production until a control method was found. The story goes that a grower just outside the town of Bordeaux was having trouble with townspeople stealing grapes as they walked by his vineyard (Prial, 1987). He decided to paint clusters near the road with a greenish blue paste made by mixing copper sulphate and lime, which gave the leaves and fruit an unappetizing, splotchy, green–blue colour. It is not reported whether this deterred the pilfering, but a passing scientist observed that the grapes did not develop downy mildew. A few enquiries and experiments later, in 1885, he published results that showed the effectiveness of this *bouilli bordelaise* (Bordeaux mixture), and it is still used around the world for control not only of downy mildew but also other disease organisms.

DISEASES

Disease in grapevines can be caused by a wide range of organisms, including those of fungal, bacterial and viral origin. The most common diseases of grapes will be discussed in this section.

Fungal diseases

On a global scale the diseases caused by fungi that are most of a concern are botrytis (bunch rot), powdery mildew, downy mildew and phomopsis. A discussion of each follows, but an expanded view of botrytis control is taken. Many aspects of control for botrytis can be applied to other grapevine diseases, so when considering a disease management strategy look for areas where these ideas have multiple uses.

Botrytis and general fungal disease

OVERVIEW Botrytis bunch rot or grey mould (caused by *Botrytis cinerea*) (see Fig. 9.1) is ubiquitous and exists in all vineyards of the world. Its genus name was coined in the early 18th century and taken from the Greek term for a cluster of grape berries (*botrus*), so it is likely to have been associated with grapes for some time (Rosslenbroich and Stuebler, 2000).

The fungus favours cooler temperatures than many others, but warm and damp conditions are best for rapid development (Nair and Allen, 1993). Although generally detrimental to grape quality by causing rotten berries in table grapes and off-flavours in wine (Loinger *et al.*, 1977), in specific cultivars and climatic conditions it is known as 'noble rot' (Müller-Thurgau (1888), cited in Rosslenbroich and Stuebler, 2000), where the berry skins retain their integrity and shrivel, resulting in a concentration of sugars and flavours inside (see Plate 41). The rot contributes to the production of the exceptional sweet wines of Hungary (Tokay), France (Sauterne) and Germany (Auslese, Beerenauslese and Trockenbeerenauslese). *B. cinerea* is not restricted to grapevines: it attacks many cultivated and wild plants and can live on dead tissue as a saprophyte (see Plate 42) or on living tissue. Because of its wide host range and adaptability to different situations, it is hard to avoid in a typical vineyard situation.

Fig. 9.1. Badly affected botrytis cluster, demonstrating how it acquired its alternative name of grey mould.

In addition, just as there are different clones of 'Pinot noir' or 'Gewürztraminer', botrytis is made up of a population of slightly different organisms (van der Vlugt-Bergmans *et al.*, 1993), each of which may have a different capacity to grow, reproduce and survive attempts at eradication. Because there is a potentially diverse population of individuals rather than millions of identical clones, attempts to control 'botrytis' through a single method are usually met with only limited success.

Just as there is a choice between single-site (those that act on a particular point in the target's physiology) and multiple-site (those that act on more than one particular point) fungicides, there is also the choice between single-idea and multiple-idea strategies to control botrytis. At every possible step, practices in the vineyard and how they may influence fungal progression must be considered.

Under harsh conditions botrytis can overwinter in hard, dark-coloured structures called sclerotia, which can resist extreme drought, heat and cold, and can be found on canes, rachis material or mummified grapes. In mild climactic areas, however, the fungus can easily survive between growing seasons as dormant hyphae inside buds, under bark, etc. and resume growing when environmental conditions improve (Pearson and Goheen, 1988).

Hyphae of botrytis grow under a wide range of conditions (even quite cool ones) and produce conidiophores bearing conidia, which are spores that can be spread by water, wind, and mechanical means (see Fig. 9.2). Ideal conditions for infection are 23.7 and 20.8°C for leaves and berries, respectively, with free water or a high RH (Nair and Allen, 1993). The time required for infection (the infection period) is from approximately 1 h, depending on the tissue type, temperature and wetness, with drier conditions requiring longer amounts of time (Nair and Allen, 1993). Berries appear to be more resistant to infection than leaves due to the resistance of the intact skin, although infection can happen through natural openings, such as stomata, and even through the cuticle itself (see Plate 43; Northover, 1987), but any breach of its integrity results in rapid colonization (Coertze *et al.*, 2001). Mechanical injury is the most frequent origin of botrytis infection, and can be caused by hail, insects, birds or even powdery mildew infections. Mechanical leaf removers may also cause physical damage to berries in the process of removing the leaves.

Spring infections of leaves and young clusters are probably caused by spores produced from sclerotia in debris on the vineyard floor or in the vine (Nair *et al.*, 1995; Seyb, 2004). Under certain conditions, infections during bloom remain latent until véraison or later (Nair *et al.*, 1995), causing damage that appears shortly before harvest.

Grape cultivars differ greatly in their susceptibility to botrytis bunch rot. Cluster architecture, microclimate around the berry, chemical composition (Hill *et al.*, 1981) and the possible intervention of phytoalexins (Langcake and McCarthy, 1979; Creasy and Coffee, 1988) are involved in varietal differences.

Fig. 9.2. Close-up of an infected 'Semillon' berry, showing the conidiophore structures that release spores.

Cultural practices – primarily aeration and exposure of the clusters (leaf removal) – can reduce the severity of the disease, although chemical protection is generally required on susceptible cultivars. Chemical treatments frequently include applications during bloom, at the time of berry touch (closure of the cluster) and just before harvest. Table grapes are frequently protected from postharvest rot by sulphur dioxide fumigation (Smilanick *et al.*, 1990) and storage at low temperatures (down to 0°C).

A coordinated approach is necessary to minimize the risk of infection by botrytis. The goals for this are to (i) minimize potential infection periods; (ii) achieve high vine health; (iii) reduce sources of inoculum; and (iv) monitor the weather and the vines.

MINIMIZING POTENTIAL INFECTION PERIODS Site selection is the first point of call for many important management concerns, botrytis included. Consideration of the macro or mesoclimatic conditions and how they would favour botrytis growth may alter plans, including the possibility of planting elsewhere rather than have ongoing costs and increased risk of infection. Cultivar and rootstock choice is important to the potential growth of the vines and thus density of the canopies, and should be matched to the site, e.g. choosing a shallow-rooting rootstock for a deep soil. Loose-clustered varieties or clones are desirable over tight-clustered types, as the fruit dries out more quickly, leading to fewer infection periods. If the weather consistently turns wet in the harvest period, an earlier-maturing variety or variety/rootstock combination should be considered.

Trellising and vine spacing should also be matched to the site, as spacing that is too close or trellising that is inadequate for the amount of growth at the site can result in vigorous and crowded canopies and, potentially, a greater number of infection periods. If vigour is expected to be high, a divided canopy should be considered, which can reduce individual shoot growth and open up the fruiting zone.

Row orientation has an impact as well. North–south oriented rows have east- and west-facing sides, and therefore the sun will spend half the day shining directly on one side and half the day shining on the other, leading to even light distribution and early drying of the eastern side in the mornings. East–west-oriented rows have a north and a south side to the canopy and only the one side (which one depends on which hemisphere the vineyard is in) gets direct sunlight. Fruit from the continually shaded side of canopies may be more susceptible to disease because of the lack of sunlight and greater fruit wetness that results.

Prevailing winds during the cropping season, particularly when the grapes are at their most susceptible (pre-bloom up to fruit set and again after véraison), can also be an important consideration in determining row orientation. Breezes down the row can aid in air circulation within the fruiting zone, reducing fruit wetness, so one consideration for row orientation would be the prevailing wind direction during the season.

Inter-row crop choice and irrigation and fertilization regimes influence vegetative growth of the vine. Vines will grow best with luxurious levels of water and fertilizer and with bare soils, so limiting the amount of water to which the vines have access is a good way of managing this. Managing water inputs can be done directly and/or indirectly, by altering between-row cover crops. Soil moisture levels can be controlled to change vine growth (see Chapter 6). In soils with high water availability cover crops such as chicory can be very effective in drawing water away from vines and thus reducing vegetative growth (Caspari *et al.*, 1997), with concomitant effects on reducing the frequency of botrytis infection periods.

One major goal of leaf removal is to reduce the incidence of botrytis. As such, there can often be significant benefit from removing leaves from just the east side of a N–S-oriented row. This exposes the fruit to the morning sun, helping to dry out the area more quickly. On an E–W-oriented row it is usually best to take leaves from the shaded side of the canopy, to improve air movement where the sun has little influence.

Shoot thinning to remove non-count shoots and open up the head area is invaluable for ensuring the canopy you planned on at pruning actually develops in the summer. Shoot positioning is equally important, to prevent crowding in the fruit zone. If your canes are lying horizontally along the row (or there are excessive numbers of laterals developing in the fruiting zone), the leaf density will be greatly increased, which will enhance conditions conducive to botrytis development.

A wide canopy presents a problem in disease control as it decreases the amount of light reaching inside, reduces air movement, increases relative humidity inside and decreases spray penetration. All of these contribute to increased risk of botrytis incidence via a greater likelihood of infection periods occurring.

ACHIEVING GOOD VINE HEALTH Other ways to combat botrytis infection are to take advantage of what nature has to offer. Aside from the physical properties that vines have to repel infection attempts (such as the cuticle (Kolattukudy, 1985)), vines possess chemical mechanisms to slow or stop botrytis development. There are also other microorganisms that can prevent botrytis from becoming established in the grapes.

If the fungal spore's hypha has successfully overcome the physical barriers, chemical barriers then come into play. Plants – particularly woody ones like grapevines – produce a number of secondary metabolites that may contribute to a passive resistance capability. These pre-formed phenolic compounds may have a significant influence on the success or failure of an infection attempt (Jersch *et al.*, 1989). For example, tannins found in grapes may bind to enzymes produced by the fungi during the infection process (Goetz *et al.* 1999), thus preventing them from acting on the plant cells.

Grapevine leaves generally are more resistant to infection as they age – perhaps because they accumulate lignin and other polyphenolic compounds

(Prins *et al.*, 2000). Grape berry resistance varies according to developmental stage and the type of pathogen attempting infection (as well as variety).

Part of the reason for this could be the active responses of the vines to attempted disease infection. There are a number of these, and it must be emphasized that one or more may be occurring at the same time. One visible example of an active response is hypersensitivity (Kombrink and Schmelzer, 2001). In this case the plant isolates the infected tissue by producing impermeable substances, like lignin, around the area and/or by accumulating toxic substances that kill the affected tissue, and thereby contain the infection.

Pathogenesis-related proteins (PRPs) are induced upon infection to a variety of pathogens including viruses, bacteria and fungi. In the grapevine, *B. cinerea* induces the production of a number of different PRPs (Renault *et al.*, 1996), which appear to act upon the fungal cells. Chitin and glucans are structural components of most fungal cells, and chitin in particular is not found in plant tissues. The production of PRPs that have enzymatic activity and break down fungal cell walls may be an important defence response by the plant (Robinson *et al.*, 1997). It appears that fungal hyphal growth is inhibited by the enzymes, supporting this theory (Giannakis *et al.*, 1998).

Developmental stage of the tissue may also influence its capacity to produce PRPs. For instance, their production in grape berries was greater after véraison once the berries had begun the ripening process (Robinson *et al.*, 1997). This may change the susceptibility of berries to botrytis during development.

While the presence of pre-formed phenolics may be a barrier to infection, the synthesis of more phenolic compounds following infection can also slow or stop the pathogen. One advantage to this class of response is that it can happen very quickly, as the compounds don't need to be evolved from scratch, but rather can be modified versions of molecules that are being made already, such as from primary metabolites. The effect of this is that cell walls near the infection site are rapidly 'reinforced' against fungal penetration and ready for more permanent formation of lignin, which can take in the order of days to occur.

Phytoalexins are compounds that inhibit microbial growth and are produced rapidly upon attempted infection by a pathogen. Many phytoalexins are phenolic in nature, and the known examples in grapevines belong to a class called stilbenes. Resveratrol is the first stilbene to be synthesized by grapevines and the first to be discovered as a fluorescing compound made along the margins of botrytis infections on leaves (Langcake and Pryce, 1976). Modifications to this molecule produce a variety of other compounds such as piceid, pterostilbene and ε-viniferin, which accumulate more slowly than resveratrol. The speed at which the vine can accumulate antifungal substances like resveratrol may influence the establishment of the fungus (Creasy and Coffee, 1988), and therefore be important in determining the success or failure of the infection attempt.

Leaves and berries are able to produce resveratrol quite effectively, but for fruit this ability drops as the berries enter the ripening phase (see Fig. 9.3), possibly due to increasing amounts of UV-absorbing compounds produced in the berry skins (Creasy and Coffee, 1988). Coincidentally, this is the same phase at which PRPs start to be manufactured by the berries (Robinson *et al.*, 1997), which means that grapes have some sort of defence throughout their development.

However, botrytis may be able to combat the effect of stilbene production by the vine. One enzyme secreted by the fungus is stilbene oxidase, which can cause the degradation of resveratrol, pterostilbene and perhaps other stilbenes such that they lose their antifungal properties (Hoos and Blaich, 1988; Adrian *et al.*, 1998).

Induced resistance and systemic acquired resistance (SAR) are concepts based on the notion that plants can be 'immunized' against future infection by pathogens (Hunt *et al.*, 1996). This is similar in concept to human immunization, where we can be inoculated in the arm and then, for some time afterward, be resistant throughout our bodies. This response can occur in plants as a result of an infection, where an active defence response occurs remote from the infection site, or it can be induced by application of a chemical or other signal. For example, treating vines with fosetyl-aluminum (Aliette®) increases the ability of those

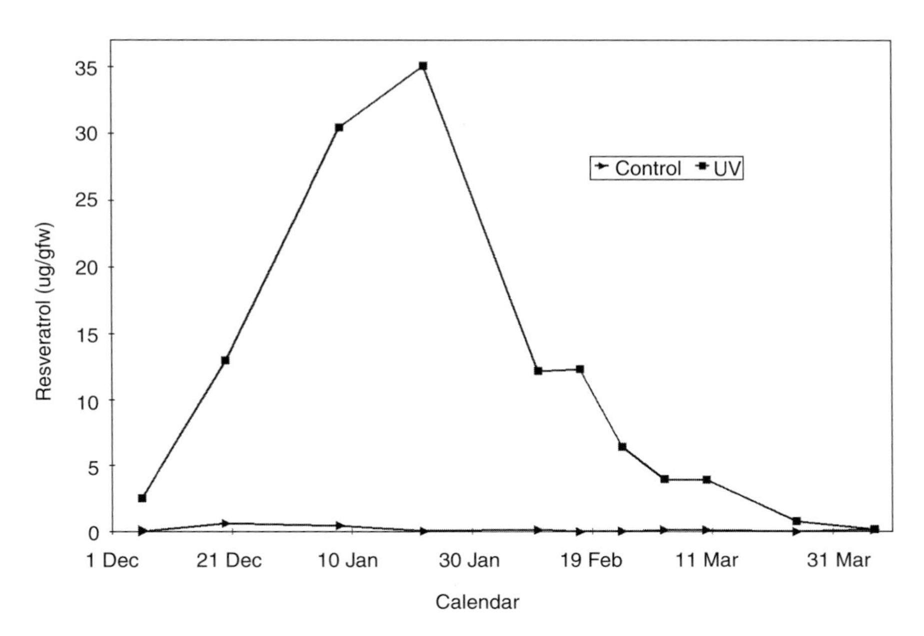

Fig. 9.3. Potential of grape berries to produce resveratrol in response to UV irradiation through the season (the difference between the UV and control lines), which is taken to be indicative of the berry's response to botrytis infection.

grapevines to produce resveratrol if an elicitation event occurs after the application of fosetyl-Al (Creasy and Qi, 2005). There is much interest in the area of induced resistance, as environmentally benign substances could be substituted for fungicides that have a direct action on botrytis.

Development of a fungus may also be inhibited by existing antagonistic organisms such as *Trichoderma* spp., though others have been tested and found to be successful. The presence of these non-pathogenic fungi living on plant surfaces inhibits the growth of a number of plant pathogens, either by taking up the space that botrytis would use, interacting directly to prevent it from getting established or even inducing a systemic response from the plant tissue (Bélanger *et al.*, 1995; De Meyer *et al.*, 1998; Elad and Kapat, 1999). Advantages to using *Trichoderma* are in short withholding periods and their eco-friendly nature. Disadvantages are in finding a strain of the fungus that will survive on grapes in the field, getting high enough populations of them on the berries and possible interactions with any use of fungicides for other reasons. Efficacy of this type of control is highly dependent on environmental factors, such as temperature and relative humidity, and so what works in the laboratory often does not translate well to commercial situations (Hjeljord *et al.*, 2000), so finding the right strain for the job is extremely important.

Despite potential problems, there remains a lot of interest in biocontrol. Many trials indicate that these methods can be just as, or even more, effective than a regular spray regime. In some cases, it is also tied in with induced resistance or SAR – the biocontrol agent eliciting some protective response in the plant.

REDUCING SOURCES OF INOCULA Cleanliness in the vineyard can contribute to reducing the chance of infection. Up to 90% of leaf trash found in some New Zealand vineyards was found to be infested with botrytis (Seyb, 2004). Old tendrils, petioles, berries, canes, etc. have been found to harbour botrytis, which sporulates in the following growing season. Machine-harvested vines can carry over a large amount of botrytis if cluster remnants are not removed at pruning.

MONITORING THE WEATHER AND THE VINES A weather station can provide more than just a record of rainfall events, wind direction, etc. The data collected can be used in botrytis infection forecasting models, which take into account the weather conditions that are needed for botrytis spores to germinate and note when an infection period has occurred. Rather than a using a preventative spray programme, fungicides are applied only when needed, thus saving money on purchases of the chemicals as well as their application, especially in low-disease pressure seasons.

However, the best way to stay on top of what's happening in a vineyard is to spend time in it! There is no substitute for getting out there and getting into the canopy. Weather stations and disease forecasting models are wonderful innovations, but they cannot actually look through the vines and see that botrytis is developing in the clusters. They also can't tell how effective the spray coverage has been, so remaining vigilant and proactive is still a must.

Keeping records of the location of trouble spots in the vineyard will also pay off. If there is one area that is particularly prone to botrytis infection, it can be used as a indicator. If disease starts developing there, then the rest of the vineyard is probably at risk.

Powdery mildew

This disease is also called oidium and is caused by *Uncinula necator* (alternate name *Erysiphe necator*) (see Plate 44). It will develop at the first instance that conditions are favourable during the season. Initial spring infections are due to spores produced by overwintering fungal bodies (cleistothecia) in the buds or on the surface of the vine (see Plate 45). The fungus spreads to new green tissue after the production of asexual spores, which can be wind blown or splash transported. Powdery mildew is a fungus that grows on the epidermis of green vine tissues. Special fungal organs called haustoria enter epidermal cells to nourish the fungus (Pearson and Goheen, 1988), and the cells must be living for the fungus to survive and reproduce. Temperatures of 20–27°C and dry conditions are optimal for infection and disease development, unlike those for some other common fungal diseases.

Powdery mildew looks like a white powder on the surface of the plant. Leaf powdery mildew is usually seen on the upper surface, but can appear on both abaxial (lower) and adaxial (upper) sides. All green parts of the vine are susceptible. Infected leaves suffer from reduced photosynthesis (Lakso *et al.*, 1982), which can debilitate the whole vine. The most economic damage, however, is from cluster infections, where infected berries fail to ripen, are prone to splitting and the infected rachis becomes brittle (see Plate 46). Once berries start to ripen, however, they are not susceptible to new infections. *Vitis* species vary in their susceptibility to powdery mildew, and selection of resistant varieties might be part of a management scheme if the grower is willing to grow the resistant variety. The desirable *V. vinifera* grapes are susceptible, the *V. vinifera* x American hybrids vary from susceptible to resistant and most native North American species are resistant.

Cultural practices can control the disease under low disease pressure and improve the control success when chemicals are necessary. These strategies rely on keeping the canopy open so that there is good air movement; however, if spores are present in the area (for example, if a neighbouring vineyard has powdery mildew), some form of chemical control will be necessary due to the continued exposure to inoculum.

Powdery mildew is typically controlled by either a programme that includes one or more modern fungicides or by elemental sulphur. Sulphur is active both as a vapour and through direct contact with the fungus. If spray coverage is not complete, control is not as good at lower temperatures due to lower volatilization of the sulphur (Wicks *et al.*, 2003). Sulphur causes phytotoxicity in some varieties of grapes (e.g. those of *V. labrusca* origin) and is thought to be phytotoxic to most grapes at high (> 30°C) temperatures, though some research suggests that this is not so (Magarey *et al.*, 2003). For effective control, any tissue to be protected must be covered with sulphur, so thorough spray coverage is necessary, as well as reapplication to newly emerging leaves or surfaces where the sulphur has been washed off. Sulphur can also disrupt the life cycles of beneficial insects and arthropods (spiders, mites, etc.) in the vineyard, and sulphur residues on berries are associated with off-flavours in wines (Thomas *et al.*, 1993).

Modern organic fungicides (such as demethylation inhibitors, or DMIs) have the advantage of being functional over a wider range of temperatures and with less phytotoxicity (Pearson and Goheen, 1988), and of course must be used on sulphur-sensitive varieties and/or when high temperatures are expected. However, the development of resistance to DMIs is a serious concern due to their specificity in action (Köller and Scheinpflug, 1987), which has resulted in recommendations to limiting their use in spray programmes. Potassium and sodium bicarbonates have been used in viticulture and reported to control powdery mildew as effectively as sulphur (Kauer *et al.*, 2001). Mono-potassium phosphate fertilizer has been used successfully to manage powdery mildew, and its alternation with DMI fungicides has resulted in control equivalent to the use of DMIs alone (Reuveni and Reuveni, 2002), which thus reduced the amount of DMI used.

Certain oils of low phytotoxicity (such as JMS Stylet-Oil® (JMS Flowers Farms, Inc., Florida), which is a refined form of paraffin) can provide adequate control (Dell *et al.*, 1998). However, these oils also appear to inhibit vine photosynthesis (see Fig. 9.4) which, if they were used over a season, was severe enough to affect grape composition (Finger *et al.*, 2002). There appear to be cultivar differences of this effect, with a *V. labrusca* grape ('Catawba') being much more affected than the *V. vinifera* grapevine 'Chardonnay' (Baudoin *et al.*, 2006) and the latter more seriously affected than 'Cabernet Sauvignon' (Finger *et al.*, 2002). The presence of trichomes (fine hairs) on the underside of leaves appears to be related to the severity of photosynthetic decrease (Baudoin *et al.*, 2006).

Bicarbonate sprays, which have relatively low impact on the environment, can provide a level of control, but were not found to be as effective as the use of either phosphates (Reuveni and Reuveni, 1995) or DMI fungicides. Other alternatives to sulphur and synthetic fungicides include milk or whey (Crisp *et al.*, 2002, 2006), which were effective at controlling powdery mildew, possibly in part due to antimicrobial activity in the milk products.

Biocontrol of the fungus is another possibility, using micophagous mites (Falks *et al.*, 1995; English-Loeb *et al.*, 1999, 2007), which feed on the fungi. Interestingly, one study reported that these mites were present on wild examples of grapes, but not commercially grown ones, which was thought to

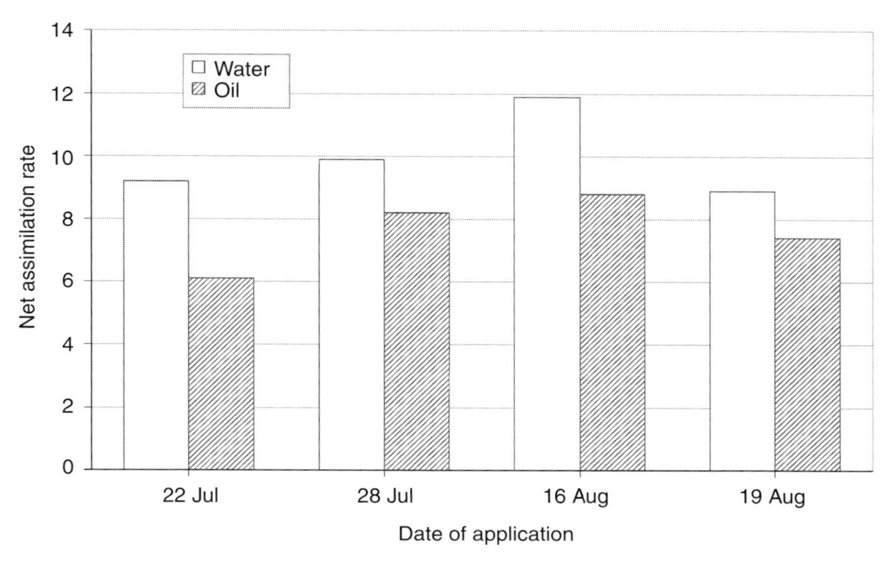

Fig. 9.4. The effect of application of a horticultural oil on 'Chardonnay' vine net assimilation rate at four different dates during one season (adapted from and used with permission, Finger *et al.*, 2002).

be due to the use of sulphur and other fungicides (English-Loeb *et al.*, 1999). This provides food for thought when it comes to rethinking disease control programmes, where beneficial insect populations can be reduced as a result of controlling the target organism, increasing the reliance on the control substance.

Nitrogen and UV radiation were also found to modify grapevine susceptibility to powdery mildew, with higher UV and lower N resulting in less disease (Keller *et al.*, 2003). This is thought to be due to the higher levels of phenolics, particularly flavonols and hydroxy-cinnamic acid derivatives, synthesized in the epidermal cells under these conditions.

Downy mildew

Downy mildew (*Plasmopara viticola*) causes leaf, shoot and cluster damage. The upper surface of the leaf will have a characteristic 'oil spot' appearance and sporulation on the lower surface, which looks downy white (see Plate 47). Berries and infected shoots have a similar downy white appearance (see Plate 48). Downy mildew is primarily a disease of warm, humid growing regions (e.g. much of Europe and eastern North America, China and Japan). The absence of rainfall and high humidity in spring and summer reduces the spread of disease in Mediterranean climates such as Chile and California (Pearson and Goheen, 1988).

Overwintering spores of downy mildew, called oospores, germinate in the presence of water as soon as the temperature reaches 11°C and, shortly thereafter, produce a sporangium (Pearson and Goheen, 1988). This asexual reproduction organ produces clonal spores dispersed, usually, by rain splash. The fungus enters the plant through the stomates, its hyphae developing inside the leaf and spores released back through the stomates to disperse to other sites by wind. Because free water is so important to spore germination, rain and high humidity are the most important factors in promoting epidemics of downy mildew. The sporangia are sensitive to desiccation and sunlight (Kast and Stark-Urnau, 1999; Rumbolz *et al.*, 2002), so the first point of control is canopy management to hasten the drying-out of the leaves and fruit, along with ensuring there is as little shade within the vine as possible.

However, in those areas where the climate is conducive to downy mildew sporulation and growth, control is primarily by fungicides, with copper (Bordeaux mixture) the primary preventive measure. Copper is phytotoxic to plants but the addition of lime reduces this, hence the creation of Bordeaux mixture. Some cultivars are more sensitive to copper than others, and application at high temperatures is more likely to cause phytotoxicity in any grapevine (Pertot *et al.*, 2006). Leaf damage (as with sulphur) results in reduced photosynthesis and low vine and berry quality. Some curative (ability to kill fungi after they have infected the plant tissue) and systemic (taken up into the tissue and transported for some distance into other, non-treated cells) fungicides, such as phenylamides (e.g. metalaxyl or oxadixyl), have been developed specifically for downy mildew.

Copper-containing pesticides have been used for many years in some vineyards, resulting in high copper concentration in the surface soil layer (Parat *et al.*, 2002; Pietrzak and McPhail, 2004). In a French study, for example, some vineyard soils were found to have up to 1500 mg Cu/kg soil (Flores-Veles *et al.*, 1996), which is many, many times higher than that considered normal (Pietrzak and McPhail, 2004). This has implications for the growth of more shallow-rooting plants than grapevines, as copper accumulates in the surface layers of the soil or, if the vineyard is removed and other plants are to be grown there. Fortunately for those who cannot use synthetic compounds, other forms of copper – which are more fungitoxic so less has to be used – are available, though they are not without their own challenges in terms of vine phytotoxicity (Pietrzak and McPhail, 2004).

Phomopsis

Phomopsis cane and leaf spot (*Phomopsis viticola*) has been reported in Africa, Asia, Australia, Europe and North America (Pearson and Goheen, 1988; Plate 49). It was formerly thought to be the sole cause of a disorder called dead-arm of grape, but later this was confirmed to be due to two separate organisms that occurred at the same time, phomopsis and eutypa dieback (see below). Phomopsis prefers dampness for infection and, if this occurs during budbreak, there is a greater chance of developing serious problems in the vineyard. Although it can the fungus rarely causes a fruit rot, unless conditions are wet enough for it to become established (Nicholas, 2004).

As for other the diseases of note, phomopsis has been carried into new growing areas by propagation material. The fungus overwinters in bark (which is frequently described as having a bleached appearance, though DNA examination of non-bleached canes also found phomopsis present (Melanson *et al.*, 2002)) and dormant buds. Spores are spread by rain splash, and any green tissue can be infected (Cucuzza and Sall, 1982; Pscheidt and Pearson, 1989). Historically, younger tissue was thought to be more susceptible than older, but more recent work suggests that tissue of any age is equally susceptible (Erincik *et al.*, 2001). Dead wood is a rich source of spores, and therefore any infected material should be removed from the vineyard. For example, the pruning stubs at the base of the vine are a source of spores that can infect the basal shoots needed for trunk renewal. Machine or minimal pruning leaves lots of dead material in close proximity to susceptible tissue, which can be a major source of infection for many shoots. Infected shoots are weak and can be more easily broken, but the major loss is caused by infection of the rachis, which can kill part of the cluster and prevent further berry development. If weather conditions are sufficiently wet, phomopsis can infect the fruits and contribute toward significant rot before harvest. The rot may be due to latent infections established shortly after bloom but, although latent infections have been identified in other phomopsis species (Williamson *et al.*, 1991; Velicheti *et al.*, 1993), it has not yet been reported as occurring in grapevine.

Control involves the use of sanitation methods to remove infected material and chemical sprays, such as those containing dithianon (a quinone fungicide), during infection periods. There are no eradicant fungicides available at this time.

Other fungal diseases of importance

EUTYPA DIEBACK (*EUTYPA LATA*) This is a disease of the woody tissues of grape vines and many other woody plants (see Plate 50). It is abundant in regions of high rainfall, and found in humid regions of Europe, North America, Australia and South Africa (Pearson and Goheen, 1988). Its symptoms include weak and stunted shoot growth, particularly noticeable early in the season with a zig-zag growth pattern, with small, cupped and frequently yellow leaves. The organism affects the plant's vascular system, and causes a slow vine decline, where the vigour of the plants reduces each year. In some cases this results in greater fruit quality, as the amount of vegetative growth is lessened leading to more fruit exposure.

Infected trunks have a pie slice-shaped brown area that shows that the wood is infected and no longer functioning. Spores of the fungus are released under wet conditions in the winter and spring and generally infect through injured areas, such as pruning cuts. Hence, reducing the chance of infection can be accomplished by avoiding pruning under these conditions. If infected material is found, it should be removed and disposed of by burning or burial. Since the same organism also infects many stone and pip fruit crops, it is important to remember that inoculum may come from outside the vineyard.

BLACK ROT (*GUIGNARDIA BIDWELLII*) Native to north-eastern North America, but is also an introduced pest problem in Europe and South America (Pearson and Goheen, 1988). Although shoot and leaf symptoms occur, they are not usually an economically significant problem and appear not to be correlated with fruit infection (Jermini and Gessler, 1996). Severe economic damage occurs from infection and death of rachis tissue or fruit rot. Rot in the latter case progresses to shrunken, black, mummified berries, unsuitable for any use. Fortunately, fruit becomes resistant to new infections about five weeks after bloom (Hoffman *et al.*, 2002), though this may vary with species and cultivar. As with many other diseases, wet conditions favour infection, with warmer temperatures needing less time for this to occur. Control is through removal of infected material or the application of either mancozeb or azoxystrobin fungicides (Nita *et al.*, 2007) as a protectant spray or myclobutanil, which has some eradicant capability (Hoffman *et al.*, 2004).

ANTHRACNOSE (*ELSINOË AMPELINA*) A disease of European origin and a particular problem in rainy, humid regions. After the introduction of Bordeaux mixture in 1885 and its extensive use to control downy mildew, the severity of the disease was greatly reduced (Pearson and Goheen, 1988). It infects *V. vinifera*, *V. labrusca* and muscadine grapes (Schilder *et al.*, 2005; Yun *et al.*, 2007). Modern organic fungicides (e.g. prochloraz or mancozeb) appear capable of controlling anthracnose (Masahiro and Shinro, 2001; Tai *et al.*, 2005), although there is concern about resistance (Thind *et al.* 2001).

BLACKFOOT DISEASE This affects roots and trunks and is caused by fungi in the genus *Cylindrocarpon*. It becomes apparent in young vines by a loss of vigour and symptoms resembling water stress (see Plate 51). If the lower part of the vine looks as if part of it is growing and part is not, then uprooting the vine may reveal death of the root system and new roots growing above that as a way of the vine compensating (see Fig. 9.5). The dead roots are sometimes in a 'J' shape from when the vine was planted, the roots having been turned upward from being placed into a hole that was too small.

Blackfoot and other trunk pathogens can cause significant losses (Scheck *et al.*, 1998), but these can be almost eliminated by using clean planting stock, digging holes for vines sufficiently large that their roots are free to point downward when planted, ensuring vines have access to adequate nutrients and water and avoiding early cropping (Halleen *et al.*, 2006).

Fig. 9.5. Blackfoot disease-affected part of the vine showing 'J'-shaped rooting habit of the original roots and the new root growth above.

Bacterial diseases

The major bacterial diseases of grape vines are Pierce's disease (*Xylella fastidiosa*) (Varela *et al.*, 2001) – found in North and South America and in Europe (Boubals, 1989; Berisha *et al.*, 1998) and crown gall (*Rhizobium vitis*, formerly *Agrobacterium tumefaciens*). California grapes were struck by a disease first seen in the Napa Valley in 1887 which, by 1906, had destroyed 14,000 ha (35,000 acres) of grapes. Named Pierce's disease after a plant pathologist, it wasn't until the 1970s that its cause was established as a bacterium. It has recently been identified in Europe, apparently introduced with nursery stock (Boubals, 1989; Berisha *et al.*, 1998). Pierce's disease is transmitted by leafhoppers (the glassy-winged sharpshooter in California), which can transmit the bacteria between grapes and the many alternate hosts (Raju *et al.*, 1980; Costa, *et al.*, 2004).

Crown gall is found wherever grapes are grown at the limits of their cold tolerance. The bacterium infects cells and alters their DNA so that they produce plant growth regulators and special amino acids. The former cause the occurrence of masses of undifferentiated cells (callus) that look like a gall (see Fig. 9.6). The disease proliferates on trunks damaged by winter freezes or through physical means, and can eventually girdle and kill the vine. It is a ubiquitous soil organism and cannot be eradicated from the environment. Though crown gall-free plants can be produced, they are easily infected in the

Fig. 9.6. Crown gall disease on grapevine cane (left) and base of trunk (right) growing in New York State, USA caused by cold damage during the dormant season.

field. Control has relied on trying to prevent the kind of damage that encourages infection. Failing this, removing the affected plants is the only option but, as the disease is so prevalent, it is likely that, if a clean vine is planted as a replacement, it, too, will become infected. Biological control methods are being investigated (Chen *et al.*, 2007), but a commercial means to manage crown gall in this way is not yet available.

Viral diseases

Viral diseases are transmissible diseases for which (at least originally) no obvious pathogen could be found. These could be spread by propagation, by (in)vertebrate vectors (generally man, insects or nematodes) or by sap inoculation. The particles identified in diseased plants by electron microscopy were determined to be the disease agents and were called viruses. These microscopic particles are composed of genetic material inside a protein coat. Viruses are much smaller than bacteria and fungi and, once inside a living plant, they have the ability to reprogramme plant cells into making more virus particles. Once a plant is infected, the virus will spread to all parts of the plant and the plant will remain infected for life. Depending on the virus involved, the vine may or may not show visible symptoms, but general loss of vine health and reduced fruit yield and quality are associated with the most significant virus diseases. Modern detection methods range from the enzyme-linked immunosorbent assay (ELISA) and DNA fingerprinting through the polymerase chain reaction (PCR) method, which are much quicker than the bioassays using indicator plants that previously were used.

Grapevine fanleaf virus (GFLV, or fanleaf degeneration)

One of the oldest described grapevine virus diseases, causing a variety of symptoms (Pearson and Goheen, 1988). One is a characteristic leaf shape, which has an open petiolar sinus, more closely spaced radial veins and exaggerated marginal points. Another is the expression of yellow banding along veins on the leaf, or more splotchy yellow areas. Additionally, there may be a proliferation of shoot fasciation.

In the field, the virus is transmitted from grapevine to grapevine by *Xiphinema index*, the dagger nematode. Rapid movement of the virus is, however, by transfer of infected propagation material. Control is limited to purchasing clean planting material and preventing the vector from entering the vineyard. There does seem to be natural resistance to feeding by the dagger nematode by some rootstocks, which could also be exploited in a management programme (Kunde *et al.*, 1968).

Grapevine leafroll

This is found in all countries where grapes are grown, and probably originated in the Near East with *V. vinifera* (Pearson and Goheen, 1988). There are several virus types within the leafroll family that have been identified, but GLRaV-1 and GLRaV-3 are most discussed, with the latter seeming to cause the most severe symptoms (Charles *et al.*, 2006). Leafroll has serious impacts on both vine health and grape quality, with growth and yield significantly reduced. Ripening is affected by delayed maturity, reduced sugar content and inhibited fruit coloration. Leaf symptoms include rolling of the leaf margins, reddening between major veins in red varieties and yellowing of leaves between the veins in white varieties (Goheen and Cook, 1959; Cabaleiro *et al.*, 1999; Borgo and Angelini, 2002; Plate 52). Rapid movement of the virus is caused by infected propagation material, but several mealybugs have been shown to vector grapevine leafroll (Cabaleiro and Segura, 1997; Golino *et al.*, 2002). Grapevine leafroll can change fruit composition of French hybrid varieties without acute leaf symptoms (Kovacs *et al.*, 2001), which makes identification of the disease more challenging.

CONTROL OF GRAPE DISEASES

Grapes are, in general, susceptible to many diseases. If a specific disease does not yet occur in a growing region is it possible that it will at some later date. The grape grower is therefore advised to be aware of many vine disease symptoms and be alert to these appearing in the surrounding region. Most countries have rigorous controls over the movement of plant materials across their borders, to deter the introduction of unwanted pests. However, even the most diligent of border precautions cannot prevent the entry of an organism that is not yet known (recall the introduction of phylloxera and various North American diseases to Europe). All growers of grapevines should be aware of diseases that are already present in their region as well as those that are potentially a threat, in order to minimize transfer of unwanted organisms. History has shown that man is the most effective of vectors when it comes to pests and diseases!

Despite the multitude of potential diseases, grapes are resistant to most microorganisms and, therefore, disease is the exception rather than the rule. Therefore the defence mechanisms of grapevines are effective in preventing many potential diseases (refer to the earlier discussion about phytoalexins).

Many cultural practices are available that can aid in disease control. Most disease organisms favour moist conditions for spore germination and infection, so therefore leaf removal, which increases air movement through the vine and the drying rate of wet leaves and fruit, is a great help. Excessive fertilization (particularly N), irrigation or an over-vigorous rootstock can result in luxuriant growth that is conducive to conditions favouring disease. Overly compact clusters (which can be due to clonal, varietal or seasonal factors), for similar reasons, are more susceptible to fungal diseases as well.

Control of fungal diseases

Chemical control of diseases is the norm in most grape-growing areas. As an example we will discuss those used against fungal diseases. Fungicides can be regarded as organic or synthetic, and their modes of action include protectant, post-infection, antisporulant and eradicant. Protectant fungicides must be in place before the fungi attack the plant part. Since the vine grows and produces new tissues throughout the season, this new growth must be covered with the protectant fungicide in order to prevent infection. These fungicides may also wash off in rains, and therefore previously covered surfaces must be recoated.

Post-infection fungicides can be applied after the infection has initiated, as they kill the fungus after infection has started and will prevent further progression of the infection to the disease stage. The time following infection when these types of fungicides can be applied varies, but may be only a few days. In the case of myclobutanil, application up to 6 days post-infection provided control of black rot, whereas the efficacy of azoxystrobin decreased linearly after 4 days post-infection (Hoffman and Wilcox, 2003; Fig. 9.7).

Antisporulants reduce spore production of infected tissue, reducing the spread of the disease to more tissue. Hoffman and Wilcox (2003) also found that myclobutanil was effective at reducing spore production of the black rot fungus providing it was applied before symptoms appeared.

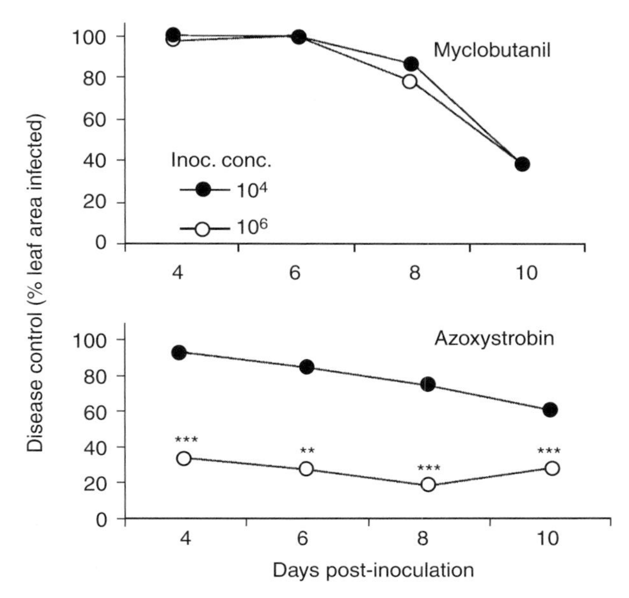

Fig. 9.7. Effect of myclobutanil and azoxystrobin as post-infection eradicants of black rot. The former does a better job of eradicating the disease if applied after infection has occurred. Inoculation concentration refers to the number of conidia of *Guignardia bidwellii* (causal organism of black rot disease) applied to the leaves before fungicidal treatment (reproduced with permission, Hoffman and Wilcox, 2003).
*indicates the degree of significant difference between the two concentrations.

Eradicants kill all or most of the fungal colony when applied after symptoms have appeared. They are rare, but an example is the use of lime sulphur to kill overwintering powdery mildew inoculum (Gadoury *et al.*, 1994).

Control of viral diseases

Prevention is the only available control for viruses because there is no cure for virus-infected vines. Some delay in damage is possible if you plant only virus-tested vines and control insects known to act as vectors for the viruses. Plant quarantines are used to slow the movement of virus-infected propagation materials from known infected areas into growing regions free of the virus. Infected vines can also be removed (rogued) from otherwise healthy vineyards to slow the spread of the disease. However, if the vector is not controlled, there is little benefit to doing so, as eventually it will spread throughout the vineyard. This is another instance that highlights the importance of sourcing clean planting material and continued monitoring for disease symptoms.

OTHER DISORDERS

There are many causes of unusual appearances of grapevine parts that are not due to pathogens or nutritional deficiencies/toxicities. Shoots and leaves are well exposed to the potentially damaging effects of environment and man-made hazards. Wayward herbicides are one cause of many leaf problems, as in some cases only very minute amounts of these can make a significant difference to the vine's appearance.

Contact herbicides, which only affect tissue they contact, can create localized leaf symptoms ranging from leaf discoloration (see Plate 53) to necrotic regions on leaves and shoots. These types of herbicides are safer to use around actively growing vines since, if some chemical contacts the plant tissue, the damage is localized.

In some cases pre-emergent herbicides are used to discourage weed growth for longer periods of time. As the chemical persists in the soil, it is important not to let it contact the vine tissue, or damage can ensue (see Plate 54).

Systemic herbicides move through the plant tissues and usually result in leaf and shoot distortions (see Plate 55), which can appear well beyond the point of contact. Therefore, their use around vines should be quite cautious, in order to prevent potential catastrophe.

Environmental hazards include hail – which can damage or strip leaves, break shoots and wound canes (see Plate 56).

Vines brought from a nursery shade house can be damaged by heat stress. This comes about from a combination of higher temperatures, increased sunlight intensity and small root systems that cannot supply enough water to the leaves, and results in necrosis of the leaf tissue (see Plate 57).

INSECT PESTS

In all grape-growing regions, roots, wood, shoots, leaves and fruits are subject to potential insect attack. Although the specific species of insect causing the damage may change in different grape regions, there is considerable similarity in their form. Some major insect pests for grapevines include phylloxera, leafhoppers, borers, mealybugs, leaf folders, cutworms and aphids. Grapes are also susceptible to infestation by many species of mites, which are not insects, but are included in the spider (arachnid) family.

Phylloxera (*Daktulosphaira vitifoliae*)

Phylloxera is an aphid-like insect (see Plate 58) of worldwide distribution. It is only known to attack grapes. When there is a point origin for phylloxera entering a vineyard, only isolated vines may be affected at first, with the area of dead and declining vines enlarging in a lens shape (long axis down the rows, Plate 59), sometimes quite rapidly (King and Buchanan, 1986) in subsequent years as the insect spreads. In the case of vines infested as planting stock, the entire vineyard may show decline after a few years of growth, though soil type was found to have an influence on phylloxera populations (Chitkowski and Fisher, 2005). High-sand content soils support smaller populations of phylloxera (Nougaret and Lapham, 1928; Galet, 1982; Pearson and Goheen, 1988; Reisenzein *et al.*, 2007), which may explain why some vineyards in infested areas have not succumbed to its effects.

The foliar form of phylloxera causes spherical galls on the underside of the leaf, as well as distortions and an exit hole on its upper side (see Plate 60; Granett *et al.*, 2001). The foliar form is known only on some cultivars in humid areas, such as in the north-eastern USA, where it originally evolved. The most serious damage is caused by the insect feeding on grape roots, the materials injected into the tissue causing the formation of nodosities (galls on younger roots) and tuberosities (galls on older roots) (see Fig. 9.8). Feeding by large numbers of phylloxera will kill a vine through the reduced capacity of its root system and potential infection of root wounds by pathogens (Omer *et al.*, 1995).

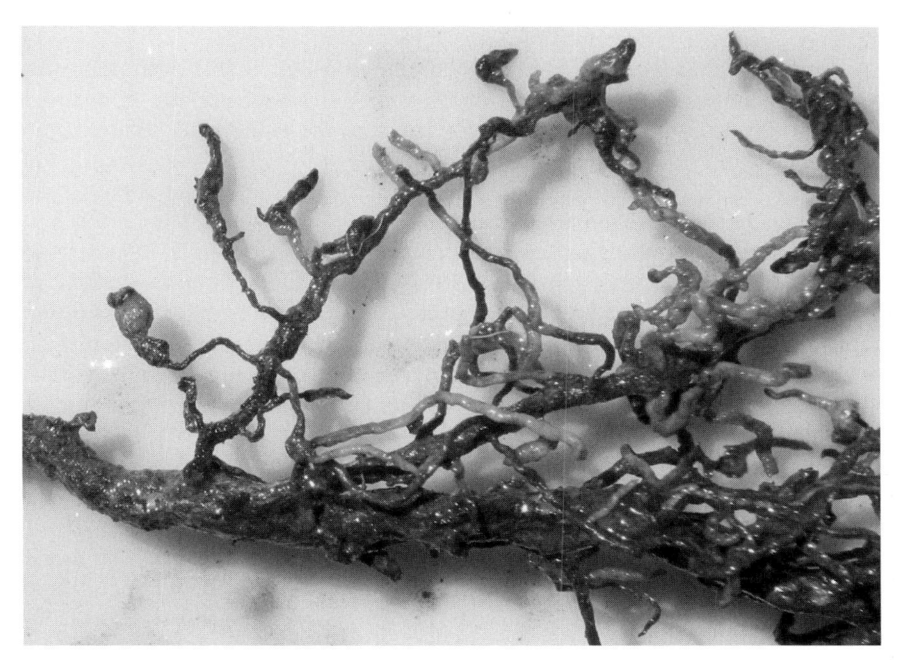

Fig. 9.8. Nodosities in grapevine roots, which are the result of phylloxera feeding. The club root or shepherd's crook shape is typical of phylloxera-infected roots.

Phylloxera spread is primarily through the action of man. The crawler form of phylloxera can be carried on vineyard machinery, although it can also be carried some distance by wind (King and Buchanan, 1986). The winged form of the insect is rarely found, but is capable of movement over much greater distances (Granett *et al.*, 2001).

Control of phylloxera in infested vines is not yet feasible, though some soil drenches have been tried. One problem with this approach is that the insect can live on roots far down into the soil profile and out of reach of applied chemicals, which are also difficult to apply in the clay soils that phylloxera prefer (Granett *et al.*, 2001). Affected vines can be managed, through additional water and fertilizer inputs, to continue producing for some additional years but, in most cases, they will succumb to the pest and need to be replaced. Some biocontrol agents are being investigated, such as an insect pathogenic fungus (Kirchmair *et al.*, 2004) and nematodes (English-Loeb *et al.*, 1999), but it is unlikely that these methods will work in more than very specific situations.

Successful management is therefore accomplished by avoiding the pest in the first place or grafting susceptible vines to resistant rootstocks, which have both been discussed earlier.

Leafhoppers

Leafhoppers are pests of many cultivated (and non-cultivated) crops and, since they have many alternate hosts, eradication of these from the vineyard is difficult. Leafhoppers feed through piercing the epidermis and taking up the cell sap. During this process there is exchange of material between the bug and the plant, which contributes to these insects acting as vectors for some disease organisms, most notably Pierce's disease.

Symptoms of leafhopper feeding include fine chlorotic spots on leaves (see Plate 61), leading to more general chlorosis or virus-like symptoms such as leaf curling and angular reddening on leaves (Winkler *et al.*, 1974; Pearson and Goheen, 1988). If the honeydew excreted from the insects falls on fruit, it can cause an unsightly sooty mould appearance, which reduces the quality of table grapes (Settle *et al.*, 1986). As for most pests and diseases, a zero tolerance to their presence is not feasible, and one study has indicated that, at least from the standpoint of impact on vine health and productivity (crop loads and vine photosynthesis), a small population of leafhoppers can be tolerated (Candolfi *et al.*, 1993).

Basal leaf removal is a good first line of defence against leafhopper infestation in vineyards (Stapleton *et al.*, 1990), as this is the location where eggs are laid. Control of leafhoppers can also be through the use of insecticides, with pyrethroids being an option that has been in use for many years (Perring *et al.*, 1999), as well as *Bacillus thuringiensis*, which is a bacterium that produces a chemical that disrupts cells in the insect gut. Synthetic insecticides such as organophosphates and carbaryl have been used, but seem likely to be phased out in the longer term due to toxicity concerns (Jenkins and Isaacs, 2007). There has also been research into successful biocontrol agents, such as wasps (Kido *et al.*, 1984; Williams and Martinson, 2000), which may form a component of integrated pest management systems.

Borer insects

There are various types of borer insects that attack grapevines. In general, the adult forms are either beetles or moths that lay eggs in the soil near the vine or on the vine itself. Root and cane borer larvae hatch from the eggs and find roots or shoots, respectively, to infest. The larvae eat through the centre of the roots/canes/shoots, damaging the vascular tissue and causing water stress-type symptoms, or killing the vines outright (se Plate 62). Control is best accomplished by removing alternate plant hosts near the vineyard and sanitation of infested material. Mating of the insects can be disrupted with pheromone release into affected areas (Johnson *et al.*, 1991), which can be effective enough to negate the requirement for insecticides. At least two fungal pathogens of the larvae have also been reported (Dutcher and All, 1978), which brings with it the promise of biological control.

Mealybugs

Mealybugs (species in the genus *Pseudococcus)* are a problem in vineyards because they vector diseases but, also, through their feeding, increase botrytis infection and physically infest the clusters, causing the appearance of sooty mould, which is unacceptable for table grapes (see Plate 63). In many areas economic control can be accomplished through the use of natural predators, such as parasitic wasps (Berlinger, 1977; Trjapitzin and Trjapitzin, 1999), though there are many other types of potential control organisms (Walton and Pringle, 2004).

The grape berry moth

The grape berry moth of the eastern USA (*Endopiza viteana*) is a serious pest in smaller vineyards with wooded boundaries harbouring many wild grapevines; moths move from the wooded areas to the vineyard (Botero-Garcés and Isaacs, 2004). Although cultural practices that alternately bury and expose the overwintering stage of the insect are sometimes effective, chemical treatment is usually required. The sex pheromone of the grape berry moth is commercially available and has been successfully used for control (Martinson, 1995;

Witzgall *et al.*, 2000), as has *B. thuringiensis* (Ifoulis and Savopoulou-Soultani, 2004). In larger vineyards with less boundary per area, chemical control can be limited to the outside rows. In other vineyards damage is rare and no chemical control is required. The European grape berry moth is a different species (*Lobesia botrana*), causing similar grape damage and requiring equivalent methods of control, though some aspects may differ (Gabel and Renczés, 1985; Witzgall *et al.*, 2000).

Thrips

Thrips are very small insects that cause damage when the eggs are laid in fruit (sometimes called halo-spotting), or when they feed by rasping on the berry surface, which causes scarring (see Plate 64). Damage is primarily visual, which is important in the table grape industry, but berry scarring can be an origin of berry splitting (Yokoyama, 1979). Enhancing the plant diversity in vineyards can enhance natural control of thrips (Nicholls *et al.*, 2000).

Beetles

Beetles can also cause damage, either through the larvae feeding on vine roots or from adults feeding on the leaves (see Fig. 9.9). The bark of the grapevine trunk is often host to a variety of insects, some of whom only come out to feed after dark.

Mites

Mites are not insects, but cause similar leaf symptoms. Spider mites, grape erineum mites, grape rust mites and false spider mites are found in most grape-growing areas. Mites cause leaf damage by feeding (see Fig. 9.10), the leaves may brown or drop and reduce vine photosynthesis or buds can also become infested (see Fig. 9.11). Many problem mites are controlled naturally by predatory mites (Karban *et al.*, 1995), so chemical applications to control mites must be carefully selected to prevent decimating the beneficial mites (English-Loeb *et al.*, 1986). For example, sulphur applications for disease control can disrupt the life cycles of both damaging and beneficial predatory mites (Hanna *et al.*, 1997).

Chemical Control

In general, chemical controls should not be applied unless the insect is present and in a form that is susceptible to the chemical. Insect life cycles vary for

Fig. 9.9. Vine weevil damage on leaves of a mature grapevine, giving a notched appearance to the leaf margins.

Fig. 9.10. Erinium (or blister) mite colonies on grapevine leaves cause these deformations on their upper sides.

different insects, and the appearance of the vulnerable form may depend on macro or microclimate. Scouting, or monitoring, can be used to determine the presence and population of relevant insects. Vines should be examined in detail for the number and type of insects and, if the number seen is known to result in economic damage (through experience or published information), a chemical or other control should be applied to reduce the risk.

Fig. 9.11. Example of shoot growth (left) and leaf (right) affected by bud mite damage. Note the zig-zag growth pattern of the shoot.

ANIMAL PESTS

Nematodes

Nematodes are generalized feeders present in most soils and many are serious pests of grapes. Root-knot nematodes (*Meloidogyne* spp.) cause general decline in grapevines, not usually killing them but rendering them more susceptible to other diseases or environmental stresses. Dagger nematodes (*Xiphinema* spp.) also cause vine decline but are also a vector of grapevine fanleaf virus (GFLV). Lesion nematodes (*Pratylenchus* spp.) also cause vine decline by root death (Nicol *et al.*, 1999). Nematode populations tend to increase with time in vineyards (Ferris and McKenry, 1974), and this has been associated with replant disorders. If a vineyard was to be planted into a soil with high nematode populations, this was traditionally corrected through soil fumigation (McKenry, 1999).

With recent and strict regulations regarding the use of methyl bromide, a commonly used pre-planting soil fumigant, alternatives such as the planting of crops that discourage nematode infestations can be tried (Nicol *et al.*, 1999). The use of soil drenches to control nematode populations suffers the same problems as for phylloxera – the organism can live quite deep into the soil profile and persist in the soil long after vines have been pulled out (Raski *et al.*, 1965). Residual roots in the soil can carry the pests, so killing roots with systemic herbicide and removal of as many roots as possible is helpful.

Vine-grazing pests

Vine-grazing pests include many species of deer and rabbits. They do not usually consume significant amounts of plant in a healthy vineyard, but are

45

46

47

Plate 45. Powdery Mildew cleistothecia, the tiny black specks, on the underside of a Sultana grapevine leaf.
Plate 46. Powdery Mildew infected berries.
Plate 47. Downy Mildew fungus on the upper and lower surfaces of a grapevine leaf.

Plate section supported by the Corporate Research Sponsors of L.L. Creasy.

49

48

50

Plate 48. Berries and rachis affected by Downy Mildew.

Plate 49. Phomopsis symptoms on the bases of shoots and on leaves.

Plate 50. Eutypa-affected shoots on a grapevine, showing short internodes with a zig-zag growth pattern and tattered leaves.

Plate 51. A vine with red foliage and failing early in the season. Closer examination (inset) shows that part of the trunk leading to the roots is dead.
Plate 52. Pinot noir leaves showing Grapevine Leafroll virus symptoms.
Plate 53. Example of contact herbicide damage on grapevine leaves: paraquat and diquat mixture.

54

55

Plate 54. The effect of a pre-emergent herbicide spray of flumioxizen on grape shoots that had been splashed with soil particles after rainfall.

Plate 55. Examples of systemic herbicide damage on grapevine shoots and leaves. **A** diuron and bromacil mixture; **B** 2,4-Dichlorophenoxyacetic acid (2,4-D) on a shoot tip; **C** 2,4-D on leaves; **D** mild glyphosate on leaves.

56

58

57

Plate 56. Left; hail damage on shoots, with one that was on the leeward side protected. Right; old wound caused by hail on the cane below the shoot. Such wounds can be entry points for disease organisms.

Plate 57. Necrosis of leaves on a recently transplanted vine, caused by heat, intense sunlight and lack of water supply.

Plate 58. Phylloxera nymph on a grapevine root.

Plate 59. Weak vines in a row, the first potential signs of Phylloxera infestation.
Plate 60. The aerial form of Phylloxera lives in these galls that appear on infested vines in some areas of the world. The upper side of leaf is shown on the left and the lower side on the right.
Plate 61. Leaf symptoms typical of those caused by leafhopper feeding, in this case, on Sauvignon blanc vines.

62

63

Plate 62. Left; the base of the trunk of a young vine killed by a Lepidopteran root borer. Right; the root borer larva becomes visible when the base of the trunk is cut open.

Plate 63. Mealybugs nestled inside a late harvest Semillon cluster.

64

65

Plate 64. Scarring on Pinot noir berries as a result of thrip feeding earlier in berry development. Most damage is usually near the stylar end of the berry as the insects feed on pollen and tender ovary tissue.

Plate 65. Grape drying racks in Australia. If necessary, covers can be set over these racks to protect the grapes from rainfall later in the season.

particularly damaging in newly planted vineyards. They can defoliate vines in a new planting, eating the tender leaf blades while leaving the petioles still attached to the shoots (see Figure 9.12). This can greatly delay the establishment of young vines through the loss of photosynthetic capacity. Rabbits, turkeys and other animals can also eat the bark from the young trunks, damaging or even killing vines (see Fig. 9.13).

Berry-eating pests

Berry eating pests include many mammals such as raccoons, skunks, opossums, rodents and others. Birds also are voracious berry consumers and can cause severe economic damage to grapes (see Fig. 9.14). Birds can be residents nesting in the vines, flocking (such as starlings) or migratory. Migratory birds travelling in large flocks are particularly dangerous, as they can consume a large amount of fruit in a short amount of time. There can be large differences in the extent of damage, depending on whether the vineyard is on a migration route or not.

Fig. 9.12. A young vine that has been defoliated by deer. This can be more of a problem in dry areas or seasons, where irrigated grapevines have the lushest growth available.

Fig. 9.13. Rabbit (left) and goose (right) damage on vine trunks. The geese have completely girdled the vine trunk, resulting in the vine pushing up shoots from the base. In severe cases the trunk can be severed by animals.

Fig. 9.14. Some bird species tend to peck at, rather than take, whole berries, which can result in secondary infections (left). Particularly in dry seasons, birds can take fruit quite early, even before it colours (right).

Although there are many possibilities for managing animal damage in vineyards, exclusion methods are most dependable for both mammalian and avian predation. Electric fences or barrier fences are both effective for mammals, though costly. Netting is effective for birds (see Fig. 9.15), although stationary noise (see Fig. 9.16) and visual scare methods may suffice temporarily for moderate bird pressure (Aubin, 1990). Many commercial operators in high-pressure areas use a combination of nets, propane cannons and frequent movement/noise in the vineyard (e.g. motorbikes), with or without actual shooting of birds. Some growers find that shooting of birds is necessary for devices such as propane cannons to be effective.

Bird netting can be applied directly on the vines and tied under to prevent entry, or applied on supports over the vines. It must be applied before birds become habituated to feeding from the grapevines, and removed just before or soon after harvest. The latter is to prolong the life of the nets, which can last up to 10 years with care in application and storage. Though effective, nets require significant labour to apply and recover (note the shoot growth through the nets

Fig. 9.15. Examples of bird netting used in vineyards. Overhead nets (upper), over-the-row nets, which can cover multiple or single rows (left) and side-netting (right).

Fig. 9.16. A stationary propane cannon situated in a vineyard; these are usually set to go off a variable intervals, sometimes with two reports in close succession.

in Fig. 9.15, which requires more effort to remove before the nets are stored), as well as space to store when not on the vines. New research on learning more about the behaviour of birds in vineyards may yield more effective and holistic management methods (Saxton *et al.*, 2004 a,b).

Fences are expensive to erect, but can last a long time with continuing maintenance. Fences for berry-eating mammals must be close enough to the ground so the animal cannot crawl under, and must extend high enough to discourage those animals that can jump. Fences to keep out deer-like animals should be at least 3 m high if they are erected on uneven terrain (VerCauteren *et al.*, 2006).

An issue worthy of serious consideration when it comes to disease and pest control is the potential effects on neighbouring properties. Spray drift and noise used to manage these problems can often be objectionable to others, so it is best to communicate well with others in your area to ensure minimal disturbance. In some cases there are local regulations covering these types of activities, which should be taken into account when developing a vineyard.

WEEDS

By definition, a weed is a plant growing where it isn't wanted. Weeds in vineyards use both nutrients and water needed for vine growth. Excessive weed growth can block light and hinder airflow as well. In some floor management systems no plants (weeds) are allowed to grow under the vines or between the rows. More machine travel in the vineyards has increased the need for a planted middle, which supports such activity, so vegetation is allowed to grow (at least for some of the year). These volunteer crops can also be called a cover crop, and may be suppressed during the highest water use season for the vines. Rather than use the volunteer plants as a cover crop, sowing a specific grass or other plant seed that has attributes better suited to vineyards is also practised. Several strategies for cover crop management are presented in Chapter 6.

Weeds can be prevented from growing, or be destroyed at some growth stage, by either mechanical (Chapter 8) or chemical methods. Chemical herbicides are used to kill established weeds, suppress weed growth in row middles and inhibit germination of weed seeds. It is possible completely to control weeds with chemical applications. Herbicides can be sprayed, wiped or broadcast. Some herbicides (contact) kill only exposed green tissue and can be used near the base of growing vines, although because only the part of the weed that was in contact with the herbicide is killed, it usually recovers. Other herbicides are translocated (systemic) by green tissues to roots and kill the whole plant. They must be applied when the vines are dormant or physically protected from contact, as they can also be taken up by the vine. Germination inhibitors are applied in conjunction with other herbicides to prevent weeds from growing rapidly in the space created by dying weeds. Weed control strategies must be developed that control the weeds without damaging the grapevines.

PESTICIDE RESISTANCE

Disease-causing organisms, weeds and insects frequently become resistant to specific pesticides. Resistance to traditional insecticides in insect control has been shown with grape berry moth (Nagarkatti *et al.*, 2002), suggesting the need for alternative methods. There are several weeds already resistant to glyphosate (Heap, 1997; Powels *et al.*, 1998). Many fungicides have become useless because of resistance development by their target pest organisms. Resistance develops because of the biology of the pesticide – usually its specificity to the pest target and how frequently it is used. The greater the specificity, generally the greater the chance that a mutation will develop that overcomes it. It is very important not to overuse new and effective pesticides, which will encourage resistance development and possibly render them ineffective. Several strategies are involved, including limiting the number of applications per season, alternating use with pesticides of different modes of action or use in combination with a second pesticide of a different mode of action (Staub, 1991; Clarke *et al.*, 1997). Despite these actions, disease management using chemicals will be dependent on new pesticides being developed as either resistance develops or regulations force their withdrawal from the market.

PEST AND DISEASE CONTROL IN ORGANIC GRAPE PRODUCTION

Organic grape production has attracted attention from consumers, and growers are responding to their desires. The principle of organic grape production is the avoidance of synthetic pesticides or fertilizers. A truer goal of organic production would be to conserve and build healthy soils, maintain diversity and reduce dependence on non-renewable resources (Henderson, 1995). These principles have often been replaced by an economic opportunity and multiple instances of official certification, which may be very different between growing regions.

The organic grape producer is faced with the same mix of pests as the non-organic producer. Organic production does not mean no-spray: organic pesticides are frequently not as long-lasting as non-organic alternatives and must be applied more frequently. Diseases, insects and weeds can be controlled by organic methods but with some limitations, particularly with regard to climate or the amount of pest/disease pressure. The more suitable the climate is for a pest, the more difficult it will be to control through organic means.

Insect damage can be tolerated to some degree, particularly for juice or wine grape production, though less so for other end uses. Insecticides for organic production include toxic botanicals extracted from plants, such as rotenone, pyrethrum, ryania and sabadilla. Some oils and soaps are effective insecticides (Sams and Deyton, 2002), such as those made from the neem or tea tree. Biologically derived products like those from the bacterium *B. thuringiensis* can be as effective as synthetic insecticides. Behavioural control agents such as repellants, antifeedants, attractants and sex pheromones can be used for insect control (Martinson, 1995). For example, biological control of lepidopteran insects (such as the grape berry moth) is possible with sex pheromones (Ifoulis and Savopoulou-Soultani, 2004). Release of the parasitic wasp *Trichogramma minutum*, which targets the moths' eggs, has resulted in acceptable control of the North American grape berry moth (Nagarkatii *et al.*, 2003).

Organic disease control depends on these major factors: climate, location, cultivar and pesticides. In dry climates, downy mildew, phomopsis and anthracnose are not problems; however they can present considerable control problems in humid growing regions. Locations away from wild grapes or abandoned vineyards reduce pressure from powdery mildew. Exposure to wind for rapid drying reduces pressure from many fungal diseases as well. Cultivar selection is important in avoiding disease damage, but frequently the cultivars required by the market are susceptible to those diseases most difficult to control, e.g. all *V. vinifera* cultivars are susceptible to powdery mildew, downy mildew and black rot. There is an available range of resistance to these diseases among French hybrids, but frequently these are not desirable cultivars. Organic pesticides available for disease control include sulphur and copper, but some cultivars that have disease resistance to one or more diseases are also sensitive to sulphur or copper, which may be necessary for control of other diseases to which the cultivar is susceptible.

Weed control in organic vineyards is more difficult than insect or disease control. Under-row weed management depends primarily on mechanical cultivation. Propane burning is permitted under some organic regulations, but is fuel intensive and difficult to carry out safely (Pinel *et al.*, 1999). Both cultivation and burning must be performed frequently when weeds are small, to prevent potential damage to lower grapevine parts.

Mulching to prevent or slow weed growth under the row shows promise. Plastic mulches are effective but expensive, may not be considered suitable for organic production and create disposal problems. More sustainable possibilities are the use of waste products, such as mussel shells or crushed glass. The former has shown promise both as a way or discouraging weed growth and in having a positive effect on the quality of wine made from the 'Pinot noir' grapes growing over it (Creasy *et al.*, 2006).

Between-row weed control can be done mechanically, although it is easy to damage roots if done too deeply (Saayman and Van Huyssteen, 1983). The maintenance of continuous grass covering the between-row area is now common, although in dry years or climates this can provide excessive competition and must be sprayed out.

PESTICIDE APPLICATION

Among the goals of pesticide application are the deposition of the recommended quantity of pesticides at the sites needed for effective pest control, and minimization of waste and the escape of pesticides into the environment. Applications should also be rapid, and low cost, in terms of both the equipment needed and the chemical itself.

Most pesticides are applied by spraying, but in specialized situations solid pesticides can be broadcast or blown as dust, and solutions can be wiped on with damp rollers or flat pads. Wipers have the advantage of direct contact with weeds away from the crop plant, and avoid the dangers of spray drift to non-target locations.

Maximizing spraying efficiency

Vineyard sprayers, wipers or spreaders can be hand carried, tractor mounted or on a trailer: the difference is only one of scale (see Fig. 9.17). A vineyard needs to be covered within a time period that includes the interval when the pest is susceptible and troublesome. It would not be satisfactory to start an application at the right time, but be unable to finish before the disease

Fig. 9.17. Examples of vineyard sprayers (clockwise, from top left): airblast sprayer, fan-assisted sprayer, ATV-mounted, controlled-droplet applicator for herbicides and two-row boom sprayer.

progressed beyond control, the pest was no longer vulnerable to control or the weather no longer favourable. Most vineyards need application equipment capacity to cover the entire planting within several days, for reasons such as the traditional sprays of elemental sulphur weathering rapidly and possibly requiring repeat applications.

For a fungicide to be effective a specified dose must reach the tissue it is meant to protect. To achieve this, the grapevine canopy must be open enough for the spray to penetrate, the sprayer must be properly calibrated and maintained and the spray applied at the appropriate time. Water-sensitive papers or clay suspensions that are visible when dried are available that allow evaluation of how well the spray is covering the target (Nicholas *et al.*, 1994).

Pesticide applications use quantity–time calculations. The target area must be known, as well as the amount of pesticide required to control the pest.

Good coverage is much easier to achieve early in the season, when the canopy is smaller and less dense, as opposed to during the ripening period. Herbicides used under the row require only a percentage of the total ground surface area to be sprayed, which must be used in rate calculations to determine the amount of pesticide to apply per unit area.

Pesticides are usually applied in motion, so the speed of application is important in determining the dose of pesticide. A speedometer (if equipped), a tachometer (and a corresponding ground speed table) or a watch can be used to determine speed. Once the rate of application is determined and the total amount is calculated, the carrier is selected – commonly water or possibly a solid. In most cases water and sprayers are used, which is the basis of the example presented here, but application of solids or liquids with wipers use the same types of calculations.

Sprayers pump liquids containing suspended or dissolved pesticides through nozzles that disperse the liquid into small droplets. The size of the droplet is important with regard to how far they travel and how well they impinge and stick to the surface of the plant (Matthews, 2001). Maximum coverage (a high percentage of the surface area of the plant covered by pesticide) is a primary goal of application. The coverage depends greatly on the design of the sprayer and the positioning and density of the vine canopy. One problem with sprayers includes fine mists that float away from the vine and result in the loss of (often expensive) material; which can create hazards for neighbouring properties or crops. Modern sprayer technology uses different approaches to minimize this problem.

- Droplet size is adjustable by pressure changes and by nozzle design and type.
- Droplets can be given an electric charge to improve their attachment to leaf surfaces (Hislop, 1988).
- Hoods cover the part of the row being sprayed and reduce spray drift (Ozkana *et al.*, 1997; Derksen *et al.*, 1999).
- Hoods can be designed to collect spray not deposited on the plant and recycle it back into the spray tank (Panneton *et al.*, 2001).

The time required to spray a vineyard depends on the size of the sprayer tank, because a significant amount of time is spent refilling with water and pesticides and in travel to and from the filling station. The speed of vineyard travel and the number of rows sprayed at a time are also very important. Travel speed can be limited by the condition of the row middles, with high speeds requiring smooth row middles. Multi-row sprayers generally move more slowly than single-row sprayers, but cover more area per unit time (see Fig. 9.18).

Application rate is also influenced by environmental conditions. The temperature should not be excessively hot, because the phytotoxicity of some pesticides is temperature dependent (i.e. copper and sulphur) (Pertot *et al.*, 2006). Hot and dry conditions can result in the spray droplet evaporating before it reaches the target, and the pesticide is free to travel in the wind and is thus lost. Windy conditions favour drifting as well, particularly for droplets

Fig. 9.18. Multiple-row spray unit. For effective use, the vineyard trellising and vines must have been installed with high precision, and a skilled tractor operator is needed.

that have not hit the target (Reichard *et al.*, 1992). The most favourable time for spray applications is in the evening or early morning, when wind is usually at its most calm. With adequate lighting, spraying can continue during the night as well. Some unfavourable conditions (wind) are ameliorated by sprayer design such as hooded, over-the-row sprayers or sensors that automatically shut off spray when passing a gap in the canopy. Recirculating sprayers are an advance over hooded sprayers; these collect spray that doesn't reach the plant, filter it and return it to the tank (Panneton *et al.*, 2001).

Pesticides are toxic; they are selected to kill some organism or group of organisms. Therefore, pesticide applicators and all those working within range should be protected from exposure and have proper and relevant training. Personal protection apparatus (PPA) is available for all pesticides, and these vary for the different levels of toxicity. The recommended (or required) PPAs will be listed on the pesticide label, which also incorporates application recommendations, safety requirements and recommended uses (this may vary with country and regulatory agency).

Pesticide application equipment

Pesticide application equipment should be carefully selected to fulfil the needs of the grower. All pesticide equipment should be calibrated each season (at least) for the delivery of known amounts of pesticides to the vines (see Fig. 9.19). Calibration is not difficult and should not be neglected. All sprayers have the ability to adjust the rate of pesticide application, at least within its design limits. This is necessary because nozzles and other parts of the spray delivery system wear out and eventually require replacement. Nozzles are available with varying delivery rates, which is one method of adjusting the rate of spray application and may be necessary for different sprays. Sprayers also have the ability to change the pump pressure, with an increase in pressure increasing the flow through a given nozzle, though most nozzles are designed to operate within a specific range of pressures. Another control point is the speed at which the sprayer moves through the vineyard. Because of time limitations and surface conditions, changing of nozzles is usually a better way to make significant changes in spray volume.

Calibrating a sprayer requires knowledge of how much liquid is sprayed per minute and the speed at which it travels. A wiper is calibrated the same way, but a duster or spreader of solid uses the weight of material and ground area covered per minute.

Fig. 9.19. Equipment made to measure the output of various types of sprayers: each of the upright panels collects the spray that hits it and feeds it to containers at the bottom. The amount of liquid in each is then measured to see how evenly the spray is being applied. Here, an airblast sprayer is being evaluated.

10

Harvest and Postharvest Processing

Unlike many other fruit crops, grapes do not ripen after they have been taken off the vine. There is no conversion of starch into sugar, as the amount of starch in berries is very low: none if you remove the brush from the fruit (Amerine and Root, 1960). Sugar produced by photosynthesis in the leaves is translocated from there to the grapes only until they are harvested or until maximum Brix is reached (around 23.5°, though the concentration of sugar can rise further than this with dehydration of the fruit). Though the concentration of berry solutes can change after harvest, and some other quality-related compounds can be evolved or degraded, in general, grapes must be harvested when the target quality parameters have been reached. Ethylene, considered to be a plant growth regulator, has been implicated in the development of grape berries, particularly for the development of colour (Hale *et al.*, 1970; Weaver and Montgomery, 1974), but it does not play the same role as it does in other, climacteric fruits, such as bananas and apples (Mailhac and Chervin, 2006).

After harvest, grapes may be consumed fresh, crushed or dehydrated. Fresh table grapes are handled, stored and marketed much like other fruits consumed fresh. Crushed grapes may be directly fermented into wine, pressed into juice that may be fermented into wine, consumed fresh or concentrated for later use. Dehydrated grapes or raisins are also a major product of the vine. However, the largest use of grapes in the world is to make wine, accounting for approximately 60% of world grape production.

TABLE GRAPES

Table grape production has been increasing in recent years. China is the largest producer and consumer of table grapes: in 2006 they produced over 6 million t. In production, Turkey is second, Italy third, Chile is fourth and the USA is fifth, producing about 800,000 t. Chile is the largest exporter of table grapes, at 800,000 t in 2006 (United States Department of Agriculture Foreign Agricultural Service *World Table Grape Situation and Outlook 2006*).

Quality parameters

The quality of table grapes depends on sugar concentration, acidity, colour, texture and aroma and flavour characteristics of the variety. Though sugar concentration is a common quality measure for grapes, there is not always a good relationship between it and the palatability of grapes for fresh consumption (Winkler, 1932). The ratio of berry juice Brix to acidity can give a better measurement of the palatability of grapes to consumers than either the sugar content or acidity used alone (Winkler *et al.*, 1974), although the best measure of quality varies between cultivars (Guelfat-Reich and Safran, 1971). Climate and variety affect the Brix:acid ratio. Ripening occurring in hot weather results in lower fruit acid and, generally, a lower sugar content will be sufficient for consumer acceptance (Winkler *et al.*, 1974). In cool climates, the acid will be higher relative to the sugar, so more of the latter will be required for equal taste.

Fruit colour and its uniformity within the cluster is also important (Clydesdale, 1993), and schemas for objective measurement of this have been proposed (Carreño *et al.*, 1995). Not only are berry size and colour significant, but so also are the condition of the cluster rachis, which should be green and firm (Winkler *et al.*, 1974), and the cluster size and shape (Cliff *et al.*, 1996), which is usually modified earlier in the season, but can be adjusted at harvest, too. Fruit texture is also important, with a crisp flesh generally being desired over those that are softer (Cliff *et al.*, 1996; Sato and Yamada, 2003).

Picking

Table grapes are all picked by hand. Pickers must be trained to select well-shaped clusters of mature grapes. Because not all clusters will be suitable for harvest at the same time, frequently there must be multiple harvests from each vineyard (Winkler *et al.*, 1974). Though berries can be tasted to determine suitability for picking, this cannot be done for every cluster so visual cues, such as berry size and colour and rachis colour, must be used to determine whether the cluster meets the picking criteria. Any poor-quality berries (e.g. those damaged by sunburn, disease, insects or wind) must be trimmed off because they will be seen by the consumer. If too many berries have to be removed, the fruit may not be marketable due to changes in cluster shape. During picking and packing grape clusters should be handled only by the stem, to preserve the natural bloom (waxy covering, see Fig. 10.1) as much as possible.

Packing

Most table grapes are packed as they are picked to avoid extra handling. Storage and shipping is most common in 10 kg boxes, which minimizes potential

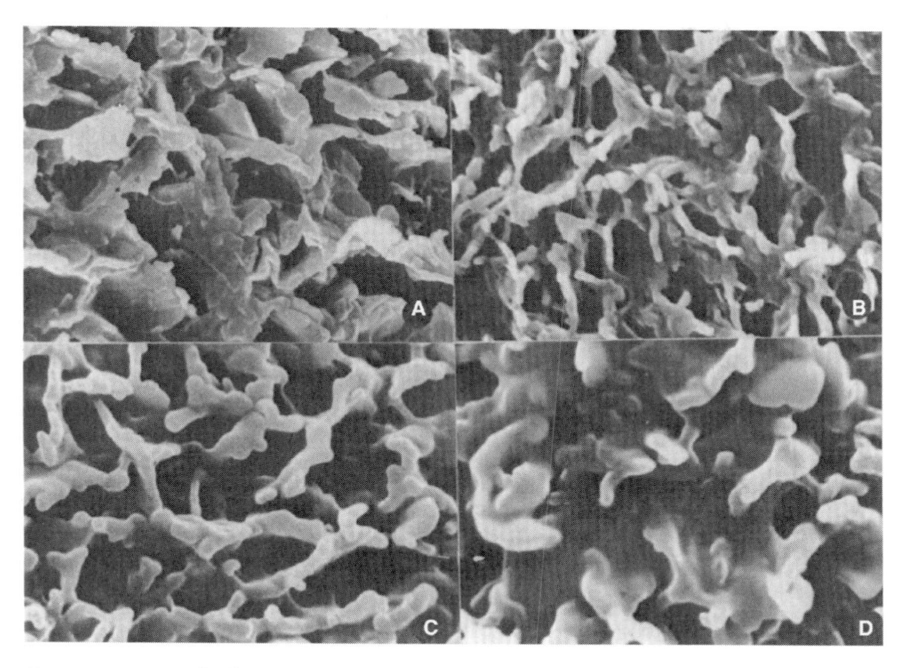

Fig. 10.1. Magnified (x 9500) scanning electron micrographs of mature grape berry epicuticular wax from 'Cabernet Sauvignon' (A), 'Pinot noir' (B), 'Grenache' (C) and 'Zinfandel' (D) fruit. Note the plate-like structure of the wax, which contributes to the bloom on grapes, and that the different cultivars have different conformations of the platelets. 'Cabernet Sauvignon' was found to have the most wax per unit of skin area (reproduced with permission from Rosenquist and Morrison, 1989).

crushing damage. Boxes, which are often wax or plastic coated to improve the container's resistance to wetting, should be kept out of the sun after (and during, if possible) filling and transported as quickly as possible to a cooling facility (see Fig. 10.2). Though grapes do not ripen after they have been removed from the vine they do continue to respire, which uses stored energy in the grapes, and warm fruit loses water more quickly than cool fruit (Szulmayer, 1971). Hence, rapid cooling of the grapes and concomitant slowing of respiration and transpiration preserves their quality.

Grapes destined for immediate delivery are sometimes shipped without pre-cooling, though they should be shipped in a refrigerated container. This is not ideal, as there is usually a considerable amount of residual heat in the grapes that is not efficiently removed in conventionally refrigerated rooms or containers (Mitchell, 1992), and high temperatures encourage disease organisms. In hot climates, delivery can be scheduled after sunset, when

Fig. 10.2. Grapes in a packing tent in Chile. From here the boxes will be transferred to a cooling and holding facility before shipment to the UK.

temperatures have dropped. Rapid cooling is extremely important for extended storage of table grapes: Nelson (1985) reported that, under Californian conditions, every hour that grapes are held at 32°C after picking they lose one week of storage at 0°C. Crisosto *et al.* (2001) noted that stem browning was associated with delays in fruit cooling, possibly due to excessive water loss during that time. Therefore, grapes should be moved to a cooling facility as soon after picking as possible and be cooled to 1°C within 4–8 h, with higher temperatures meaning less time is permissible (Thompson *et al.*, 2001).

If the fruit remains at temperatures above 0°C for too long, there are a number of disease organisms that can become established and decrease fruit quality. Among these are different types of moulds caused by botrytis or *Penicillium* and *Rhizopus* rot (Beattie and Dahlenberg, 1989). The best way to combat these problems is to bring the fruit to as close to 0°C as quickly as possible. At temperatures below this there is a risk of the berries or rachis freezing, which causes catastrophic damage. Grapes of different species have varying temperature thresholds for freezing, with *V. vinifera* grapes generally tolerating greater cold (down to −1°C) than fruit from *V. labrusca* or 'Muscadinia' cultivars (Himelrick, 2003). Rather than this being a physiological resistance to freezing, it probably has more to do with the amount of solutes in the berries (Jie *et al.*, 2003), as *V. vinifera* tend to have a higher Brix than the other grape types.

Cooling at the packing house

Grape boxes should be designed to facilitate air movement through them so that heat can be removed efficiently. Cooling using water, which is used in other crops, is not feasible with grapes due to the potential for disease spread. Moderate cooling rates can be accomplished in cold rooms (0°C) with high-velocity air movement, if boxes are placed with enough room between them so that air can move through easily (Mitchell and Crisosto, 1995). It is much more effective to stack the boxes in rows and pull cold air through the box slots and over the berries (Gentry and Nelson, 1964; Nelson, 1985), as this will result in cooling to a given low temperature in one-sixth to one-eighth the amount of time it takes for room cooling (Mitchell and Crisosto, 1995). After rapid cooling the grapes should stored at 0°C and 95–98 % humidity, which minimizes respiration, water loss and development of disease.

Storage

Holding cooled grapes at 0°C and high humidity will prolong green stem colour, reduce dehydration losses and berry shatter, and maintain fruit firmness (Crisosto *et al.*, 2001). Treatment with sulphur dioxide (SO_2) is effective in preventing grape decay and also helps prevent browning of the cluster stems. The discovery that sulphur dioxide reduces the activity of many decay organisms – even botrytis, which is active at low temperatures – was made in the 1920s, and its use was rapidly taken up by the California table grape industry (Winkler *et al.*, 1974).

Continuous room fumigation at low (30 ppm) SO_2 concentrations is the most effective treatment for preserving table grapes and has been the standard procedure for many years (Winkler *et al.*, 1974). Sulphur dioxide at high concentrations, however, can cause bleaching of the berries, with a white colour being first noticed at any breaks in the berry skin (Nelson and Ahmedullah, 1970). To avoid high concentrations of sulphur dioxide during shipping, grape boxes can be fitted with in-package, two-stage sulphur dioxide generators that release SO_2 in response to water vapour released by the grapes. The first stage is quick release, high concentration, effective in killing surface disease organisms on the fruit at packing. A slow-release second stage allows for several months of storage (Nelson, 1983), increasing opportunities for lengthening the marketing season and transport to more distant markets.

Some disadvantages of using SO_2 are that it is a colourless gas with a very irritating, pungent odour and it tends to corrode steel pipes and other equipment. Because some consumers (and workers) have sensitivities to sulphur, there is interest in finding alternative decay-controlling treatments. Those being considered include the use of acetaldehyde vapours (Avissar and Pesis, 1991), carbon dioxide-enriched atmospheres (controlled atmospheres) (Crisosto *et al.*, 2002), ozone gas (Francisco Artés-Hernandez *et al.*, 2007) and natural volatile plant components. Examples of the latter include (E)-2-hexenal, which has been shown to reduce mould on stored table grapes (Archbold *et al.*, 1999) and volatiles from 'Isabella' grapes (*V. labrusca*), which were shown to reduce botrytis development on *V. vinifera* grapes (Kulakiotu *et al.*, 2004). Some of these innovative approaches would be acceptable for organically grown grapes.

DRIED GRAPES

Preservation through drying is another method of storing grapes. Many fruits are dried, concentrating the sugars inside and making spoilage much less likely. Grapes can do this while on or off the vine. Raisins are any type of dried grapes, and several cultivars are used to make these.

The USA is the largest producer of raisins (400,000 t, 57% of total world production), which takes up 25% of the US grape harvest. Turkey, China, Chile, South Africa, Greece and Australia comprise most of the rest of world production. Turkey, however, is the largest exporter of raisins in the world (United States Department of Agriculture Foreign Agricultural Service *Market News Raisin Update*, August 2007).

Cultivars

Most raisins are made from the 'Sultana' (also known as 'Thompson Seedless' and other names) cultivar, with some also being made from 'Black Corinth' and 'Muscat' cultivars. The latter are usually seeded but the former are seedless, from a parthenocarpic fruit. The 'Black Corinth' has very small berries which, when dried, are called Zante Currants (or just currants) and are used primarily in baking. Because of their small size, the berries dry more rapidly than larger ones, which is an advantage that has ensured it being grown in areas that have high humidity or more frequent rain late in the harvest season (Whiting, 1992). 'Muscat of Alexandria' is a seeded grape that has a distinctive fruity flavour, but is less desired in the modern fresh grape market because of its seeds (Alston *et al.*, 1997). When they are mechanically seeded they become sticky and, although difficult to eat 'out of hand', they are frequently used in baking. Most of the raisins found in consumer markets and in world trade are made from 'Sultana'. This cultivar is desirable for raisin production for a number of reasons: a noticeable and natural bloom, meaty texture and characteristic flavour, not sticky and a low tendency to cake in storage (Winkler *et al.*, 1974).

Harvest

Harvest criteria for raisins revolve around improving the final yield of dried fruit, of which sugar concentration in the fresh fruit is a major component. As the concentration of sugar in the grapes rises, so does the yield of finished raisins, according to the drying ratio, which is the mass of fresh grapes it takes to produce 1 kg of raisins at 15% moisture (Christensen and Peacock, 2000b). Because gross returns are based on the amount of raisins produced, starting with higher-Brix fruit provides more money per tonne of fresh grapes. Raisin quality also improves with increasing Brix (Winkler *et al.*, 1974), though achieving higher sugars carries with it a greater risk of rainfall late in the season.

Traditionally, clusters of grapes were harvested by hand and laid onto trays between the grapevine rows. The exposure of the fruit to the sun hastens the loss of water, and berries shrivelled and browned to the point where they no longer expressed juice easily and could be handled neatly. More recent improvements to this method have been the adoption of techniques to minimize handling, such as laying the grapes on continuous sheets of paper that could be laid down mechanically as well as picked up for harvesting of the dried fruit (Christensen, 2000). Soil between the rows can also be moved to create a slope that angles toward the sun, which enhances drying rates (Winkler *et al.*, 1974; Christensen and Peacock, 2000). Part of this phenomenon may be that warmer grapes lose water at a faster rate than cooler grapes (Szulmayer, 1971). Turning of the grapes, once practised to improve the drying rate, appears not to be necessary in most situations (Christensen and Peacock, 2000), saving much in labour costs. Only when fruit is damaged or clusters are unusually large is this practice worth the extra cost.

The beginnings of dried-on-vine (DoV) production of raisins were with a technique devised to allow mechanical harvesting and laying out of grapes on to continuous drying sheets (Studer and Olmo, 1971). Canes of vines bearing grapes were cut and a mechanical harvester run through approximately one week later, which allowed partial drying of the grapes such that they would be removed from the rachis by the harvester, then conveyed on to the sheets of paper. More modern DoV techniques perform the drying completely on the vine, with no need to dry grapes on trays placed between rows (Peacock and Swanson, 2005). Aside from the significant savings in production costs, this method results in higher grades of raisins due to the different drying environment of the grapes, though consumer preference appears to be largely unaffected (Angulo *et al.*, 2007).

Rain is a serious threat to raisin production, and was found to be responsible for a significant amount of the variation in production in Australia (Considine, 1973). Those sites suitable for raisin production must be good grape-growing sites and have a very low probability of rain during and after the harvest season. Raisins can be harvested or picked up from the aisles if sufficient warning of impending rain is available. They can finish drying in a dehydrator if necessary, though this is an expensive process. Because of the impact weather events have on production, forecasting is extremely important to raisin growers (Lave, 1963). In the raisin-producing regions of California, Turkey and Iran, rain is very rare during the drying period. In countries where there is even a slightly greater chance of rain, frequently raisins are dried by placing the grapes on stacked wooden racks (see Plate 65). The intensity of the sun, temperature and air movement are satisfactory for production of raisins but, in case of occasional rain, the stacks can be covered. In most areas the probability of rain increases as autumn approaches, and the harvest decision for raisin grapes is a compromise between obtaining the most mature grapes and getting them dried economically before rain.

Raisins dried in the sun are invariably dark coloured due to the oxidation of phenolic compounds. However, lighter-coloured raisins (often called sultanas or sultana raisins) can also be produced. For 'Sultana' (the variety) this used to be performed by dipping the fruit in hot caustic soda, which caused tiny cracks in the cuticle of the berries, rinsing them in cool water and then exposing them to sulphur fumes for several hours to prevent oxidation. These golden raisins are finally dried in a dehydrator at 60–70°C to reduce the water content to approximately 15%. The increased rate of drying inhibits the browning reactions (Grncarevic and Radler, 1971). More recently, cold emulsions have been used to hasten water loss, a technique originally used by the Greeks and Romans, where grapes are dipped in a mixture of olive oil and wood ash (Whiting, 1992). The drying rates of grapes so treated approximate those of peeled berries (Grncarevic, 1963), which in some ways is astonishing! Therefore, the use of emulsions brings the drying time for grapes down from 20 to 8–10 days in the sun, or in a dehydrator from 3–5 days to 24–30 h (Grncarevic and Radler, 1971).

Golden raisins, which are very light in colour, are produced by dipping the fresh fruit (usually 'Sultana') in a very hot solution of caustic soda and sodium sulphite for just a few seconds (Winkler *et al.*, 1974). This process creates fine fissures in the berry skin, enhancing the drying rate. Following the dip, fruit are subjected to additional gaseous sulphiting and then forced-air dried. The action of the sulphites produces the yellow golden colour, which is preferable for some raisin uses. Modification to this procedure, using sulphite solutions, has been investigated for the purpose of reducing sulphite emissions and increasing the efficiency of sulphite transfer to the fruit (Lydakis *et al.*, 2003).

Processing

Once the average moisture of raisins has been reduced to 15%, they are conditioned to allow the moisture content to equilibrate, with low-moisture berries gaining some water from the higher-moisture berries. This reduces the variation in the raisins, making sorting and quality grading more efficient. Put into large boxes, they are then taken to a processing plant (see Fig. 10.3).

Raisins are usually graded in an airstream sorter (Baranek *et al.*, 1970): they are fed into a stream of air flowing upward and, due to their volume:density ratio, either fall downstream to be collected in a bin or are blown upwards to a different bin. Since raisin production often results in close proximity of the soil to the fruit, contamination with soil and sand is a possibility and the amount of grit mixed with the raisins is sometimes used as another quality parameter (Winkler *et al.*, 1974).

Raisins (unlike grapes) do not respire and can be stored and shipped under less controlled conditions. However, they are still subject to insect infestation and disease, which must be managed mainly through exclusion, as the product is to be consumed directly.

JUICE/PRESERVES

Fresh grape juice includes purple grape juice, traditionally made from the 'Concord' grape and familiar in the USA. Purple 'Concord' juice was introduced in 1869 and made possible by the commercialization of the experiments of Louis Pasteur (pasteurization), which killed spoilage organisms and allowed the juice to keep much longer. The impetus behind pasteurized grape juice at that time came with the Temperance movement, as some people wished to find a substitute for

Fig. 10.3. Raisins entering a processing plant. Note the material other than grapes (MOG).

wine used in religious ceremonies (Bailey, 1912). The imposition of national prohibition in the USA in 1920 created a great demand for 'Concord' grapes, which were planted heavily and still dominate viticulture in the eastern states of the country. Today, 'Concord' grapes are grown in the US states of New York, Pennsylvania, Ohio and Michigan, with the largest producer being Washington (http://www.welchs.com).

Juice is also made from neutrally flavoured *V. vinifera* varieties, most notably 'Sultana', and also in much smaller quantities from muscadine grapes (Olien, 1990). While the juice of these and other cultivars of grapes is sometimes sold on its own, it is more often used as an ingredient in other products.

Uses

Grape juice concentrate is used frequently as a food ingredient and has the advantage of being a means of adding sugar to other products in a natural form. Generally, white grape concentrates are used as sweeteners and juice stock while red grape concentrates are used to provide colour and as stock for red grape juice, although pigments from 'Concord' grape juice have been tested as colorants for other beverages and also for gelatin products (Clydesdale *et al.*, 1978). Grape juices sweetened and diluted with grape juice concentrate can be labelled as 100% grape juice, while other fruit juices can be labelled as 100% juice (Tipton *et al.*, 1999). California is a major producer of juice concentrate, with about 18% of the total grape crush in California (3.6 million t in 2004) being made into juice concentrate. The major variety is 'Sultana' (65%), with some table grapes and some red wine grapes (Paggi and Yamazaki, 2005).

Grape juice is a normal product of international trade; however, accurate data for the production of juices are difficult to find as there is uncertainty in terms of the data collected over its eventual end use (OIV *Situation Report and Statistics for the World Vitivinicultural Sector in 2002*). Much of it is used as a food additive or as fresh juice. Argentina, Chile, Brazil and Mexico account for more than 90% of the juice/concentrate imported into the USA. Argentina produced 180,000 t of concentrated grape juice in 2006 and exported 160,000 t, about 50% of that to the USA (Pirovano, 2006) Given the amount of grape juice produced in the world it is surprising that more accurate data have not yet been found.

Processing

Red and purple grape juices are made with a hot press process that extracts many of the skin components, most importantly the pigments (anthocyanins), as well as sugar and flavour compounds (Pederson, 1971). Juice colour and flavour can be affected by the maximum temperature to which the juice is heated, so provided that the process regulates this temperature closely, unwanted organisms can be killed and juice quality largely unaffected (Flora, 1976).

White grape juices are made with a cold press, leaving many skin components behind. This can be advantageous if a neutrally flavoured juice is desired, as many flavour components are sequestered in the skin cells as well as in the berry pulp (Wilson *et al.*, 1986).

Sulphur dioxide can be used to prolong the quality of wine or juice grapes waiting to be processed. This is particularly useful with mechanically harvested grapes that are susceptible to decay (Morris *et al.*, 1972). It is notable that the storage temperature of 'Concord' juice was found to have an influence on its quality – at temperatures of 35°C the loss of colour and desirable flavours was greater than if the juice had been stored at 24°C (Morris *et al.*, 1986). This demonstrates the importance of postharvest factors that are sometimes out of the control of the producer or marketer.

WINE

Total world wine production is well in excess of 240 million hl, with EU countries producing about 70% of this. The major producing countries are France (55 million hl), Italy (50 million hl), Spain (40 million hl), the USA (28 million hl), Argentina (15 million hl), South Africa (12 million hl), Germany (10 million hl), Chile (8 million hl) and Portugal (6 million hl) (United States Department of Agriculture Foreign Agricultural Service *World Wine Situation and Outlook, Aug 2006*). There is little doubt that wine is an important part of many cultures, leading to its production in virtually every area that can support vine growth.

Harvest

Grape maturity for wines depends on the wine being produced. Grapes for sparkling wine production, for example, are harvested at a lower Brix than those for still table wines (Zoecklein, 2002), and grapes for late-harvest wines are picked when the sugar levels are higher than for table wines. Grapes for most wines, however, require sufficient acid and sugar to result in the desired pH, alcohol content and flavour, and also acceptable varietal aroma, to produce a quality wine.

Most grapes intended for wine are harvested mechanically, although grapes for fine wines may be picked by hand to select for even maturity and to avoid poor-quality grapes. For example, mechanical harvesting cannot distinguish between the primary crop and second set crop that may develop on shoot laterals, resulting in a lower-quality must (Clary *et al.*, 1990). Thus, in some cases, pickers may go through a vineyard to remove immature or diseased clusters before mechanical picking. As there is mechanical damage to the fruit during the process, SO_2 may be added to the harvesting bin to slow oxidation while the fruit is transported to the winery (Ough and Berg, 1971). In some areas where the wineries are distant from the vineyards, grapes are harvested, crushed and the juice pressed out near the vineyard and then transported by tanker truck to the winery, which can take upwards of 24 h or more (Pocock *et al.*, 1998). Adding SO_2 to the juice at crushing prevents oxidation and microbial spoilage during the long transport times.

Hand-harvested grapes are picked into small baskets or bins that are carried to larger bins of up to 500 kg capacity – the size of these being limited by how much weight the grape clusters at the bottom can carry before crushing of the berries (Hamilton and Coombe, 1992). Bins are usually lined with plastic to retain any juice that may escape from the fruit, e.g. during transport.

Hand-harvested fruit should arrive as whole clusters at the winery, which can be transferred directly to a press (see Fig. 10.4) and juice extracted from the undamaged fruit. This yields a lighter juice, usually of more delicate quality and with fewer phenolics, though there is less volume per tonne of grapes obtained. This process is used for sparkling wine grapes and some white wine styles. The juice is then fermented with yeasts, either added or naturally occurring, which convert the sugar to alcohol and carbon dioxide.

Fig. 10.4. Left, a hydraulic, basket-type press, which is often used for whole-bunch fruit. Right, a close-up of the side of a basket press, showing some seeds escaping from the slotted basket.

In some cases hand-harvested grapes are sorted before they are put into the press or crusher/de-stemmer. This is a last point of quality control, where material other than grapes, immature fruit and diseased fruit can be removed from the crop.

Processing

Fruit entering the winery is usually sampled to acquire information about Brix, acidity, pH, MOG and disease. Payment for grapes is mainly dependent on these factors, so obtaining an accurate and representative sample is important for both grower and winery. In large bins of fruit a remote-controlled sampling arm can be used, which can take a sample spanning the top to the bottom of the load (see Fig. 10.5).

Clusters of red wine grape cultivars are usually crushed and de-stemmed (in a machine with rotating tines and a grating that removes the berries from the rachides, followed by closely placed rollers that break open the berries),

Fig. 10.5. An example of a remote sampling arm, which is used to take a representative sample of grapes from a large bin on arrival at the winery.

resulting in a must, which is the soupy mix of the skins, seeds and pulp of the fruit (see Fig. 10.6). Fermentation is carried out with yeasts as for white wines, but this time the skins and seeds are included. When the yeast fermentation has finished the alcoholic wine is drained away, leaving the skins and seeds to be transferred to a press where the remaining wine is extracted.

The cake of pressed grape skins, either those from fresh grapes destined for white wine or those for red wine that have been through fermentation, are a source of many antioxidant compounds (Larrauri *et al.*, 1996). As winemaking does not extract everything from the grapes, sugar and water can be added back to the pressed grapes, which are then re-fermented and the resulting wine distilled to make grappa, or marc. These high-alcohol spirits often contain flavours quite reminiscent of the grape itself.

Fig. 10.6. A crusher/de-stemmer. Grapes are fed into the chute at top right and the rachides are removed in the cylindrical section at middle left. Below the de-stemmer are rollers that crush the fruit, which is then emptied out into the vessel below, where the must is pumped into a press or temporary holding tank.

OTHER WINE STYLES

Passito wines, or wine made from dried grapes as the term is used in Italy, are another category of wine types. It is not unusual to make wine from raisined grapes, as is the case for wines made from botrytis-affected 'Noble Rot' fruit, but it is rarely made using red grapes. The exceptions to the rule are recioto (sweet)- and amarone (dry)-style wines made in Italy, where red-skinned grapes are harvested at ripeness and laid out on trays (much like those in Fig. 10.5, but under shelter) to lose water and concentrate sugars and flavour compounds at moderate temperatures (Versari *et al.*, 2001). Because the grapes are under shelter and do not dry out as quickly as in raisin production, the flavour profile is said to be unique (Belfrage and Loftus, 1993). Under conditions like these, botrytis is likely to become established in the fruit, which is fine for white grapes as long as it is of the 'Noble Rot' type, but it is not desirable in red grapes (Belfrage and Loftus, 1993).

Icewine is another style of sweet wine that is out of the ordinary. The special characteristics of this product come from the fact that, as water freezes in a berry, it concentrates the remaining solutes. So, if grapes are picked and pressed at very cold temperatures, such as −8°C or below (Pickering, 2006), a very high-Brix juice can be extracted and fermented into a very sweet wine. Understandably, there are few areas in the world that have the right combination of climate and cultivars, as well as hardy people available to work under such conditions. Traditionally, these wines were made in Germany and Austria, but more recently Canada has made it a focus and is producing more and more (Pickering, 2006), and China has areas with appropriate climate, too, and is ramping up production (Patton, 2006).

References

Abel, N., Baxter, J., Campbell, A., Cleugh, H., Fargher, J., Lambeck, R., Prinsley, R., Prosser, M., Reid, R., Revell, G., Schmidt, C., Stirzaker, R. and Thorburn, P. (1997) *Design Principles for Farm Forestry*. Rural Industries Research & Development Corporation/Land and Water Resources R&D Corporation/Forest and Wood Products R&D Corporation Joint Venture Agroforestry Program, Canberra, Australia.

Adams, L.D. (1985) *The Wines of America*. McGraw-Hill Book Company, New York.

Adrian, M., Rajaei, H., Jeandet, P., Veneau, J. and Bessis, R. (1998) Resveratrol oxidation in *Botrytis cinerea* conidia. *Phytopathology* 88, 472–476.

Aguero, C.B., Uratsu, S.L., Greve, C.L., Powell, A.L.T., Labavitch, J.M., Meredith, C.P. and Dandekar, A.M. (2005) Evaluation of tolerance to Pierce's disease and Botrytis in transgenic plants of *Vitis vinifera* L. expressing the pear PGIP gene. *Molecular Plant Pathology* 6(1), 43–51.

Aguín, O., Mansilla, J.P., Vilarino, A. and Sainz, M.J. (2004) Effects of mycorrhizal inoculation on root morphology and nursery production of three grapevine rootstocks. *American Journal of Enology and Viticulture* 55(1), 108–111.

Ahmad, M.F. and Zargar, G.H. (2005) Effect of trunk girdling, flower thinning, GA3 and ethephon application on quality characteristics in grape cv. Perlette under temperate Kashmir valley conditions. *Indian Journal of Horticulture* 62(3), 285–287.

Allen, M., Lacey, M.J. and Boyd, S.J. (1997) Methoxypyrazines of grapes and wines: new insights into their biosynthesis and occurrence. *Proceedings of the Fourth International Symposium for Cool Climate Viticulture and Enology*, 16–20 July, 1996, Rochester, New York. Agricultural and Experimental Station, Geneva, New York, pp. V, 36–39.

Alleweldt, G. and Possingham, J.V. (1988) Progress in grapevine breeding. *TAG Theoretical and Applied Genetics* 75(5), 669–673.

Alston, J.M., Chalfant, J.A., Christian, J.E., Meng, E. and Piggott, N. (1997) *The California Table Grape Commission's Promotion Program: an Evaluation*. Paper 43, Giannini Foundation of Agricultural Economics, University of California, 133 pp.

Amerine, M.A. and Roessler, E.B. (1976) *Wines. Their Sensory Evaluation*. W.H. Freeman and Company, San Francisco, California.

Amerine, M.A. and Root, G.A. (1960) Carbohydrate content of various parts of the grape cluster. II. *American Journal of Enology and Viticulture* 11(3), 137–139.

Amerine, M.A. and Winkler, A.J. (1944) Composition and quality of musts and wines of California grapes. *Hilgardia* 15, 493–675.

Amerine, M.A., Roessler, E.B. and Filipello, F. (1959) Modern sensory methods of evaluating wine. *Hilgardia* 28, 477–567.

Amiri, M.E. and Fallahi, E. (2007) Influence of mineral nutrients on growth, yield, berry quality, and petiole mineral nutrient concentrations of table grape. *Journal of Plant Nutrition* 30(3), 463–470.

Anderson, L.J., Comas, L.H., Lakso, A.N. and Eissenstat, D.M. (2003) Multiple risk factors in root survivorship: a 4-year study in Concord grape. *New Phytologist* 158, 489–501.

Angulo, O., Fidelibus, M.W. and Heymann, H. (2007) Grape cultivar and drying method affect sensory characteristics and consumer preference of raisins. *Journal of the Science of Food and Agriculture* 87(5), 865–870.

Annabell, R. (2001) *Novel Way to Fight Frost Threat*. Wairarapa Times-Age, Masterton, New Zealand.

Anon. (2007) *Phylloxera-resistant Rootstocks for Grapevines*. Hellman, E., Northwest Berry and Grape Information Network, 2008.

Anon. (2008) *Yalumba Nursery Rootstock Selector*, Yalumba Nursery, Australia, 2008.

Anstey, T.H. (1966) Prediction of full bloom date for apple, pear, cherry, peach, and apricot from temperature data. *Proceedings of the American Horticultural Society* 88, 57–66.

Antcliff, A.J. and Webster, W.J. (1955) Studies on the Sultana vine I. Fruit bud distribution and bud burst with reference to forecasting potential crop. *Australian Journal of Agricultural Research* 6(4), 565–588.

Antcliff, A., Webster, W. and May, P. (1957) Studies on the Sultana vine. V. Further studies on the course of bud burst with reference to time of pruning. *Australian Journal of Agricultural Research* 8(1), 15–23.

Arbabzadeh, F. and Dutt, G. (1987) Salt tolerance of grape rootstocks under greenhouse conditions. *American Journal of Enology and Viticulture* 38(2), 95–99.

Archbold, D.D., Hamilton-Kemp, T.R., Clements, A.M. and Collins, R.W. (1999) Fumigating 'Crimson Seedless' table grapes with (E)-2-hexenal reduces mold during long-term postharvest storage. *HortScience* 34(4), 705–707.

Ascard, J. (1998) Comparison of flaming and infrared radiation techniques for thermal weed control. *Weed Research* 38(1), 69–76.

Aubert, C., Baumes, R., Gunata, Z., Lepoutre, J.P., Cooper, J.F. and Bayonove, C. (1998) Effects of sterol biosynthesis inhibiting fungicides on the aroma of grape. *Vitis* 18(1), 41–58.

Aubin, T. (1990) Synthetic bird calls and their application to scaring methods. *Ibis* 132(2), 290–299.

Augsburger, H.K.M. (2000) Frost control in temperate climates through dissipation of cold air. *Aspects of Applied Biology* 61, 201–204.

Avissar, I. and Pesis, E. (1991) The control of postharvest decay in table grapes using acetaldehyde vapours. *Annals of Applied Biology* 118(1), 229–237.

Bailey, L.H. (1912) *Cyclopedia of American Agriculture: Vol. II – crops*. Ayer Company Publishers, Manchester, New Hampshire.

Bailey, L.H. (1933) The species of grapes peculiar to North America. *Gentes Herbarium* 3, 150–244.

Bangerth, F. (2000) Abscission and thinning of young fruit and their regulation by plant hormones and bioregulators. *Plant Growth Regulation* 31(1–2), 43–59.

Baranek, P., Miller, M.W., Kasimatis, A.N. and Lynn, C.D. (1970) Influence of soluble solids in 'Thompson Seedless' grapes on airstream grading for raisin quality. *American Journal of Enology and Viticulture* 21(1), 19–25.

Barnes, P.W., Flint, S.D. and Caldwell, M.M. (1990) Morphological responses of crop and weed species of different growth forms to ultraviolet-B radiation. *American Journal of Botany* 77(10), 1354–1360.

Bates, T.R., Dunst, R.M., Taft, T. and Vercant, M. (2002) The vegetative response of 'Concord' grapevines to soil pH. *HortScience* 37(6), 890–893.

Baudoin, A.B., Finger, S.A. and Wolf, T.K. (2006) Factors affecting reductions in photosynthesis caused by applying horticultural oil to grapevine leaves. *HortScience* 41(2), 346–351.

Beattie, B.B. and Dahlenberg, A.P. (1989) Grapes. In: Beattie, B.B., McGlasson, W.B. and Wade, N.L. (eds) *Postharvest Diseases of Horticultural Produce Vol. 1. Temperate Fruit*. CSIRO Publications, East Melbourne, Victoria, Australia, pp. 67–73.

Becker, N.J. (1977) Experimental research on the influence of microclimate on grape constituents and on the quality of the crop. *Proceedings of the OIV Symposium on Quality and Vintage*, Cape Town, South Africa, Oenological and Viticulture Research Institute, pp. 181–188.

Belancic, A., Agosin, E., Ibacache, A., Bordeu, E., Baumes, R., Razungles, A. and Bayonove, C. (1997) Influence of sun exposure on the aromatic composition of Chilean Muscat grape Cultivars Moscatel de Alejandria and Moscatel rosada. *American Journal of Enology and Viticulture* 48(2), 181–186.

Bélanger, R.R., Dufour, N., Caron, J. and Benhamou, N. (1995) Chronological events associated with the antagonistic properties of *Trichoderma harzianum* against *Botrytis cinerea*: indirect evidence for sequential role of antibiosis and parasitism. *Biocontrol Science and Technology* 5, 41–54.

Belfrage, N. and Loftus, S. (1993) Dried grapes: the classic wines of antiquity. *Journal of Wine Research* 4(3), 205–225.

Bennett, J. (2002) *Relationships between Carbohydrate Supply and Reserves and the Reproductive Growth of Grapevines* (Vitis vinifera *L.*). Soil, Plant and Ecological Sciences Division, Lincoln University, Christchurch, New Zealand, 204 pp.

Bennett, J., Jarvis, P., Creasy, G.L. and Trought, M.C.T. (2005) Influence of defoliation on overwintering carbohydrate reserves, return bloom, and yield of mature Chardonnay grapevines. *American Journal of Enology and Viticulture* 56(4), 386–393.

Bergström, L. and Brink, N. (1986) Effects of differentiated applications of fertilizer N on leaching losses and distribution of inorganic N in the soil. *Plant and Soil* 93(3), 333–345.

Berisha, B., Chen, Y.D., Zhang, G.Y., Xu, B.Y. and Chen, T.A. (1998) Isolation of Pierce's disease bacteria from grapevines in Europe. *European Journal of Plant Pathology* 104(5), 427–433.

Berlinger, M.J. (1977) The Mediterranean vine mealybug and its natural enemies in southern Israel. *Phytoparasitica* 5, 3–14.

Bettiga, L.J., Dokoozlian, N.K. and Williams, L.E. (1996) Windbreaks improve the growth and yield of Chardonnay grapevines grown in a cool climate. *Proceedings of the Fourth International Symposium for Cool Climate Viticulture and Enology*, New York State Agricultural Experimental Station, Rochester, New York, pp. II, 43–46.

Biricolti, S., Ferrini, F., Rinaldelli, E., Tamantini, I. and Vignozzi, N. (1997) VAM fungi and soil lime content influence rootstock growth and nutrient content. *American Journal of Enology and Viticulture* 48, 93–99.

Bisson, L.F. and Butzke, C.E. (2000) Diagnosis and rectification of stuck and sluggish fermentations. *American Journal of Enology and Viticulture* 51(2), 168–177.

Blanke, M.M. and Leyhe, A. (1987) Stomatal activity of the grape berry cv. Riesling, Müller–Thurgau and Ehrenfelser. *Plant Physiology and Biochemistry* 127(5), 451–460.

Blanke, M.M. and Leyhe, A. (1988) Stomatal and cuticular transpiration of the cap and berry of grape. *Journal of Plant Physiology* 132, 250–253.

Bodin, F. and Morlat, R. (2006) Characterization of viticultural terroirs using a simple field model based on soil depth I. Validation of the water supply regime, phenology and vine vigour, in the Anjou Vineyard (France). *Plant and Soil* 281(1–2), 37–54.

Bonilla, F., Mayen, M., Merida, J. and Medina, M. (1999) Extraction of phenolic compounds from red grape marc for use as food lipid antioxidants. *Food Chemistry* 66(2), 209–215.

Borgo, M. and Angelini, E. (2002) Influence de l'enroulement foliaire GLRaV3 sur les paramètres de la production du Merlot. *Bulletin de l'OIV* 72, 611–622.

Boss, P.K., Davies, C. and Robinson, S.P. (1996) Anthocyanin composition and anthocyanin pathway gene expression in grapevine sports differing in berry skin colour. *Australian Journal of Grape and Wine Research* 2(3), 163–170.

Botelho, R.V., Pavanello, A.P., Pires, E.J.P., Terra, M.M. and Muller, M.M.L. (2007) Effects of chilling and garlic extract on bud dormancy release in Cabernet Sauvignon grapevine cuttings. *American Journal of Enology and Viticulture* 58(3), 402–404.

Botero-Garcés, N. and Isaacs, R. (1993) Distribution of grape berry moth, *Endopiza viteana* (Lepidoptera: Tortricidae), in natural and cultivated habitats. *Environmental Entomology* 32(5), 1187–1195.

Botero-Garcés, N. and Isaacs, R. (2004) Movement of the grape berry moth, *Endopiza viteana*: displacement distance and direction. *Physiological Entomology* 29(5), 443–452.

Boubals, D. (1989) La maladie de Pierce arrive dans les vignobles d'Europe. *Bulletin de l'OIV* 62, 309–314.

Bowers, J.E., Bandman, E.B. and Meredith, C.P. (1993) DNA fingerprint characterization of some wine grape cultivars. *American Journal of Enology and Viticulture* 44(3), 266–274.

Brenchley, W.E. and Warington, K. (1933) The weed seed population of arable soil: II. Influence of crop, soil and methods of cultivation upon the relative abundance of viable seeds. *Journal of Ecology* 21(1), 103–127.

Bridle, P. and Garcia-Viguera, C. (1996) A simple technique for the detection of red wine adulteration with elderberry pigments. *Food Chemistry* 55(2), 111–113.

Brossaud, F., Cheynier, V., Asselin, C. and Moutounet, M. (1999) Flavonoid compositional differences of grapes among site test plantings of Cabernet franc. *American Journal of Enology and Viticulture* 50(3), 277–284.

Brun, L.A., Maillet, J., Hinsinger, P. and Pepin, M. (2001) Evaluation of copper availability to plants in copper-contaminated vineyard soils. *Environment and Pollution* 111(2), 293–302.

Bruwer, J. (2003) South African wine routes: some perspectives on the wine tourism industry's structural dimensions and wine tourism product. *Tourism Management* 24(4), 423–435.

Buttery, R.G., Seifert, R.M., Guadagni, D.G. and Ling, L.C. (1969) Characterization of some volatile constituents of bell peppers. *Journal of Agricultural and Food Chemistry* 17, 1322–1327.

Buttrose, M.S. (1969) Fruitfulness in grapevines: effects of light intensity and temperature. *Botanical Gazette* 130(6), 166–173.

Cabaleiro, C. and Segura, A. (1997) Field transmission of grapevine leafroll associated virus 3 (GLRaV-3) by the mealybug *Planococcus citri*. *Plant Disease* 81(3), 283–287.

Cabaleiro, C., Segura, A. and Garcia-Berrios, J.J. (1999) Effects of grapevine leafroll-associated Virus 3 on the physiology and must of *Vitis vinifera* L. cv. Albarino following contamination in the field. *American Journal of Enology and Viticulture* 50(1), 40–44.

Cakmak, I., Hengeler, C. and Marschner, H. (1994) Partitioning of shoot and root dry matter and carbohydrates in bean plants suffering from phosphorus, potassium and magnesium deficiency. *Journal of Experimental Botany* 45(9), 1245–1250.

Camp, C.R. (1998) Subsurface drip irrigation: a review. *Transactions of the ASAE* 41(5), 1353–1367.

Campbell, C. (2004) *Phylloxera: how Wine was Saved for the World*. Harper Perennial, London.

Campbellclause, J.M. (1998) Stomatal response of grapevines to wind. *Australian Journal of Experimental Agriculture* 38, 77–82.

Candolfi, M., Jermini, M., Carrera, E. and Candolfi-Vasconcelos, M. (1993) Grapevine leaf gas exchange, plant growth, yield, fruit quality and carbohydrate reserves influenced by the grape leafhopper, *Empoasca vitis*. *Entomologia Experimentalis et Applicata* 69(3), 289–296.

Carbonneau, A. and Casteran, P. (1987) Optimization of vine performance by the lyre training systems. *Procedings of the 6th Australian Wine Industry Technical Conference*, Adelaide, Australia. Winetitles, Underdale, South Australia pp. 194–204.

Carreño, J., Martínez, A., Almela, L. and Fernández-López, J.A. (1995) Proposal of an index for the objective evaluation of the colour of red table grapes. *Food Research International* 28(4), 373–377.

Caspari, H.W., Neal, S., Naylor, A., Trought, M.C.T. and Tannock, S. (1997) Use of cover crops and deficit irrigation to reduce vegetative vigour of 'Sauvignon blanc' grapevines in a humid climate. *Proceedings of the Fourth International Symposium for Cool Climate Viticulture and Enology*, Rochester, New York, pp. II, 63–66.

Cavalieri, D., McGovern, P., Hartl, D., Mortimer, R. and Polsinelli, M. (2003) Evidence for *S. cerevisiae* fermentation in ancient wine. *Journal of Molecular Evolution* 57, S226–S232.

Cawthon, D.L. and Morris, J.R. (1982) Relationship of seed number and maturity to berry development, fruit maturation, hormonal changes and uneven ripening of 'Concord' (*Vitis labrusca* L.) grapes. *Journal of the American Society for Horticultural Science* 107, 1097–1104.

Cervera, M.-T., Cabezas, J.A., Sancha, J.C., Toda, F.M.D. and Martínez-Zapater, J.M. (1998) Application of AFLPs to the characterization of grapevine *Vitis vinifera* L. genetic resources. A case study with accessions from Rioja (Spain). *Theoretical Applied Genetics* 97(1–2), 51–59.

Chalmers, D.J., Mitchell, P.D. and van Heek, L. (1981) Control of peach tree growth and productivity by regulated water supply, tree density and summer pruning. *Journal of the American Society for Horticultural Science* 106, 307–312.

Charles, J.G., Cohen, D., Walker, J.T.S., Forgie, S.A., Bell, V.A. and Breen, K.C. (2006) A review of the ecology of grapevine leafroll associated virus type 3(GLRaV-3). *New Zealand Plant Protection* 59, 330–337.

Chaves, M.M. and Pereira, J.S. (1992) Water stress, CO_2 and climate change. *Journal of Experimental Botany* 43(8), 1131–1139.

Chen, F., Guo, Y.B., Wang, J.H., Li, J.Y. and Wang, H.M. (2007) Biological control of grape crown gall by *Rahnella aquatilis* HX2. *Plant Disease* 91(8), 957–963.

Chen, L.S. and Cheng, L. (2003) Carbon assimilation and carbohydrate metabolism of 'Concord' grape (*Vitis labrusca* L.) leaves in response to nitrogen supply. *Journal of the American Society for Horticultural Science* 128(5), 754–760.

Cheynier, V. and Rigaud, J. (1986) HPLC separation and characterization of flavonols in the skins of *Vitis vinifera* var. Cinsault. *American Journal of Enology and Viticulture* 37, 248–252.

Cheynier, V., Fulcrand, H., Sarni, P. and Moutounet, M. (1998) Progress in phenolic chemistry in the last ten years. *Proceedings of the ASVO Seminar Phenolics and Extraction*, pp. 12–17.

Chitkowski, R.L. and Fisher, J.R. (2005) Effect of soil type on the establishment of grape phylloxera colonies in the Pacific Northwest. *American Journal of Enology and Viticulture* 56(3), 207–211.

Choné, X., Van Leeuwen, C., Dubourdieu, D. and Gaudillére, J.P. (2001) Stem water potential is a sensitive indicator of grapevine water status. *Australian Journal of Experimental Agriculture* 87(4), 477–483.

Christen, E.W., Moll, J.L. and Muirhead, W.A. (1995) Furrows that trickle? *Proceedings of the 9th Australian Wine Industry Technical Conference*, July 16–19, Adelaide, South Australia. Winetitles, Underdale, South Australia, pp. 118.

Christensen, L.P. (2000) Current developments in harvest mechanization and DOV. *Raisin Production Manual*, University of California, Agricultural and Natural Resources, Oakland, California, pp. 252–263.

Christensen, L.P. and Peacock, W.L. (2000a) Raisin grape varieties. *Raisin Production Manual*. University of California, Agricultural and Natural Resources, Oakland, California, pp. 38–47.

Christensen, L.P. and Peacock, W.L. (2000b) Harvesting and handling. *Raisin Production Manual*. University of California, Agricultural and Natural Resources, Oakland, California, pp. 193–206.

Christensen, L.P., Boggero, J. and Bianchi, M. (1990) Comparative leaf tissue analysis of potassium deficiency and a disorder resembling potassium deficiency in Thompson Seedless grapevines. *American Journal of Enology and Viticulture* 41(1), 77–83.

Christensen, L.P., Leavitt, G.M., Hirschfelt, D.J. and Bianchi, M.L. (1994) The effects of pruning level and post-budbreak cane adjustment on Thompson Seedless raisin production and quality. *American Journal of Enology and Viticulture* 45(2), 141–149.

Christensen, P. (1984) Nutrient level comparisons of leaf petioles and blades in twenty-six grape cultivars over three years (1979 through 1981). *American Journal of Enology and Viticulture* 35(3), 124–133.

Clarke, J.H., Clark, W.S. and Hancock, M. (1997) Strategies for the prevention of development of pesticide resistance in the UK – lessons for and from the use of herbicides, fungicides and insecticides. *Pesticide Science* 51(3), 391–397.

Clary, C.D., Steinhauer, R.E., Frisinger, J.E. and Peffer, T.E. (1990) Evaluation of machine- vs. hand-harvested Chardonnay. *American Journal of Enology and Viticulture* 41(2), 176–181.

Cleugh, H. (1998) Effects of windbreaks on airflow, microclimates and crop yields. *Agroforestry Systems* 41(1), 55–84.

Cliff, M.A., Dever, M.C. and Reynolds, A.G. (1996) Descriptive profiling of new and commercial British Columbia table grape cultivars. *American Journal of Enology and Viticulture* 47(3), 301–308.

Clingeleffer, P.R. (1993a) Vine response to modified pruning practices. *Proceedings of the 2nd New Jersey Shaulis Grape Symposium on Pruning Mechanization and Crop Control*, Fredonia, New York. New York State Agriculture Experiment Station, Geneva, New York, pp. 20–30.

Clingeleffer, P.R. (1993b) Development of management systems for low cost, high quality wine production and vigour control in cool climate Australian vineyards. *Die Wein-Wissenschaft* 48, 130–134.

Clingeleffer, P.R. (1999) Developments in Australian winegrape production. *Proceedings du Groupe d' Étude des Systèmes de Conduite de la Vigne, 11th Meeting*, University of Palermo, Marsala, Italy, Vol. I, pp. 56–69.

Clingeleffer, P.R. and May, P. (1981) The swing-arm trellis for Sultana grapevine management. *South African Journal of Enology and Viticulture* 2, 37–44.

Clore, W.J., Wallace, M.A. and Fay, R.D. (1974) Bud survival of grape varieties at sub-zero temperatures in Washington. *American Journal of Enology and Viticulture* 25(1), 24–29.

Clydesdale, F.M. (1993) Color as a factor in food choice. *Critical Reviews in Food Science and Nutrition* 33(1), 82–101.

Clydesdale, F.M., Main, J.H., Francis, F.J. and Damon, R.A. (1978) Concord grape pigments as colorants for beverages and gelatin desserts. *Journal of Food Science* 43(6), 1687–1692.

Coertze, S., Holz, G. and Sadie, A. (2001) Germination and establishment of infection on grape berries by single airborne conidia of *Botrytis cinerea*. *Plant Disease* 85(6), 668–677.

Coipel, J., Lovelle, B.R., Sipp, C. and Van Leeuwen, C. (2006) 'Terroir' effect, as a result of environmental stress, depends more on soil depth than on soil type (*Vitis vinifera* L. cv. Grenache noir, Côtes du Rhône, France, 2000). *Journal International des Sciences de la Vigne et du Vin* 40(4), 177–185.

Collins, R.M., Bertram, A., Roche, J.-A. and Scott, M.E. (2002) Preliminary studies in the comparison of hot water and hot foam for weed control. *Proceedings of the 5th EWRS Workshop on Physical Weed Control*, Pisa, Italy, 11–13 March, pp. 207–215.

Comas, L.H., Eiseeenstat, D.M. and Lakso, A.N. (2000) Assessing root death and root system dynamics in a study of grape canopy pruning. *New Phytologist* 147(1), 171–178.

Conradie, W.J. (1980) Seasonal uptake of nutrients by Chenin blanc in sand culture: I. Nitrogen. *South African Journal of Enology and Viticulture* 1, 59–65.

Conradie, W.J. (1990) Distribution and translocation of nitrogen absorbed during late spring by two-year-old grapevines grown in sand culture. *American Journal of Enology and Viticulture* 41(3), 241–250.

Conradie, W.J. (2005) Partitioning of mineral nutrients and timing of fertilizer application for optimum efficiency. *Proceedings of the Soil, Environment, Vine and Mineral Nutrition Symposium*, American Society for Enology and Viticulture, Davis, California, pp. 69–81.

Conradie, W.J. and Saayman, D. (1989) Effects of long-term nitrogen, phosphorus, and potassium fertilization on Chenin blanc vines. II. Leaf analyses and grape composition. *American Journal of Enology and Viticulture* 40(2), 91–98.

Considine, J.A. (1973) A statistical study of rain damage of grapes grown for drying in Victoria. *Australian Journal of Experimental Agriculture and Animal Husbandry* 13, 604–611.

Considine, J.A. and Knox, R.B. (1979) Development and histochemistry of the cells, cell walls, and cuticle of the dermal system of fruit of the grape, *Vitis vinifera* L. *Protoplasma* 99(4), 347–365.

Cook, J.A., Lynn, C.D. and Kissler, J.J. (1960) Boron deficiency in California vineyards. *American Journal of Enology and Viticulture* 11(4), 185–194.

Coombe, B.G. (1960) Relationship of growth and development to changes in sugars, auxins, and gibberellins in fruit of seeded and seedless varieties of *Vitis vinifera*. *Plant Physiology* 35, 241–250.

Coombe, B.G. (1973) The regulation of set and development of the grape berry. *Acta Horticulturae* 34, 261–273.

Coombe, B.G. (1976) The development of fleshy fruits. *Annual Review of Plant Physiology* 27, 507–528.

Coombe, B.G. (1995) Adoption of a system for identifying grapevine growth stages. *Australian Journal of Grape and Wine Research* 1(2), 104–110.

Coombe, B.G. and McCarthy, M.G. (1997) Identification and naming of the inception of aroma development in ripening grape berries. *Australian Journal of Grape Wine Research* 3(1), 18–20.

Coombe, B.G. and McCarthy, M.G. (2000) Dynamics of grape berry growth and physiology of ripening. *Australian Journal of Grape and Wine Research* 6(2), 131–135.

Cornford, C.E. (1938) Katabatic winds and the prevention of frost damage. *Quarterly Journal of the Royal Meteorological Society* 64, 553–587.

Costa, H.S., Raetz, E., Pinckard, T.R., Gispert, C., Hernandez-Martinez, R., Dumenyo, C.K. and Cooksey, D.A. (2004) Plant hosts of *Xylella fastidiosa* in and near southern California vineyards. *Plant Disease* 88(11), 1255–1261.

Creasy, G.L. (1991) *Xylem Discontinuity in* Vitis vinifera *L. Berries*. Department of Horticulture, Oregon State University, Corvallis, Oregon, 89 pp.

Creasy, G.L. (1996) *Inflorescence Necrosis, Ammonium, and Evidence for Ferredoxin-Glutamate Synthase Activity in Grape (*Vitis vinifera *L.)*. Department of Horticulture, Oregon State University, Corvallis, Oregon, 112 pp.

Creasy, G.L. (2004) Soil and air temperatures as affected by soil cultivation in a Canterbury vineyard. *Proceedings of the 12th Australian Wine Industry Technical Conference*, Melbourne, Victoria, Australia. Australian Wine Industry Technical Conference, Inc, pp. 245.

Creasy, G.L. and Qi, G. (2005) Grapevine leaf and fruit tissue responses to fosetyl-Al and UV radiation. *American Journal of Enology and Viticulture* 56(3), 303A.

Creasy, G.L., Crawford, M., Ibbotson, L., Gladstone, P., Kavanaugh, J. and Sutherland, A. (2006) Mussel shell mulch alters Pinot noir grapevine development and wine qualities. *Australian Grapegrower and Winemaker* 509a, 12–18.

Creasy, L.L. and Coffee, M. (1988) Phytoalexin production potential of grape berries. *Journal of the American Society of Horticultural Science* 113, 230–234.

Crisosto, C.H., Smilanick, J.L. and Dokoozlian, N.K. (2001) Table grapes suffer water loss, stem browning during cooling delay. *California Agriculture* 55(1), 39–42.

Crisosto, C.H., Garner, D. and Crisosto, G. (2002) Carbon dioxide-enriched atmospheres during cold storage limit losses from Botrytis but accelerate rachis browning of 'Redglobe' table grapes. *Postharvest Biology and Technology* 26(2), 181–189.

Crisp, P., Scott, E.S. and Wicks, T.J. (2002) Novel control of grapevine powdery mildew. *Proceedings of the 4th International Workshop on Grapevine Powdery and Downy Mildew*, Napa, California. University of California, Davis, California, pp. 78–79.

Crisp, P., Wicks, T.J., Troup, G. and Scott, E.S. (2006) Mode of action of milk and whey in the control of grapevine powdery mildew. *Australian Plant Pathology* 35(5), 487–493.

Cristofolini, F. and Gottardini, E. (2000) Concentration of airborne pollen of *Vitis vinifera* L. and yield forecast: a case study at S. Michele all'Adige, Trento, Italy. *Aerobiologia* 16(1), 125–129.

Cucuzza, J.D. and Sall, M.A. (1982) Phomopsis cane and leaf spot disease of grape vine: effect of chemical treatments on inoculum level, disease severity, and yield. *Plant Disease* 66, 794–797.

Cunha, M., Abreu, I., Pinto, P. and Castro, R.D. (2003) Airborne pollen samples for early-season estimates of wine production in a Mediterranean climate area of northern Portugal. *American Journal of Enology and Viticulture* 54(3), 189–194.

Dadic, M. and Belleau, G. (1973) Polyphenols and beer flavor. *Proceedings of the American Society of Brewing Chemists* 4, 107–114.

Davenport, J.R. and Stevens, R.G. (2006) High soil moisture and low soil temperature are associated with chlorosis occurrence in Concord grape. *HortScience* 41(2), 418–422.

Deal, D.R., Boothroyd, C.W. and Mai, W.F. (1972) Replanting of vineyards and its relationship to vesicular–arbuscular mycorrhiza. *Phytopathology* 62, 172–175.

De Boubée, D.R. (2003) *Research on 2-Methoxy-3-Isobutylpyrazine in Grapes and Wines*. Amorim Academy competition, Paris, 21 pp.

De Boubée, R. (2004) Recherches sur le caractère végétal-poivron vert dans les raisins et dans les vins. *Revue des Oenologues* 110, 6–10.

Deckers, T., Daemen, E., Lemmens, K. and Missotten, C. (1997) Influence of foliar applications of Mn during summer on the fruit quality of Jonagold. *Acta Horticulturae* 448, 467–473.

De Kok, L.J. (1990) Sulfur metabolism in plants exposed to atmospheric sulfur. In: Rennenberg, H., Brunold, C., De Kok, L.J. and Stulen, I. (eds) *Sulfur Nutrition and Sulfur Assimilation in Higher Plants: Fundamental, Environmental and Agricultural Aspects*. SPB Academic Publishing, The Hague, Netherlands, pp. 111–130.

Delcroix, A., Gunata, Z., Sapis, J.-C., Salmon, J.–M. and Bayonove, C. (1994) Glycosidase activities of three enological yeast strains during winemaking: effect on the terpenol content of Muscat wine. *American Journal of Enology and Viticulture* 45(3), 291–296.

Dell, K.J., Gubler, W.D., Krueger, R., Sanger, M. and Bettiga, L.J. (1998) The efficacy of JMS stylet-oil on grape powdery mildew and botrytis bunch rot and effects on fermentation. *American Journal of Enology and Viticulture* 49(1), 11–16.

De Meyer, G., Bigirimana, J., Elad, Y. and Hafte, M. (1998) Induced systemic resistance in *Trichoderma harzianum* T39 biocontrol of *Botrytis cinerea*. *European Journal of Plant Pathology* 104(3), 279–286.

Derksen, R.C., Coffman, C.W., Jiang, C. and Gulyas, S.W. (1999) Influence of hooded and air-assist vineyard applications on plant and worker protection. *Transactions of the American Society of Agricultural Engineers* 42(1), 31–36.

Dermen, H. (1954) Colchiploidy in grapes. *Journal of Heredity* 45(4), 159–172.

Dermen, H. (1964) Cytogenetics in hybridization of bunch and muscadine-type grapes. *Economic Botany* 18(2), 137–148.

Dethier, B.E. and Shaulis, N. (1964) Minimizing the hazard of cold in New York vineyards. *New York Agricultural Experiment Station Bulletin*, Geneva, New York, 8 pp.

De Toda, F.M. and Sancha, J.C. (1999) Long-term effects of simulated mechanical pruning on Grenache vines under drought conditions. *American Journal of Enology and Viticulture* 50(1), 87–90.

Dirr, M.A. and Heuser, C.W. (1987) *The Reference Manual of Woody Plant Propagation*. Varsity Press, Athens, Georgia.

Dobrowski, S.Z., Ustin, S.L. and Wolpert, J.A. (2002) Remote estimation of vine canopy density in vertically shoot positioned vineyards: determining optimal vegetation indices. *Australian Journal of Grape and Wine Research* 8(2), 117–125.

Dokoozlian, N.K. and Hirschfelt, D.J. (1995) The influence of cluster thinning at various stages of fruit development on Flame Seedless table grapes. *American Journal of Enology and Viticulture* 46(4), 429–436.

Doran, J.W. (1980) Soil microbial and biochemical changes associated with reduced tillage. *Soil Science Society of America Journal* 44(4), 765–771.

Downey, M.O., Harvey, J.S. and Robinson, S.P. (2003) Analysis of tannins in seeds and skins of Shiraz grapes throughout berry development. *Australian Journal of Grape and Wine Research* 9(1), 15–27.

Downton, W.J.S. (1985) Growth and mineral composition of the Sultana grapevine as influenced by salinity and rootstock. *Australian Journal of Agricultural Research* 36(3), 425–434.

Dry, P.R. (1997) *The Response of Grapevines to Partial Drying of the Rootzone*. University of Adelaide, Adelaide, South Australia.

Dry, P.R., Reed, S. and Potter, G. (1989) The effect of wind on the performance of Cabernet Franc grapevines. *Acta Horticulturae* 240, 143–146.

Due, G. (1990) The use of polypropylene shelters in grapevine establishment – a preliminary trial. *Australian Grapegrower and Winemaker* 318, 29–33.

Dutcher, J.D. and All, J.N. (1978) Survivorship of the grape root borer in commercial grape vineyards with contrasting cultural practices. *Journal of Economic Entomology* 71, 751–754.

Ebadi, A., Coombe, B.G. and May, P. (1995) Fruit-set on small Chardonnay and Shiraz vines grown under varying temperature regimes between budburst and flowering. *Australian Journal of Grape and Wine Research* 1(1), 3–10.

Ebadi, A., May, P. and Coombe, B.G. (1996) Effect of short-term temperature and shading on fruit-set, seed and berry development in model vines of *V. vinifera*, cvs Chardonnay and Shiraz. *Australian Journal of Grape and Wine Research* 2, 2–9.

Edson, C.E., Howell, G.S. and Flore, J.A. (1993) Influence of crop load on photosynthesis and dry matter partitioning of Seyval grapevines I. Single leaf and whole vine response pre- and post-harvest. *American Journal of Enology and Viticulture* 44(2), 139–147.

Eijkhoff, P. (2003) *Wine in China. Its History and Contemporary Developments*. Wine Guild, Netherlands, 191 (http://www.eykhoff.nl/Wine_in_China_UK.htm).

Einset, J. and Pratt, C. (1975) Grapes. In: Janick, J. and Moore, J.N. (eds) *Advances in Fruit Breeding*. Purdue University Press, West Lafayette, Indiana, pp. 130–153.

Elad, Y. and Kapat, A. (1999) The role of *Trichoderma harzianum* protease in the biocontrol of *Botrytis cinerea*. *European Journal of Plant Pathology* 105(2), 177–189.

Ellis, R.J. (1979) The most abundant protein in the world. *Trends in Biochemical Science* 4, 241–244.

English-Loeb, G.M., Flaherty, D.L., Wilson, L.T., Barnett, W.W., Leavitt, G.M. and Settle, W.H. (1986) Pest management affects spider mites in vineyards. *California Agriculture* 40, 28–30.

English-Loeb, G., Norton, A.P., Gadoury, D.M., Seem, R.C. and Wilcox, W.F. (1999a) Control of powdery mildew in wild and cultivated grapes by a tydeid mite. *Biological Control* 14(2), 97–103.

English-Loeb, G., Villani, M., Martinson, T., Forsline, A. and Consolie, N. (1999b) Use of entomopathogenic nematodes for control of grape *Phylloxera* (Homoptera: Phylloxeridae): a laboratory evaluation. *Environmental Entomology* 28(5), 890–894.

English-Loeb, G., Norton, A.P., Gadoury, D., Seem, R. and Wilcox, W. (2007) Biological control of grape powdery mildew using mycophagous mites. *Plant Disease* 91(4), 421–429.

Epstein, E. (1961) The essential role of calcium in selective cation transport by plant cells. *Plant Physiology* 36(4), 437–444.

Erincik, O., Madden, L.V., Ferree, D.C. and Ellis, M.A. (2001) Effect of growth stage on susceptibility of grape berry and rachis tissues to infection by *Phomopsis viticola*. *Plant Disease* 85(5), 517–520.

Ernst, B. (2007) Frutas Argentinas. *Top Info Marketing Southern Hemisphere Congress*, Buenos Aires, 28–30 November 2007.

Escalona, J., Flexas, J. and Medrano, H. (2002) Drought effects on water flow, photosynthesis and growth of potted grapevines. *Vitis* 41(2), 57–62.

Evans, R.G. (2000) The art of protecting grapevines from low temperature injury. *Proceedings of the American Society for Enology and Viticulture 50th Annual Meeting, Seattle, Washington*, pp. 60–72.

Ewart, A.J. (1987) Influence of vineyard site and grape maturity on juice and wine quality of *Vitis vinifera*, cv. Riesling. *Proceedings of the 6th Australian Wine Industry Technical Conference*, Adelaide, South Australia. Australian Industrial Publishers, Australia, pp. 71–74.

Falks, S.P., Gadoury, D.M., Pearson, R.C. and Seem, R.C. (1995) Partial control of grape powdery mildew by the mycoparasite *Ampelomyces quisqualis*. *Plant Disease* 75, 483–490.

FAO (2006) *FAOSTAT ProdSTAT Agricultural Production Data, 2008*, Rome.

FAS (2006) *World Table Grape Situation and Outlook*. USDA FAS Horticultural & Tropical Products Division, Washington, DC, 6 pp.

FAS (2007) *Raisin Production, Supply and Distribution Data*. USDA FAS Office of Global Analysis, Washington, DC, 3 pp.

Feil, H. and Purcell, A.H. (2001) Temperature-dependent growth and survival of *Xylella fastidiosa in vitro* and in potted grapevines. *Plant Disease* 85(12), 1230–1234.

Fennell, A. and Hoover, E. (1991) Photoperiod influences growth, bud dormancy, and cold acclimation in *Vitis labruscana* and *V. riparia*. *Journal of the American Society for Horticultural Science* 116, 270–273.

Fereres, E. and Soriano, M.A. (2007) Deficit irrigation for reducing agricultural water use. *Journal of Experimental Botany* 58(2), 147–159.

Ferris, H. and McKenry, M.V. (1974) Seasonal fluctuation in the spatial distribution of nematode populations in a California vineyard. *Journal of Nematology* 6, 203–210.

Finger, S.A., Wolf, T.K. and Baudoin, A.B. (2002) Effects of horticultural oils on the photosynthesis, fruit maturity, and crop yield of winegrapes. *American Journal of Enology and Viticulture* 53(2), 116–124.

Fisher, K.H., Piott, B. and Barkovic, J. (1997) Adaptability of labrusca and French Hybrid grape varieties to mechanical pruning and mechanical thinning. *Proceedings of the Fourth International Symposium for Cool Climate Viticulture and Enology*, Rochester, New York. New York State Agricultural and Experimental Station, Geneva, New York, pp. IV, 33–39.

Flora, L.F. (1976) Time–temperature influence on Muscadine grape juice quality. *Journal of Food Science* 41(6), 1312–1315.

Flores-Velez, L.M., Ducaroir, J., Jaunet, A.M. and Robert, M. (1996) Study of the distribution of copper in an acid sandy vineyard soil by three different methods. *European Journal of Soil Science* 47(4), 523–532.

Forsline, P.L., Musselman, R.C., Dee, R.J. and Kender, W.J. (1983) Effects of acid rain on grapevines. *American Journal of Enology and Viticulture* 34(1), 17–22.

Francisco Artés-Hernández, E.A., Francisco Artés, Francisco A Tomás-Barberán, (2007) Enriched ozone atmosphere enhances bioactive phenolics in seedless table grapes after prolonged shelf life. *Journal of the Science of Food and Agriculture* 87(5), 824–831.

Freeman, B.M., Kliewer, W.M. and Stern, P. (1982) Influence of windbreaks and climatic region on diurnal fluctuation of leaf water potential, stomatal conductance, and leaf temperature of grapevines. *American Journal of Enology and Viticulture* 33, 233–236.

Freeman, B.M., Tassie, E. and Rebbechi, M.D. (1992) Training and trellising. In: Coombe, B.G. and Dry, P.R. (eds) *Viticulture Volume 2. Practices*. Winetitles, Underdale, South Australia, pp. 42–65.

Friend, A.P. (2005) *Berry Set and Development in* Vitis vinifera L. Food and Wine Group, Agriculture and Life Sciences Division, Lincoln University, Christchurch, New Zealand, 193 pp.

Friend, A.P. and Trought, M.C.T. (2007) Delayed winter spur-pruning in New Zealand can alter yield components of Merlot grapes. *Australian Journal of Grape and Wine Research* 13(3), 157–164.

Friend, A.P., Trought, M.C.T. and Creasy, G.L. (2000) Pruning to increase yield: the influence of delayed pruning time on yield components of *Vitis vinifera* L. cv. Merlot. *Proceedings of the Fifth International Symposium for Cool Climate Viticulture and Oenology*, Melbourne, Australia, 16–20 January 2000.

Friend, A.P., Creasy, G.L., Trought, M.C.T. and Lang, A. (2003) Use of tagging to trace capfall and development of individual *Vitis vinifera* L. cv. Pinot noir flowers. *American Journal of Enology and Viticulture* 54(4), 313–317.

Gabel, B. and Renczés, V. (1985) Factors affecting the monitoring of flight activity of *Lobesia botrana* and *Eupoecilia ambiguella* (Lepidoptera: Tortricidae) by pheromone traps. *Acta Entomologica Bohemoslovaca* 82, 269–277.

Gadoury, D.M., Pearson, R.C., Riegel, D.G., Seem, R.C., Becker, C.M. and Pscheidt, J.W. (1994) Reduction of powdery mildew and other diseases by over-the-trellis applications of lime sulfur to dormant grapevines. *Plant Disease* 78(1), 83–87.

Galet, P. (1982) Phylloxera. *Les Maladies et le Parasites de la Vigne, Tome II. Les Parasites Animaux*. Paysan du Midi, Montpellier, France, pp. 1059–1313.

Gamon, J.A. and Pearcy, R.W. (1989) Leaf movement, stress avoidance and photosynthesis in *Vitis californica*. *Oecologia* 79(4), 475–481.

Gao, L., Girard, B., Mazza, G. and Reynolds, A.G. (1997) Simple and polymeric anthocyanins and color characteristics of Pinot Noir wines from different vinification processes. *Journal of Agricultural and Food Chemistry* 45, 2003–2008.

Gardea, A.A. (1987) *Freeze Damage of Pinot Noir (*Vitis vinifera *L.) as Affected by Bud Development, INA-Bacteria, and a Bacterial Inhibitor*. Department of Horticulture, Oregon State University, Corvallis, Oregon, 86 pp.

Gawel, R. (1998) Red wine astringency: a review. *Australian Journal of Grape and Wine Research* 4, 74–95.

Gawel, R., Oberholster, A. and Francis, I.L. (2000) A 'mouth-feel wheel': terminology for communicating the mouth-feel characteristics of red wine. *Australian Journal of Grape and Wine Research* 6(3), 203–207.

Gentry, J.P. and Nelson, K.E. (1964) Conduction cooling of table grapes. *American Journal of Enology and Viticulture* 15(1), 41–46.

Giannakis, C., Buchell, C.S., Skene, K.G.M., Robinson, S.P. and Scott, N.S. (1998) Chitinase and beta-1,3-glucanase in grapevine leaves: a possible defence against powdery mildew infection. *Australian Journal of Grape and Wine Research* 4(1), 14–22.

Giovannoni, J. (2001) Molecular biology of fruit maturation and ripening. *Annual Review of Plant Physiology and Plant Molecular Biology* 52(1), 725–749.

Glad, C., Regnard, J.L., Querou, Y., Brun, O. and Morot-Gaudry, J.F. (1992) Flux and chemical composition of xylem exudates from Chardonnay grapevines: temporal evolution and effect of recut. *American Journal of Enology and Viticulture* 43(3), 275–282.

Gladstone, J. (1992) *Viticulture and Environment*. Winetitles, Underdale, South Australia, 320 pp.

Goetz, G., Fkyerat, A., Métais, N., Kunz, M., Tabacchi, R., Pezet, R. and Pont, V. (1999) Resistance factors to grey mould in grape berries: identification of some phenolics inhibitors of *Botrytis cinerea* stilbene oxidase. *Phytochemistry* 52, 759–767.

Goffinet, M.C. (2004) *Anatomy of Grapevine Winter Injury and Recovery*. New York State Agricultural Experiment Station, Geneva, New York, 21 pp.

Goffinet, M.C. and Martinson, T. (2007) Burying canes in the Finger Lakes: grower experience and impact on buds. *Proceedings of Viticulture 2007*, 7–9 February, Rochester, New York, pp 54–55.

Goheen, A.C. and Cook, J.A. (1959) Leafroll (red-leaf or rougeau) and its effects on vine growth, fruit quality, and yields. *American Journal of Enology and Viticulture* 10(4), 173–181.

Goldhamer, D. and Fereres, E. (2001) Irrigation scheduling protocols using continuously recorded trunk diameter measurements. *Irrigation Science* 20(3), 115–125.

Golino, D.A., Sim, S.T., Gill, R. and Rowhani, A. (2002) California mealybugs can spread grapevine leafroll disease. *California Agriculture* 56, 196–201.

González-Neves, G., Gómez-Cordovés, C. and Barreiro, L. (2001) Anthocyanic composition of Tannat, Cabernet Sauvignon and Merlot young red wines from Uruguay. *Journal of Wine Research* 12(2), 125–133.

Goodess, C.M., Hanson, C., Hulme, M. and Osborn, T.J. (2003) Representing climate and extreme weather events in integrated assessment models: a review of existing methods and options for development. *Integrated Assessment* 4(3), 145–172.

Gossett, D.R., Egli, D.B. and Leggett, J.E. (1977) The influence of calcium deficiency on the translocation of photosynthetically fixed 14C in soybeans. *Plant and Soil* 48(1), 243–251.

Goto-Yamamoto, N., Mouri, H., Azumi, M. and Edwards, K.J. (2006) Development of grape microsatellite markers and microsatellite analysis including oriental cultivars. *American Journal of Enology and Viticulture* 57(1), 105–108.

Grando, M.S., Bellin, D., Edwards, K.J., Pozzi, C., Stefanini, M. and Velasco, R. (2003) Molecular linkage maps of *Vitis vinifera* L. and *Vitis riparia* Mchx. *Theoretical and Applied Genetics* 106(7), 1213–1224.

Granett, J., Timper, P. and Lider, L.A. (1985) Grape phylloxera (*Daktulosphaira vitifoliae*) (Homoptera: Phylloxeridae) biotypes in California. *Journal of Economic Entomology* 78(6), 1463–1467.

Granett, J., Walker, M.A., Kocsis, L. and Omer, A.D. (2001) Biology and management of grape phylloxera. *Annual Review of Entomology* 46, 387–412.

Granett, J., Walker, M.A. and Fossen, M.A. (2007) Association between grape phylloxera and strongly resistant rootstocks in California: bioassays. *Acta Horticulturae* 733, 25–31.

Green, S.R. and Clothier, B.E. (1988) Water use of kiwifruit vines and apple trees by the heat-pulse technique. *Journal of Experimental Botany* 39(1), 115–123.

Grncarevic, M. (1963) Effect of various dipping treatments on the drying rate of grapes for raisins. *American Journal of Enology and Viticulture* 14(4), 230–234.

Grncarevic, M. and Radler, F. (1971) A review of the surface lipids of grapes and their importance in the drying process. *American Journal of Enology and Viticulture* 22(2), 80–86.

Groot Obbink, J. and Alexander, D.M. (1973) Response of six grapevine cultivars to a range of chloride concentrations. *American Journal of Enology and Viticulture* 24(2), 65–68.

Gu, S., Du, G., Zoldoske, D., Hakim, A., Cochran, R., Fugelsang, K. and Jorgensen, G. (2004) Effects of irrigation amount on water relations, vegetative growth, yield and fruit composition of Sauvignon blanc grapevines under partial rootzone drying and conventional irrigation in the San Joaquin Valley of California, USA. *Journal of Horticultural Science and Biotechnology* 79(1), 26–33.

Guarga, R., Scaglione, G., Supino, E. and Mastrangelo, P. (2003) Evaluation of the SIS – a new frost protection method applied in a citrus orchard. *Proceedings of the 9th International Citrus Congress*, Orlando, Florida, pp. 583.

Gubler, W.D., Marois, J.J., Bledsoe, A.M. and Bettiga, L.J. (1987) Control of Botrytis bunch rot of grape with canopy management. *Plant Disease* 71(7), 599–601.

Guelfat-Reich, S. and Safran, B. (1971) Indices of maturity for table grapes as determined by variety. *American Journal of Enology and Viticulture* 22(1), 13–18.

Guilford, T., Nicol, C., Rothschild, M. and Moore, B. (1987) The biological roles of pyrazines: evidence for a warning odour function. *Biological Journal of the Linnean Society* 31, 113–128.

Gunkle, W.W. and Throop, J.A. (1993) Robotic pruning of grapevines: development of intelligent pruning and thinning machines. *Proceedings of the 2nd New Jersey Shaulis Grape Symposium on Pruning Mechanization and Crop Control*, Fredonia, New York. New York State Agriculture Experiment Station, Geneva, New York, pp. 51–54.

Haeseler, C.W., Smith, C.B., Kardos, L.T. and Fleming, H.K. (1980) Response of mature vines of *Vitis labrusca* L. cv Concord to applications of phosphorus and potassium over an eight-year span in Pennsylvania. *American Journal of Enology and Viticulture* 31(3), 237–244.

Hale, C.R. and Weaver, R.J. (1962) The effect of developmental stage on direction of translocation of photosynthate in *Vitis vinifera*. *Hilgardia* 33, 89–131.

Hale, C.R., Coombe, B.G. and Hawker, J.S. (1970) Effects of ethylene and 2-chloroethylphosphonic acid on the ripening of grapes. *Plant Physiology* 45(5), 620–623.

Halleen, F., Fourie, P.H. and Crous, P.W. (2006) A review of black foot disease of grapevine. *Phytopathology Mediterranea* 45, S55–S67.

Hamilton, R.P. and Coombe, B.G. (1992) Harvesting of winegrapes. In: Coombe, B.G. and Dry, P.R. (eds) *Viticulture Vol. 2. Practices*. Winetitles, Underdale, South Australia, pp. 302–327.

Hamman Jr., R.A., Dami, I.E., Walsh, T.M. and Stushnoff, C. (1996) Seasonal carbohydrate changes and cold hardiness of Chardonnay and Riesling grapevines. *American Journal of Enology and Viticulture* 47, 31–36.

Hanna, R., Zalom, F.G., Wilson, L.T. and Leavitt, G.M. (1997) Sulfur can suppress mite predators in vineyards. *California Agriculture* 51, 19–21.

Harbertson, J.F., Kennedy, J.A. and Adams, D.O. (2002) Tannin in skins and seeds of Cabernet Sauvignon, Syrah, and Pinot noir berries during ripening. *American Journal of Enology and Viticulture* 53(1), 54–59.

Hardie, W.J. and Cirami, R.M. (1988) Grapevine rootstocks. In: Coombe, B.G. and Dry, P.R. (eds) *Viticulture Vol. 1. Resources*. Winetitles, Underdale, South Australia. pp. 154–176.

Hardie, W.J. and Considine, J.A. (1976) Response of grapes to water-deficit stress in particular stages of development. *American Journal of Enology and Viticulture* 27(2), 55–61.

Harrell, D.C. and Williams, L.E. (1987) The influence of girdling and gibberellic acid application at fruitset on Ruby Seedless and Thompson Seedless grapes. *American Journal of Enology and Viticulture* 38(2), 83–88.

Harris, J.M., Kriedemann, P.E. and Possingham, J.V. (1968) Anatomical aspects of grape berry development. *Vitis* 7, 106–119.

Hartwig, N.L. and Ammon, H.U. (2002) Cover crops and living mulches. *Weed Science* 50(6), 688–699.

Hashizume, K. and Samuta, T. (1999) Grape maturity and light exposure affect berry methoxypyrazine concentration. *American Journal of Enology and Viticulture* 50, 194–198.

Haslam, E. (1980) In vino veritas: oligomeric procyanidins and the ageing of red wines. *Phytochemistry* 19(12), 2577–2582.

Heap, I.M. (1997) The occurrence of herbicide-resistant weeds worldwide. *Pesticide Science* 51(3), 235–243.

Henderson, E. (1995) Why grow organically? *Organic Grape and Wine Production Symposium*, Geneva, New York, Communications Services, New York State Agricultural Experiment Station, Cornell University, Geneva, New York.

Herderich, M.J. and Smith, P.A. (2005) Analysis of grape and wine tannins: methods, applications and challenges. *Australian Journal of Grape and Wine Research* 11(2), 205–214.

Hill, G.K., Stellwaag-Kittler, F., Huth, G. and Schlösser, E. (1981) Resistance of grapes in different development stages to *Botrytis cinerea*. *Phytopathology Zeitschrift* 102, 329–338.

Himelrick, D.G. (1991) Growth and nutritional responses of nine grape cultivars to low soil pH. *HortScience* 26, 269–271.

Himelrick, D.G. (2001) Economic consequences of Phylloxera in cold climate wine grape production areas of eastern Washington. *Small Fruits Review* 1(4), 3–15.

Himelrick, D.G. (2003) Handling, storage and postharvest physiology of muscadine grapes: a review. *Small Fruits Review* 2(4), 45–62.

Hislop, E.C. (1988) Electrostatic ground-rig spraying: an overview. *Weed Technology* 2(1), 94–105.

Hjeljord, L.G., Stensvand, A. and Tronsmo, A. (2000) Effect of temperature and nutrient stress on the capacity of commercial *Trichoderma* products to control *Botrytis cinerea* and *Mucor piriformis* in greenhouse strawberries. *Biological Control* 19(2), 149–160.

Hoag, J.C., Simonson, B., Cornforth, B. and St John, L. (2001) Waterjet Stinger: a tool for planting dormant nonrooted cuttings. *Native Plants Journal* 2, 84–89.

Hoddle, M.S. (2004) The potential adventive geographic range of glassy-winged sharpshooter, *Homalodisca coagulata* and the grape pathogen *Xylella fastidiosa*: implications for California and other grape growing regions of the world. *Crop Protection* 23(8), 691–699.

Hoffman, L.E. and Wilcox, W.F. (2003) Factors influencing the efficacy of myclobutanil and azoxystrobin for control of grape black rot. *Plant Disease* 87(3), 273–281.

Hoffman, L.E., Wilcox, W.F., Gadoury, D.M. and Seem, R.C. (2002) Influence of grape berry age on susceptibility to *Guignardia bidwellii* and its incubation period length. *Phytopathology* 92(10), 1068–1076.

Hoffman, L.E., Wilcox, W.F., Gadoury, D.M., Seem, R.C. and Riegel, D.G. (2004) Integrated control of grape black rot: influence of host phenology, inoculum availability, sanitation, and spray timing. *Phytopathology* 94(6), 641–650.

Hollman, P.C.H. and Arts, I.C.W. (2000) Review. Flavonols, flavones and flavanols – nature, occurrence and dietary burden. *Journal of Science in Food and Agriculture* 80, 1081–1093.

Hoos, G. and Blaich, R. (1988) Metabolism of stilbene phytoalexins in grapevines: oxidation of resveratrol in single-cell cultures. *Vitis* 27, 1–12.

Hoskins, N., Bonfiglioli, R. and Wright, C. (2003) *The Riversun Rootstock Project – Part 1*. Riversun Nurseries Ltd., Gisborne, New Zealand, 1 p.

Howard, K.L., Mike, J.H. and Riesen, R. (2005) Validation of a solid-phase micro-extraction method for headspace analysis of wine aroma components. *American Journal of Enology and Viticulture* 56(1), 37–45.

Howell, G.S. (2001) Sustainable grape productivity and the growth–yield relationship: a review. *American Journal of Enology and Viticulture* 52(3), 165–174.

Howell, G.S., Stergios, B.G. and Stackhouse, S.S. (1978) Interrelation of productivity and cold hardiness of Concord grapevines. *American Journal of Enology and Viticulture* 29, 187–191.

Hrazdina, G., Parsons, G.F. and Mattick, L.R. (1984) Physiological and biochemical events during development and maturation of grape berries. *American Journal of Enology and Viticulture* 35(4), 220–227.

Huang, H.T. (2000) Fermentations and food science. In: Needham, J. (ed.) *Science and Civilisation in China. Volume 6: Biology and Biological Technology*. Cambridge University Press, Cambridge, UK.

Huang, P.-D., Cash, J.N. and Santerre, C.R. (1988) Influence of stems, petioles and leaves on the phenolic content of Concord and Aurora blanc juice and wine. *Journal of Food Science* 53(1), 173–175.

Huang, S. and Gale, F. (2006) *China's Rising Fruit and Vegetable Exports Challenge U.S. Industries*. USDA, Washington, DC, p. 21.

Huggett, J.M. (2006) Geology and wine: a review. *Proceedings of the Geological Association* 117(2), 239–247.

Huisman, J.A., Hubbard, S.S., Redman, J.D. and Annan, A.P. (2003) Measuring soil water content with ground penetrating radar: a review. *Vadose Zone Journal* 2(4), 476–491.

Hunt, M.D., Neuenschwander, U.H., Delaney, T.P., Weymann, K.B., Friedrich, L.B., Lawton, K.A., Steiner, H.-Y. and Ryals, J.A. (1996) Recent advances in systemic acquired resistance research – a review. *Gene* 179, 89–95.

Ibacache, A.E. (1990) *Girdling and Shading Effects on Inflorescence Necrosis and Rachis Composition of Pinot Noir Grapevine*. Department of Horticulture, Oregon State University, Corvallis, Oregon.

Ifoulis, A.A. and Savopoulou-Soultani, M. (2004) Biological control of *Lobesia botrana* (Lepidoptera: Tortricidae) larvae by using different formulations of *Bacillus thuringiensis* in 11 vine cultivars under field conditions. *Journal of Economic Entomology*, 340–343.

Ingalsbe, D.W., Neubert, A.M. and Carter, G.H. (1963) Concord grape pigments. *Journal of Agricultural and Food Chemistry* 11, 263–268.

Intrieri, C. and Filippetti, I. (2000) Innovations and outlook in grapevine training systems and mechanization in North-Central Italy. *Proceedings of the American Society for Enology and Viticulture 50th Anniversary Annual Meeting*, Seattle, Washington, pp. 170–184.

Intrigliolo, D. and Castel, J. (2007) Evaluation of grapevine water status from trunk diameter variations. *Irrigation Science* 26(1), 49–59.

Jackson, D.I. (1991) Environmental and hormonal effects on development of Early Bunch Stem Necrosis. *American Journal of Enology and Viticulture* 42(4), 290–294.

Jackson, D.I. (1997) *Monographs in Cool Climate Viticulture 1. Pruning and Training*. Lincoln University Press, Canterbury, New Zealand, 69 pp.

Jackson, R.S. (2000) *Wine Science. Principles Practices Perception*. Academic Press, San Diego, California, 654 pp.

Jackson, D.I. and Cherry, N.J. (1988) Prediction of a district's grape-ripening capacity using a Latitude–Temperature Index (LTI) *American Journal of Enology and Viticulture* 39(1), 19–28.

Jackson, D.I. and Coombe, B.G. (1988) Early bunchstem necrosis – a cause of poor set. *Proceedings of the 2nd International Cool Climate Viticulture and Oenolology Symposium*, Auckland, New Zealand, pp. 72–75.

Jackson, D.I. and Lombard, P.B. (1993) Environmental and management practices affecting grape composition and wine quality – a review. *American Journal of Enology and Viticulture* 44(4), 409–430.

Jackson, R.B. and Caldwell, M.M. (1993) The scale of nutrient heterogeneity around individual plants and its quantification with geostatistics. *Ecology* 74(2), 612–614.

Janick, J. (2002) Ancient Egyptian agriculture and the origins of horticulture. *Acta Horticulturae* 582, 23−39.

Jansen, M.A.K., Gaba, V. and Greenberg, B.M. (1998) Higher plants and UV-B radiation: balancing damage, repair and acclimation. *Trends in Plant Science* 3(4), 131–135.

Jenkins, P.E. and Isaacs, R. (2007) Reduced-risk insecticides for control of grape berry moth (Lepidoptera: Tortricidae) and conservation of natural enemies. *Journal of Economic Entomology* 100(3), 855–865.

Jensen, F., Swanson, F., Peacock, W. and Leavitt, G. (1975) The effect of width of cane and trunk girdles on berry weight and soluble solids in table 'Thompson Seedless' vineyards. *American Journal of Enology and Viticulture* 26(2), 90–91.

Jensen, F., Andris, H. and Beede, R. (1981) A comparison of normal girdles and knife-line girdles on Thompson Seedless and Cardinal grapes. *American Journal of Enology and Viticulture* 32(3), 206–207.

Jermini, M. and Gessler, C. (1996) Epidemiology and control of grape black rot in southern Switzerland. *Plant Disease* 80(3), 322–325.

Jersch, S., Scherer, C., Huth, G. and Schlösser, E. (1989) Proanthocyanidins as basis for quiescence of *Botrytis cinerea* in immature strawberry fruits. *Journal of Plant Disease Protection* 96(4), 365–378.

Jie, W., Lite, L. and Yang, D. (2003) The correlation between freezing point and soluble solids of fruits. *Journal of Food Engineering* 60(4), 481–484.

Johnson, D.T., Lewis, B.A. and Snow, J.W. (1991) Control of grape root borer (Lepidoptera: Sesiidae) by mating disruption with two synthetic sex pheromone compounds. *Environmental Entomology* 20(3), 930–934.

Johnson, W. and Grgich, M. (1975) Another look at machine harvesting/wine quality. *Wines and Vines* 57, 40.

Johnston, W.H. (2000) Calibration of gypsum blocks and data loggers and their evaluation for monitoring soil water status. *Australian Journal of Experimental Agriculture* 40(8), 1131–1136.

Jordan, D. (1993) Cool climate effects on fruit set in grapevines through inflorescence necrosis. *Die Wein-Wissenschaft* 48, 3–6.

Jordan, T.D., Pool, R.M., Zabadal, T.J. and Tomkins, J.P. (1981) *Cultural Practices for Commercial Vineyards.* Miscellaneous Bulletin, New York State College of Agriculture and Life Sciences, Cornell University, Ithaca, New York, 69 pp.

Kaiser, B.N., Gridley, K.L., Brady, J.N., Phillips, T. and Tyerman, S.D. (2005) The role of molybdenum in agricultural plant production. *Australian Journal of Experimental Agriculture* 96(5), 745–754.

Karban, R., English-Loeb, G., Walker, M.A. and Thaler, J. (1995) Abundance of phytoseiid mites on *Vitis* species: effects of leaf hairs, domatia, prey abundance and plant phylogeny. *Experimental and Applied Acarology* 19(4), 189–197.

Kasimatis, A.N., Weaver, R.J., Pool, R.M. and Halsey, D.D. (1971) Response of 'Perlette' grape berries to gibberellic acid applied during bloom or at fruit set. *American Journal of Enology and Viticulture* 22(1), 19–23.

Kassemeyer, H.-H. and Staudt, G. (1981) Development of the embryo sac and fertilization in grapevine. *Vitis* 20, 202–210.

Kast, W.K. and Stark-Urnau, M. (1999) Survival of sporangia from *Plasmopara viticola*, the downy mildew of grapevine. *Vitis* 38(4), 185–186.

Kato, A.E. and Fathallah, F.A. (2002) Ergonomic evaluation of California winegrape trellis systems. *Proceedings of the Human Factors and Ergonomics Society Annual Meeting.* Pittsburgh, Pennsylvania, 23–27 September 2002.

Kauer, R., Gaubatz, B., Kornitzer, U., Kirchner, B., Wöhrle, M. and Schultz, H.R. (2001) *Organic Viticulture without Sulphur? Three Years of Experience with Sodium and Potassium Bicarbonate.* GESCO Compte Rendu, Montpellier, France, pp. 213–218.

Keller, M. and Torres-Martinez, N. (2004) Does UV radiation affect winegrape composition? *Acta Horticulturae* 640, 313–319.

Keller, M., Arnink, K.J. and Hrazdina, G. (1998) Interaction of nitrogen availability during bloom and light intensity during veraison. I. Effects on grapevine growth, fruit development, and ripening. *American Journal of Enology and Viticulture* 49(3), 333–340.

Keller, M., Rogiers, S.Y. and Schultz, H.R. (2003) Nitrogen and ultraviolet radiation modify grapevines' susceptibility to powdery mildew. *Vitis* 42(2), 87–94.

Keller, M., Mills, L.J., Wample, R.L. and Spayd, S.E. (2005) Cluster thinning effects on three deficit-irrigated *Vitis vinifera* cultivars. *American Journal of Enology and Viticulture* 56(2), 91–103.

Kido, H., Flaherty, D.L., Bosch, D.F. and Valero, K.A. (1984) French prune trees as overwintering sites for the grape leafhopper egg parasite. *American Journal of Enology and Viticulture* 35(3), 156–160.

Kimura, P.H., Okamoto, G. and Hirano, K. (1998) The mode of pollination and stigma receptivity in *Vitis coignetiae* Pulliat. *American Journal of Enology and Viticulture* 49(1), 1–5.

King, P.D. and Buchanan, G.A. (1986) The dispersal of phylloxera crawlers and spread of phylloxera infestations in New Zealand and Australian vineyards. *American Journal of Enology and Viticulture* 37(1), 26–33.

Kirchmair, M., Huber, L., Porten, M., Rainer, J. and Strasser, H. (2004) *Metarhizium anisopliae*, a potential agent for the control of grape phylloxera. *BioControl* 49(3), 295–303.

Klepper, B., Douglas, V. and Taylor, H.M. (1971) Stem diameter in relation to plant water status. *Plant Physiology* 48, 683–685.

Kliewer, W.M. (1968) Changes in the concentration of free amino acids in grape berries during maturation. *American Journal of Enology and Viticulture* 19(3), 166–174.

Kliewer, W.M. (1977) Influence of temperature, solar radiation and nitrogen on coloration and composition of Emperor grapes. *American Journal of Enology and Viticulture* 28(2), 96–103.

Kliewer, W.M. and Antcliff, A.J. (1970) Influence of defoliation, leaf darkening, and cluster shading on the growth and composition of Sultana grapes. *American Journal of Enology and Viticulture* 21(1), 26–36.

Kliewer, W.M. and Dokoozlian, N.K. (2005) Leaf area/crop weight ratios of grapevines: influence on fruit composition and wine quality. *American Journal of Enology and Viticulture* 56(2), 170–181.

Kliewer, W.M. and Lider, L.A. (1968) Influence of cluster exposure to the sun on the composition of Thompson Seedless fruit. *American Journal of Enology and Viticulture* 19(3), 175–184.

Kliewer, W.M. and Torres, R.E. (1972) Effect of controlled day and night temperatures on grape coloration. *American Journal of Enology and Viticulture* 23(2), 71–77.

Kliewer, W.M., Bowen, P. and Benz, M. (1989) Influence of shoot orientation on growth and yield development in Cabernet Sauvignon. *American Journal of Enology and Viticulture* 40(4), 259–264.

Kluba, R.M., Mattick, L.R. and Hackler, L.R. (1978) Changes in the free and total amino acid composition of several *Vitis labruscana* grape varieties during maturation. *American Journal of Enology and Viticulture* 29(2), 102–111.

Kolattukudy, P.E. (1985) Enzymatic penetration of the plant cuticle by fungal pathogens. *Annual Review of Phytopathology* 23, 223–250.

Kolb, C.A., Kopecký, J., Riederer, M. and Pfündel, E.E. (2003) UV screening by phenolics in berries of grapevines (*Vitis vinifera*). *Functional Plant Biology* 30, 1177–1186.

Köller, W. and Scheinpflug, H. (1987) Fungal resistance to sterol biosynthesis inhibitors: a new challenge. *Plant Disease* 71, 1066–1074.

Kombrink, E. and Schmelzer, E. (2001) The hypersensitive response and its role in local and systemic disease resistance. *European Journal of Plant Pathology* 107(1), 69–78.

Kondo, K. and Ting, K.C. (1998) Robotics for plant production. *Artificial Intelligence Review* 12, 227–243.

Konrad, H., Lindner, B., Bleser, E. and Rühl, E.H. (2003) Strategies in the genetic selection of clones and the preservation of genetic diversity within varieties. *Acta Horticulturae* 603, 105–110.

Kovacs, L.G., Hanami, H., Fortenberry, M. and Kaps, M.L. (2001) Latent infection by leafroll agent GLRaV-3 is linked to lower fruit quality in French–American hybrid grapevines Vidal blanc and St. Vincent. *American Journal of Enology and Viticulture* 52(3), 254–259.

Kubota, N. and Miyamuki, M. (1992) Breaking bud dormancy in grapevines with garlic paste. *Journal of the American Society for Horticultural Science* 117(6), 898–901.

Kubota, N., Matthews, M.A., Takahagi, T. and Kliewer, W.M. (2000) Budbreak with garlic preparations: effects of garlic preparations and of calcium and hydrogen cyanamides on budbreak of grapevines grown in greenhouses. *American Journal of Enology and Viticulture* 51(4), 409–414.

Kulakiotu, E.K., Thanassoulopoulos, C.C. and Sfakiotakis, E.M. (2004) Biological control of *Botrytis cinerea* by volatiles of 'Isabella' grapes. *Phytopathology* 94(9), 924–931.

Kunde, R.M., Lider, L.A. and Schmitt, R.V. (1968) A test of *Vitis* resistance to *Xiphinema index*. *American Journal of Enology and Viticulture* 19(1), 30–36.

Lakso, A.N., Pratt, C., Pearson, R.C., Pool, R.M., Seem, R.C. and Welser, M.J. (1982) Photosynthesis, transpiration, and water use efficiency of mature grape leaves infected with *Uncinula necator* (powdery mildew). *Phytopathology* 72, 232–236.

Landis, D.A., Wratten, S.D. and Gurr, G.M. (2000) Habitat management to conserve natural enemies of arthropod pests in agriculture. *Annual Review of Entomology* 45(1), 175–201.

Lang, A. (1970) Gibberellins: structure and metabolism. *Annual Review of Plant Physiology* 21, 537–570.

Langcake, P. and McCarthy, W.V. (1979) The relationship of resveratrol production to infection of grapevine leaves by *Botrytis cinerea*. *Vitis* 18, 244–253.

Langcake, P. and Pryce, R.J. (1976) The production of resveratrol by *Vitis vinifera* and other members of the Vitaceae as a response to infection or injury. *Physiological Plant Pathology* 9, 77–86.

Larrauri, J.A., Ruperez, P. and Calixto, F.S. (1996) Antioxidant activity of wine pomace. *American Journal of Enology and Viticulture* 47(4), 369–372.

Lave, L.B. (1963) The value of better weather information to the raisin industry. *Econometrica* 31(1/2), 151–164.

Lavee, S. (1974) Dormancy and bud break in warm climates: considerations of growth regulators involvement. *Acta Horticulturae* 34, 225–232.

Lavee, S. and Nir, G. (1986) Grape. In: Monselise, S.P. (ed.) *CRC Handbook of Fruit Set and Development*. CRC Press, Boca Raton, Florida, pp. 167–191.

Lavee, S., Regev, U. and Samish, R.M. (1967) The determination of induction and differentiation in grapevines. *Vitis* 6, 1–13.

Li, X. (2004) *Potassium Nutrition of Self-rooted Pinot Noir Grapevines (*Vitis vinifera *L.) Grown in Pots.* Agriculture and Life Science Division, Lincoln University, Canterbury, New Zealand, 145 pp.

Lider, L.A. (1958) Phylloxera-resistant grape rootstocks for the coastal valleys of California. *Hilgardia* 27, 287–318.

Lin, C.H. and Wang, T.Y. (1985) Enhancement of bud sprouting in grape single bud cuttings by cyanamide. *American Journal of Enology and Viticulture* 36(1), 15–17.

Lipe, W.N., Baumhardt, L., Wendt, C.W. and Rayburn, D.J. (1992) Differential thermal analysis of deacclimating Chardonnay and Cabernet Sauvignon grape buds as affected by evaporative cooling. *American Journal of Enology and Viticulture* 43(4), 355–361.

Liu, H.-F., Wu, B.-H., Fan, P.-G., Xu, H.-Y. and Li, S.-H. (2007) Inheritance of sugars and acids in berries of grape (*Vitis vinifera* L.). *Euphytica* 153(1), 99–107.

Loinger, C., Cohen, S., Dror, N. and Berlinger, M.J. (1977) Effect of grape cluster rot on wine quality. *American Journal of Enology and Viticulture* 28(4), 196–199.

Lomkatsi, S.I., Ordzhonikidze, A.A., Mgeladze, T.A., Gurabanidze, L.R., Dzhimshitashvili, L.G., Pipiya, A.D., Chubinishvili, T.-M. and Mazanashvili, T.G. (1983) Protection of vineyards from hail damage. *Vestnik-Sel'skokhozyaistvennoi-Nauki* 2, 66–68.

Long, Z. (1997) Developing wine flavor in the vineyard. *Practical Winery & Vineyard* 18, 6–9.

Lovisolo, C. and Schubert, A. (2000) Downward shoot positioning affects water transport in field-grown grapevines. *Vitis* 39(2), 49–53.

Lydakis, D., Fysarakis, I., Papadimitriou, M. and Kolioradakis, G. (2003) Optimization study of sulfur dioxide application in processing of Sultana raisins. *International Journal of Food Properties* 6(3), 393–403.

Magarey, P., Wicks, T., Hitch, C. and Emmett, B. (2003) Effects of temperature and application rates on the phytotoxicity of sulphur on grapevines. In: Emmett, R.W. (ed.) *Strategic Use of Sulphur in Integrated Pest and Disease Management (IPM) Programs for Grapevines.* Project Number: DAV 98/1, Victoria Dept. of Primary Industries, Mildura, Victoria, Australia, pp. 110–116.

Mailhac, N. and Chervin, C. (2006) Ethylene and grape berry ripening. *Stewart Postharvest Review* (2), 5 pp.

Marais, J. (1994) Sauvignon blanc cultivar aroma – a review. *South African Journal of Enology and Viticulture* 15, 41–45.

Marais, J., Wyk, C.J.v. and Rapp, A. (1992) Effect of sunlight and shade on norisoprenoid levels in maturing Weisser Riesling and Chenin blanc grapes and Weisser Riesling wines. *South African Journal of Enology and Viticulture* 13, 23–32.

Marschner, H. (1986) *Mineral Nutrition of Higher Plants.* Academic Press, Harcourt Brace Jovanovich, London.

Marshall, J.K. (1967) The effect of shelter on the productivity of grasslands and field crops. *Field Crops Abstracts* 20, 1–14.

Martini, A., Ciani, M. and Scorzetti, G. (1996) Direct enumeration and isolation of wine yeasts from grape surfaces. *American Journal of Enology and Viticulture* 47(4), 435–440.

Martinson, T. (1995) Management of insect pests in organic vineyards. *Organic Grape and Wine Production Symposium*, Geneva, New York. Communications Services, New York State Agricultural Experiment Station, Cornell University, Geneva, New York.

Masahiro, F. and Shinro, K. (2001) Control effects of application of chemicals without mixing lime sulfur to the branches of the grape tree before sprouting against anthracnose of grape. *Research Bulletin of the Aichi-ken Agricultural Research Center* 33, 187–193.

Matthews, G.A. (2001) The application of chemicals for plant disease control. In: Waller, J.M., Lenné, J.M. and Waller, S.J. (eds) *Plant Pathologist's Pocketbook.* CAB International, Wallingford, UK, pp. 345–353.

Matthews, M.A., Cheng, G. and Weinbaum, S.A. (1987) Changes in water potential and dermal extensibility during grape berry development. *Journal of the American Society of Horticultural Science* 112(2), 314–319.

Mattick, L.R., Shaulis, N.J. and Moyer, J.C. (1972) The effect of potassium fertilization on the acid content of 'Concord' grape juice. *American Journal of Enology and Viticulture* 23(1), 26–30.

May, P. (1965) Reducing inflorescence formation by shading individual Sultana buds. *Australian Journal of Biological Science* 18, 463–473.

May, P. (1986) The grapevine as a perennial, plastic and productive plant. *Proceedings of the 6th Australian Wine Industry Technical Conference*, Adelaide. Australian Industrial Publishers, Adelaide, South Australia, Australia, pp. 40–49.

May, P. (2000) From bud to berry, with special reference to inflorescence and bunch morphology in *Vitis vinifera* L. *Australian Journal of Grape and Wine Research* 6, 82–98.

May, P., Scholefield, P.B., Clingeleffer, P.R. and Smith, L. (1974) Experiments on the mechanical harvesting of sultanas for drying. *Journal of Science and Food Agriculture* 25(5), 541–552.

Maynard, D.N., Barker, A.V. and Lachman, W.H. (1968) Influence of potassium on the utilization of ammonium by tomato plants. *Proceedings of the American Society for Horticultural Science* 92, 331–337.

McBride, R.L. and Johnson, R.L. (1987) Perception of sugar–acid mixtures in lemon juice drink. *International Journal of Food Science and Technology* 22(4), 399–408.

McCarthy, M.G., Cirami, R.M. and Furkaliev, D.G. (1997) Rootstock response of Shiraz (*Vitis vinifera*) grapevines to dry and drip-irrigated conditions. *Australian Journal of Grape and Wine Research* 3(2), 95–98.

McGovern, P.E., Zhang, J., Tang, J., Zhang, Z., Hall, G.R., Moreau, R.A., Nunez, A., Butrym, E.D., Richards, M.P., Wang, C., Cheng, G., Zhao, Z. and Wang, C. (2004) Fermented beverages of pre- and proto-historic China. *PNAS* 101(51), 17593–17598.

McKenry, M.V. (1999) *The Replant Problem and its Management*. Catalina Circle, Fresno, California, 125 pp.

Melander, B., Heisel, T. and Jørgensen, M.H. (2002) Band-steaming for intra-row weed control. *Proceedings of the 5th EWRS Workshop on Physical Weed Control*, Pisa, Italy, 11–13 March, pp. 216–219.

Mengel, K., Breininger, T. and Bubl, W. (1984) Bicarbonate, the most important factor inducing iron chlorisis in vine grapes on calcareous soil. *Plant and Soil* 81, 333–344.

Milne, G., Lang, S. and Kingston, C. (2003) Can pollen source influence grape fruit set or berry characteristics? *Australian and New Zealand Grapegrower and Winemaker* 471, 32–34.

Mirabel, M., Saucier, C., Guerra, C. and Glories, Y. (1999) Copigmentation in model wine solutions: occurrence and relation to wine aging. *American Journal of Enology and Viticulture* 50(2), 211–218.

Mitchell, F.G. (1992) Cooling horticultural commodities. In: Kader, A.A. (ed.) *Postharvest Technology of Horticultural Crops*. University of California, Division of Agriculture and Natural Resources, Oakland, California, pp. 53–63.

Mitchell, F.G. and Crisosto, C.H. (1995) The use of cooling and cold storage to stabilize and preserve fresh stone fruits. *Seminario sobre Calidad Post-Cosecha y Productos Derivados en Frutos de Hueso*, Lleida, Spain.

Moio, L. and Etievant, P.X. (1995) Ethyl anthranilate, ethyl cinnamate, 2,3-dihydrocinnamate, and methyl anthranilate: four important odorants identified in Pinot noir wines of Burgundy. *American Journal of Enology and Viticulture* 46(3), 392–398.

Mollah, M. (1997) *Practical Aspects of Grapevine Trellising*. Winetitles, Underdale, South Australia.

Moncur, M.W., Rattigan, K., Mackenzie, D.H. and McIntyre, G.N. (1989) Base temperature for budbreak and leaf appearance of grapevines. *American Journal of Enology and Viticulture* 40(1), 21–26.

Monta, M., Kondo, N. and Shibano, Y. (1995) Agricultural robot in grape production system. *Proceedings of the IEEE International Conference on Robotics and Automation*, Nagoya, Japan, IEEE, Los Alamitos, California, pp. 2504–2509.

Moore, M.O. and Giannasi, D.E. (1994) Foliar flavonoids of eastern North American *Vitis* (Vitaceae) north of Mexico. *Plant Systematics and Evolution* 193(1–4), 21–36.

Moran, W. (2006) Crafting terroir: people in cool climates, soils and markets. *Proceedings of the 6th International Symposium on Cool Climate Viticulture and Oenology*, Christchurch, New Zealand. New Zealand Society for Viticulture and Oenology, Auckland, New Zealand.

Morin, F., Fortin, J.A., Hamel, C., Granger, R.L. and Smith, D.L. (1994) Apple rootstock response to vesicular–arbuscular mycorrhizal fungi in a high phosphorus soil. *Journal of the American Society for Horticultural Science* 119, 578–583.

Morris, J.R. (1989) Producing quality grape juice. *110th Annual Meeting of the Arkansas State Horticultural Society* (http://www.uark.edu/depts/ifse/grapeprog/table.htm).

Morris, J.R. (2000) Past, present, and future of vineyard mechanization. *Proceedings of the American Society for Enology and Viticulture* 50th *Annual Meeting*, Seattle, Washington, pp. 155–167.

Morris, J. (2006) Development and incorporation of mechanisation into intensely managed grape vineyards. *Proceedings of the 6th International Cool Climate Symposium on Viticulture and Oenology*, Christchurch, New Zealand. New Zealand Society for Viticulture and Oenology, Auckland, New Zealand.

Morris, J.R., Fleming, J.W., Benedict, R.H. and McCaskill, D.R. (1972) Effects of sulfur dioxide on postharvest quality of mechanically harvested grapes. *Arkansas Farm Research* 21(2), 5.

Morris, J.R., Cawthon, D.L. and Fleming, J.W. (1975) Effect of mechanical pruning on yield and quality of 'Concord' grapes. *Arkansas Farm Research* 24(3), 12.

Morris, J.R., Sistrunk, W.A., Junek, J. and Sims, C.A. (1986) Effects of fruit maturity, juice storage, and juice extraction temperature on quality of 'Concord' grape juice. *Journal of the American Society for Horticultural Science* 111(5), 742–746.

Mullins, M.G., Bouquet, A. and Williams, L.E. (1992) *Biology of the Grapevine*. Cambridge University Press, Cambridge, UK, 251 pp.

Nagarkatti, S., Tobin, P.C., Muza, A.J. and Saunders, M.C. (2002) Carbaryl resistance in populations of grape berry moth (Lepidoptera: Tortricidae) in New York and Pennsylvania. *Journal of Economic Entomology* 95, 1027–1032.

Nagarkatti, S., Tobin, P.C., Saunders, M.C. and Muza, A.J. (2003) Release of native *Trichogramma minutum* to control grape berry moth. *Canadian Entomologist* 135(4), 589–598.

Nair, N.G. and Allen, R.N. (1993) Infection of grape flowers and berries by *Botrytis cinerea* as a function of time and temperature. *Mycological Research* 97, 1012–1014.

Nair, N.G., Guilbaud-Oulton, S., Barchia, I. and Emmett, R. (1995) Significance of carry over inoculum, flower infection and latency on the incidence of *Botrytis cinerea* in berries of grapevines at harvest in New South Wales. *Australian Journal of Experimental Agriculture* 35(8), 1177–1180.

Nakagawa, S. and Nanjo, Y. (1965) A morphological study of Delaware grape berries. *Journal of the Japanese Society for Horticultural Science* 34, 85–95.

Naylor, A.P., Creasy, G.L. and Vanhanen, L. (2003) Effects of row orientation and cluster exposure on light interception and Sauvignon blanc fruit composition. *Australian and New Zealand Grapegrower & Winemaker* 474, 97–100.

Neel, D. (2000) Windbreaks mitigate climate. *Practical Winery & Vineyard* XXI(1) (May/June), 20–22.

Neeson, R., Glasson, A., Morgan, A., Macalpine, S. and Darnley-Naylor, M. (1995) On-farm options. *Murrumbidgee Irrigation Areas and Districts Land and Water Management Plan*, NSW Agriculture, Australia.

Nelson, K.E. (1983) Effects of in-package sulfur dioxide generators, package liners, and temperature on decay and desiccation of table grapes. *American Journal of Enology and Viticulture* 34(1), 10–16.

Nelson, K.E. (1985) Harvesting and handling California table grapes for market. *University of California Agricultural and Experimental Station Research Bulletin 1913*, Davis, California, p. 72.

Nelson, K.E. and Ahmedullah, M. (1970) Effect on 'Cardinal' grapes of position of sulfur dioxide generators and retention of gas and water vapor in unvented containers. *American Journal of Enology and Viticulture* 21(2), 70–77.

Nelson, K.E., Baker, G.A., Winkler, A.J., Amerine, M.A., Richardson, H.B. and Jones, F.R. (1963) Chemical and sensory variability in table grapes. *Hilgardia* 34, 1–42.

Nelson, R.R., Acree, T.E., Lee, C.Y. and Butts, R.M. (1977) Methyl anthranilate as an aroma constituent of American wine. *Journal of Food Science* 42, 57–59.

Newman, J.L. (1986) Vines, wines, and regional identity in the Finger Lakes Region. *Geographical Review* 76(3), 301–316.

Nicholas, P. (2004) *Soil, Irrigation and Nutrition*. Winetitles, Underdale, South Australia.

Nicholas, P.R., Chapman, A.P. and Cirami, R.M. (1992) Grapevine propagation. In: Coombe, B.G. and Dry, P.R. (eds) *Viticulture. Volume 2, Practices*. Winetitles, Underdale, South Australia, pp. 1–22.

Nicholas, P., Magarey, P.A. and Wachtel, M. (1994) *Diseases and Pests*. Grape Production Series 1, Hyde Park Press, Adelaide, South Australia, 106 pp.

Nicholls, C.I., Parrella, M.P. and Altieri, M.A. (2000) Reducing the abundance of leafhoppers and thrips in a northern California organic vineyard through maintenance of full season floral diversity with summer cover crops. *Agricultural and Forest Entomology* 2(2), 107–113.

Nicol, J.M., Stirling, G.R., Rose, B.J., May, P. and Heeswijck, R.V. (1999) Impact of nematodes on grapevine growth and productivity: current knowledge and future directions, with special reference to Australian viticulture. *Australian Journal of Grape and Wine Research* 5(3), 109–127.

Nita, M., Ellis, M.A., Wilson, L.L. and Madden, L.V. (2007) Evaluation of the curative and protectant activity of fungicides and fungicide–adjuvant mixtures on Phomopsis cane and leaf spot of grape: a controlled environment study. *Crop Protection* 26(9), 1377–1384.

Noble, A.C., Ough, C.S. and Kasimatis, A.N. (1975) Effect of leaf content and mechanical harvest on wine 'quality'. *American Journal of Enology and Viticulture* 26(3), 158–163.

Noble, A.C., Arnold, R.A., Masuda, B.M., Pecore, S.D., Schmidt, J.O. and Stern, P.M. (1984) Progress towards a standardized system of wine aroma terminology. *American Journal of Enology and Viticulture* 35(2), 107–109.

Noble, A.C., Arnold, R.A., Buechsenstein, J., Leach, E.J., Schmidt, J.O. and Stern, P.M. (1987) Modification of a standardized system of wine aroma terminology. *American Journal of Enology and Viticulture* 38(2), 143–146.

Northover, J. (1987) Infection sites and fungicidal prevention of *Botrytis cinerea* bunch rot of grapes in Ontario. *Canadian Journal of Plant Pathology* 9, 129–136.

Nougaret, R.L. and Lapham, M.H. (1928) *A Study of Phylloxera Infestation in California as Related to Types of Soils.* Technical Bulletin No. 2, US Department of Agriculture, Washington, DC.

Nuttall, S.L., Kendall, M.J., Bombardelli, E. and Morazzoni, P. (1998) An evaluation of the antioxidant activity of a standardized grape seed extract, Leucoselect. *Journal of Clinical Pharmacological Therapy* 23(5), 385–389.

Ochs, E.S., Throop, J.A. and Gunkel, W.W. (1992) *Vision Control and Rough Terrain Compensation for a Robotic Grape Pruner.* ASAE Paper No. 92–7043, American Society of Agricultural Engineers, St Joseph, Michigan.

OIV (1983) *Le Code des Charactères Descriptifs des Varietés et Espèces de* Vitis. Office International de la Vigne et du Vin, Dedon, Paris, 152 pp.

OIV (2000) *Description of World Vine Varieties.* L'Organisation Internationale de la Vigne et du Vin, Paris, 542 pp.

OIV (2003) *Situation Report for the World Vitivinicultural Sector in 2003.* L'Organisation Internationale de la Vigne et du Vin, Paris, 59 pp.

OIV (2006) *State of the Vitiviniculture World Report.* L'Organisation Internationale de la Vigne et du Vin, Paris, 14 pp.

Okamoto, G. (1994) Grape. In: Konishi, K., Iwahori, S., Kitagawa, H. and Yakuwa, T. (ed.) *Horticulture in Japan.* Asakura Publishing Co., Tokyo, pp. 27–33.

Olien, W.C. (1990) The Muscadine grape: botany, viticulture, history, and current industry. *HortScience* 25(7), 732–739.

Ollat, N. (1992) Nouaison chez *Vitis vinifera* L. cv. 'Merlot Noir': rôle de l'intensité lumineuse et de la photosynthèse à la floraison. *Proceedings of the IVth International Symposium on Grapevine Physiology*, Istituto Agrario San Michele all'Adige, University of Turin, Italy, pp. 113–116.

Olmo, H.P. (1946) Correlations between seed and berry development in some seeded varieties of *Vitis vinifera*. *Proceedings of the American Society for Horticultural Science* 48, 291–297.

Olmstead, M.A. and Tarara, J.M. (2001) Physical principles of row covers and grow tubes with application to small fruit crops. *Small Fruits Review* 1(3), 29–46.

Omer, A.D., Granett, J., De Benedictis, J.A. and Walker, M.A. (1995) Effects of fungal root infections on the vigor of grapevines infested by root-feeding grape phylloxera. *Vitis* 34(3), 165–170.

Ortega-Heras, M., González-SanJosé, M.L. and Beltrán, S. (2002) Aroma composition of wine studied by different extraction methods. *Analytica Chimica Acta* 458, 85–93.

Ough, C.S. and Berg, H.W. (1971) Simulated mechanical harvest and gondola transport. 11. Effect of temperature, atmosphere, and skin contact on chemical and sensory qualities of white wines. *American Journal of Enology and Viticulture* 22(4), 194–198.

Ozkana, H.E., Mirallesb, A., Sinfortb, C., Zhuc, H. and Fox, R.D. (1997) Shields to reduce spray drift. *Journal of Agricultural Engineering Research* 67(4), 311–322.

Paggi, M.S. and Yamazaki, F. (2005) *Grape Juice Concentrate Trade Profile 2004*. California Association of Winegrape Growers, p. 9.

Palliotti, A., Cartechini, A. and Ferranti, F. (2000) Morpho-anatomical and physiological characteristics of primary and lateral shoot leaves of Cabernet Franc and Trebbiano Toscano grapevines under two irradiance regimes. *American Journal of Enology and Viticulture* 51(2), 122–130.

Palma, B.A. and Jackson, D.I. (1989) Inflorescence initiation in grapes – response to plant growth regulators. *Vitis* 28, 1–12.

Pankhurst, C., Hawke, B., McDonald, H., Kirkby, C., Buckerfield, J., Michelsen, P., O'Brien, K., Gupta, V. and Doube, B. (1995) Evaluation of soil biological properties as potential bioindicators of soil health. *Australian Journal of Experimental Agriculture* 35(7), 1015–1028.

Panneton, B., Thériault, R. and Lacasse, B. (2001) Efficacy evaluation of a new spray-recovery sprayer for orchards. *Transactions of the American Society of Agricultural Engineering* 44(3), 473–479.

Parat, C., Chaussod, R., Lévêque, J., Dousset, S. and Andreux, F. (2002) The relationship between copper accumulated in vineyard calcareous soils and soil organic matter and iron. *European Journal of Soil Science* 53, 663–670.

Parish, M.E. and Carroll, D.E. (1985) Indigenous yeasts associated with Muscadine (*Vitis rotundifolia*) grapes and musts. *American Journal of Enology and Viticulture* 36(2), 165–169.

Parr, W.V., Green, J.A., White, K.G. and Sherlock, R.R. (2007) The distinctive flavour of New Zealand Sauvignon blanc: sensory characterisation by wine professionals. *Food Quality Preference* 18(6), 849–861.

Patakas, A., Noitsakis, B. and Chouzouri, A. (2005) Optimization of irrigation water use in grapevines using the relationship between transpiration and plant water status. *Agriculture, Ecosystems & Environment* 106(2–3), 253–259.

Patrick, V. (1983) Mechanical pruning trials in Coonawarra. *Coonawarra Viticulture: Proceedings of a Seminar*, Coonawarra, South Australia. Australian Society of Viticulture and Enology.

Patton, D. (2006) *China to Start Making Ice Wine*. APFI, Singapore (http://www.APFoodTechnology.com).

Peacock, W.L. and Swanson, C.A. (2005) The future of California raisins is drying on the vine. *California Agriculture* 59(2), 70–74.

Peacock, W.L., Jensen, F., Else, J. and Leavitt, G. (1977a) The effects of girdling and ethephon treatments on fruit characteristics of Red Malaga. *American Journal of Enology and Viticulture* 28(4), 228–230.

Peacock, W.L., Rolston, D.E., Aljibury, F.K. and Rauschkolb, R.S. (1977b) Evaluating drip, flood, and sprinkler irrigation of wine grapes. *American Journal of Enology and Viticulture* 28(4), 193–195.

Pearson, R.C. and Goheen, A.C. (1988) *Compendium of Grape Diseases*. American Phytopathological Society, St. Paul, Minnesota.

Pederson, C.S. (1971) Grape juice. In: Tressler, D.K.J. (ed.) *Fruit and Vegetable Juice: Processing Technology*. Avi Publishing Co., Westport, Connecticut, pp. 234–271.

Penfold, C. (2004) Mulching and its impact on weed control, vine nutrition, yield and quality. *Australian and New Zealand Grapegrower and Winemaker* 485a, 32–38.

Perold, A.I. (1927) *A Treatise on Viticulture*. Macmillan and Co. Ltd., London.

Perring, T.M., Gruenhagen, N.M. and Farrar, C.A. (1999) Management of plant viral diseases through chemical control of insect vectors. *Annual Review of Entomology* 44(1), 457–481.

Pertot, I., El Bilali, H., Simeone, V., Vecchione, A. and Zulini, L. (2006) Efficacy evaluation and phytotoxicity assessment of copper peptidate on seven grapevine varieties and identification of the potential factors that induced copper damages on leaves. *Integrated Protection in Viticulture, IOBC/WPRS Bulletin* Vol. 29 (11), Boario Terme, Italy.

Petrie, P.R. and Clingeleffer, P.R. (2006) Crop thinning (hand versus mechanical), grape maturity and anthocyanin concentration: outcomes from irrigated Cabernet Sauvignon (*Vitis vinifera* L.) in a warm climate. *Australian Journal of Grape and Wine Research* 12(1), 21–29.

Petrie, P.R., Trought, M.C.T. and Howell, G.S. (2000a) Influence of leaf ageing, leaf area and crop load on photosynthesis, stomatal conductance and senescence of grapevine (*Vitis vinifera* L. cv. Pinot noir) leaves. *Vitis* 39, 31–36.

Petrie, P.R., Trought, M.C.T. and Howell, G.S. (2000b) Growth and dry matter partitioning of Pinot Noir (*Vitis vinifera* L.) in relation to leaf area and crop load. *Australian Journal of Grape and Wine Research* 6, 40–45.

Petrucci, V.E. and Siegfried, R. (1976) The extraneous matter in mechanically harvested wine grapes. *American Journal of Enology and Viticulture* 27(1), 40–41.

Phelps, G.W. (1999) The influence of site, crop load, and cluster light exposure on Pinot noir fruit composition in Canterbury. M Appl. Sci. thesis, Lincoln University, Christchurch, New Zealand, 100 pp.

Pickering, G. (2006) Icewine – the frozen truth. *Proceedings of the 6th International Cool Climate Symposium on Viticulture and Oenology*, Christchurch, New Zealand. New Zealand Society for Viticulture and Oenology, Auckland, New Zealand.

Pierquet, P. and Stushnoff, C. (1980) Relationship of low temperature exotherms to cold injury in *Vitis riparia* Michx. *American Journal of Enology and Viticulture* 31(1), 1–6.

Pierquet, P., Stushnoff, C. and Burke, M.J. (1977) Low temperature exotherms in stem and bud tissues of *Vitis riparia* Michx. *Journal of the American Society for Horticultural Science* 102, 54–55.

Pietrzak, U. and McPhail, D.C. (2004) Copper accumulation, distribution and fractionation in vineyard soils of Victoria, Australia. *Geoderma* 122(2–4), 151–166.

Pike, W. and Melewar, T.C. (2006) The demise of independent wine production in France: a marketing challenge? *International Journal of Wine Marketing* 18(3), 183–203.

Pillitteri, L.J., Lovatt, C.J. and Walling, L.L. (2004) Isolation and characterization of a terminal flower homolog and its correlation with juvenility in citrus. *Plant Physiology* 135(3), 1540–1551.

Pinel, M.P.C., Bond, W., White, J.G. and de Courcy Williams, M. (1999) *Field Vegetables: Assessment of the Potential for Mobile Soil Steaming Machinery to Control Diseases, Weeds and Mites of Field Salad and Related Crops*. Final Report on HDC Project FV229, UK Horticultural Development Council, East Malling, UK, 55 pp.

Pinney, T. (1989) *A History of Wine in America: from the Beginnings to Prohibition*. University of California Press, Berkeley, California.

Pirovano, F. (2006) *GAIN Report Number AR6004*. USDA Foreign Agricultural Service GAIN Report, Rojas, K., Buenos Aires, 7 pp.

Plauborg, F., Iversen, B.V. and Laerke, P.E. (2005) *In situ* comparison of three dielectric soil moisture sensors in drip irrigated sandy soils. *Vadose Zone Journal* 4(4), 1037–1047.

Pocock, K.F., Hayasaka, Y., Peng, Z., Williams, P.J. and Waters, E.J. (1998) The effect of mechanical harvesting and long-distance transport on the concentration of haze-forming proteins in grape juice. *Australian Journal of Grape and Wine Research* 4(1), 23–29.

Poling, E.B. (2007) Spring frost control. In: Poling, E.B. (ed.) *North Carolina Winegrape Grower's Guide*. North Carolina Cooperative Extension Service, Raleigh, North Carolina, 196 pp.

Pollock, J.G., Shepardson, E.S., Shaulis, N.J. and Crowe, D.E. (1977) Mechanical pruning of American hybrid grapevines. *Transactions of the American Society of Agricultural Engineers* 20, 817–821.

Poni, S., Bernizzoni, F., Briola, G. and Cenni, A. (2005) Effects of early leaf removal on cluster morphology, shoot efficiency and grape quality in two *Vitis vinifera* cultivars. *Acta Horticulturae* 689, 217–226.

Pool, R.M. (1987) Thin grapes mechanically. *Western Fruit Grower* 107, 17–19.

Pool, R.M., Dunst, R.M., Kamas, J.S., Gunkel, W.W., Lakso, A.N. and Goffinet, M.C. (1990) *Shoot Positioning Native American (Concord-type) Grapevines*. New York's Food and Life Sciences Bulletin No. 131, Cornell University Agricultural Experiment Station, Ithaca, New York, 5 pp.

Pool, R.M., Dunst, D.C., Crowe, H., Hubbard, G.E., Howard, G.E. and DeGolier, G. (1993) Predicting and controlling crop on machine or minimal pruned grapevines. *Proceedings of the 2nd New Jersey Shaulis Grape Symposium on Pruning Mechanization and Crop Control*, Fredonia, New York. New York State Agriculture Experiment Station, Geneva, New York, pp. 31–45.

Possingham, J.V. and Groot Obbink, J. (1971) Endotrophic mycorrhiza and the nutrition of grape vines. *Vitis* 10, 120–130.

Possingham, J.V., Chambers, T.C., Radler, F. and Grncarevic, M. (1967) Cuticular transpiration and wax structure and composition of leaves and fruit of *Vitis vinifera*. *Australian Journal of Biological Science* 20, 1149–1153.

Pouget, R. and Delas, J. (1989) Le choix des porte-greffes de la vigne pour une production de qualité. *Connaisance Vigne et du Vin Hors Série* 23, 27–31.

Powles, S.B., Lorraine-Colwill, D.F., Dellow, J.J. and Preston, C. (1998) Evolved resistance to glyphosate in rigid ryegrass (*Lolium rigidum*) in Australia. *Weed Science* 46(5), 604–607.

Pratt, C. (1971) Reproductive anatomy in cultivated grapes – a review. *American Journal of Enology and Viticulture* 22(2), 92–109.

Pratt, C. (1979) Shoot and bud development during the prebloom period of *Vitis*. *Vitis* 18(1), 1–5.

Pratt, C. and Coombe, B.G. (1978) Shoot growth and anthesis in *Vitis*. *Vitis* 17, 125–133.

Prial, F.J. (1987) Wine; stealing from thieves. *The New York Times* March 15, 1987.

Price, S.F., Breen, P.J., Valladao, M. and Watson, B.T. (1995) Cluster sun exposure and quercetin in Pinot noir grapes and wine. *American Journal of Enology and Viticulture* 46, 187–194.

Price, S.F., Watson, B.T. and Valladao, M. (1996) Vineyard and winery effects on wine phenolics – flavonols in Oregon Pinot noir. *Proceedings of the 9th Australian Wine*

Industry Technical Conference, Adelaide, South Australia. Winetitles, Underdale, South Australia, pp. 93–97.

Prins, T.W., Tudzynski, P., von Tiedemann, A., Tudzynski, B., ten Have, A., Hansen, M.E., Tenberge, K. and van Kan, J.A.L. (2000) Infection strategies of *Botrytis cinerea* and related necrotrophic pathogens. In: Kronstad, J. (ed.) *Fungal Pathology*. Kluwer Academic Publishers, Dordrecht, Netherlands, pp. 32–64.

Pscheidt, J.W. and Pearson, R.C. (1989) Effect of grapevine training systems and pruning practices on occurrence of Phomopsis cane and leaf spot. *Plant Disease* 73(10), 825–828.

Purcell, A.H. and Hopkins, D.L. (1996) Fastidious xylem-limited bacterial plant pathogens. *Annual Review of Phytopathology* 34(1), 131–151.

Quinlan, J.D. and Weaver, R.J. (1970) Modification of pattern of photosynthate movement within and between shoots of *Vitis vinifera* L. *Plant Physiology* 46(4), 527–530.

Raju, B.C., Nome, S.F., Docampo, D.M., Goheen, A.C. and Nyland, G. (1980) Alternative hosts of Pierce's disease of grapevines that occur adjacent to grape growing areas in California. *American Journal of Enology and Viticulture* 31(2), 144–148.

Raski, D.J., Hewitt, W.B., Goheen, A.C., Taylor, C.E. and Taylor, R.H. (1965) Survival of *Xiphinema Index* and reservoirs of fanleaf virus in fallowed vineyard soil. *Nematologica* 11(3), 349–352.

Reedy, R.C. and Scanlon, B.R. (2003) Soil water content monitoring using electromagnetic induction. *Journal of Geotechnical and Geoenvironmental Engineering* 129(11), 1028–1039.

Reichard, D.L., Zhu, H., Fox, R.D. and Brazee, R.D. (1992) Computer simulation of variables that influence spray drift. *Transactions of the American Society of Agricultural Engineers* 35(5), 1401–1407.

Reisenzein, H., Pfeffer, M., Aust, G. and Baumgarten, A. (2007) The influence of soil properties on the development of grape phylloxera populations in Austrian viticulture. *Acta Horticulturae* 733, 13–23.

Renault, A.S., Deloire, A. and Bierne, J. (1996) Pathogenesis-related proteins in grapevines induced by salicylic acid and *Botrytis cinerea*. *Vitis* 35, 49–52.

Reuveni, M. and Reuveni, R. (1995) Efficacy of foliar sprays of phosphates in controlling powdery mildews in field-grown nectarine, mango trees and grapevines. *Crop Protection* 14(4), 311–314.

Reuveni, M. and Reuveni, R. (2002) Mono-potassium phosphate fertilizer (PeaK) – a component in integrated control of powdery mildews in fruit trees and grapevines. *Acta Horticulturae* 594, 619–625.

Reynolds, A.G. (1988) Response of Okanagan Riesling vines to training system and simulated mechanical pruning. *American Journal of Enology and Viticulture* 39(3), 205–212.

Ribereau-Gayon, J. and Ribereau-Gayon, P. (1958) The anthocyans and leucoanthocyans of grapes and wines. *American Journal of Enology and Viticulture* 9(1), 1–9.

Richardson, E.A., Seeley, S.D., Walker, D.R., Anderson, J.L. and Ashcroft, G.L. (1975) Pheno-climatography of spring peach bud development. *HortScience* 10, 236–237.

Rieger, M. (1989) Freeze protection for horticultural crops. *Horticultural Reviews* 11, 45–109.

Ristic, R. and Iland, P.G. (2005) Relationships between seed and berry development of *Vitis vinifera* L. cv Shiraz: developmental changes in seed morphology and phenolic composition. *Australian Journal of Grape Wine Research* 11(1), 43–58.

Ritchie, G.A. and Hinckley, T.M. (1975) The pressure chamber as an instrument for ecological research. *Advances in Ecological Research* 9, 165–253.

Robinson, J. and Harding, J. (2006) *The Oxford Companion to Wine*. Oxford University Press, Oxford, UK.

Robinson, J.B. (1992) Grapevine nutrition. In: Coombe, B.G. and Dry, P.R. (ed.) *Viticulture Volume 2. Practices*. Winetitles, Underdale, South Australia, pp. 178–208.

Robinson, J.B., Nicholas, P.R. and McCarthy, J.R. (1978) A comparison of three methods of tissue analysis for assessing the nutrient status of plantings of *Vitis vinifera* in an irrigated area in South Australia. *Australian Journal of Experimental Agriculture* 18(91), 294–300.

Robinson, S.P., Jacobs, A.K. and Dry, I.B. (1997) A class IV chitinase is highly expressed in grape berries during ripening. *Plant Physiology* 114, 771–778.

Roby, G. and Matthews, M.A. (2004) Relative proportions of seed, skin and flesh, in ripe berries from Cabernet Sauvignon grapevines grown in a vineyard either well irrigated or under water deficit. *Australian Journal of Grape and Wine Research* 10, 74–82.

Roper, M. and Gupta, V. (1995) Management practices and soil biota. *Australian Journal of Soil Research* 33(2), 321–339.

Rosenquist, J.K. and Morrison, J.C. (1989) Some factors affecting cuticle and wax accumulation on grape berries. *American Journal of Enology and Viticulture* 40(4), 241–244.

Rosslenbroich, H.-J. and Stuebler, D. (2000) *Botrytis cinerea* – history of chemical control and novel fungicides for its management. *Crop Protection* 19(8–10), 557–561.

Rumbolz, J., Wirtz, S., Kassemeyer, H.-H., Guggenheim, R., Schäfer, E. and Büche, C. (2002) Sporulation of *Plasmopara viticola*: differentiation and light regulation. *Plant Biology* 4, 413–422.

Ryle, G.J.A. and Hesketh, J.D. (1969) Carbon dioxide uptake in nitrogen-deficient plants. *Crop Science* 9(4), 451–454.

Saayman, D. and Van Huyssteen, L. (1983) Preliminary studies on the effect of a permanent cover crop and root pruning on an irrigated Colombar vineyard. *South African Journal of Enology and Viticulture* 4, 7–12.

Sampson, B., Noffsinger, S., Gupton, C. and Magee, J. (2001) Pollination biology of the muscadine grape. *HortScience* 36, 120–124.

Sams, C.E. and Deyton, D.E. (2002) Botanical and fish oils: history, chemistry, refining, formulation and current uses. In: Beattie, G.A.C., Watson, D.M., Stevens, M.L., Rae, D.J. and Spooner-Hart, R.N. (eds) *Spray Oils Beyond 2000: Sustainable Pest and Disease Management*. University of Western Sydney, Sydney, Australia, pp. 19–28.

Santiago, J.L., Boso, S., Martinez, M.d.-C., Pinto-Carnide, O. and Ortiz, J.M. (2005) Ampelographic comparison of grape cultivars (*Vitis vinifera* L.) grown in Northwestern Spain and Northern Portugal. *American Journal of Enology and Viticulture* 56(3), 287–290.

Sartorato, I., Zanin, G., Baldoin, C. and Zanche, C. (2006) Observations on the potential of microwaves for weed control. *Weed Research* 46(1), 1–9.

Sato, A. and Yamada, M. (2003) Berry texture of table, wine, and dual-purpose grape cultivars quantified. *HortScience* 38(4), 578–581.

Saxton, V.P., Creasy, G.L., Paterson, A.M. and Trought, M.C.T. (2004a) Experimental method to investigate and monitor bird behavior and damage in vineyards. *American Journal of Enology and Viticulture* 55(3), 288–291.

Saxton, V.P., Hickling, G.J., Trought, M.C.T. and Creasy, G.L. (2004b) Comparative behavior of free-ranging blackbirds (*Turdus merula*) and silvereyes (*Zosterops lateralis*) with hexose sugars in artificial grapes. *Applied Animal Behaviour Science* 85, 157–166.

Scheck, H.J., Vasquez, S.J., Gubler, W.D. and Foggle, D. (1998) Grape growers report losses to black-foot and young vine decline. *California Agriculture* 54, 19–23.

Schilder, A.M.C., Smokevitch, S.M., Catal, M. and Mann, W.K. (2005) First report of anthracnose caused by *Elsinoë ampelina* on grapes in Michigan. *Plant Disease* 89(9), 1011.

Schnabel, B.J. and Wample, R.L. (1987) Dormancy and cold hardiness in *Vitis vinifera* L. cv. White Riesling as influenced by photoperiod and temperature. *American Journal of Enology and Viticulture* 38(4), 265–272.

Scholander, P.F., Bradstreet, E.D., Hemmingsen, E.A. and Hammel, H.T. (1965) Sap pressure in vascular plants: negative hydrostatic pressure can be measured in plants. *Science* 148(3668), 339–346.

Schreiber, M.M. (1992) Influence of tillage, crop rotation, and weed management on giant foxtail (*Setaria faberi*) population dynamics and corn yield. *Weed Science* 40(4), 645–653.

Schreiner, R.P. (2002) Seasonal dynamics of mineral uptake and allocation in whole Pinot noir vines in a Red Hill soil. *Proceedings of the Oregon Wine Industry Symposium*, Oregon State University, Corvallis, Oregon.

Schreiner, R.P., Scagel, C.F. and Baham, J. (2006) Nutrient uptake and distribution in a mature 'Pinot noir' vineyard. *HortScience* 41, 336–345.

Schubert, A., Lovisolo, C. and Peterlunger, E. (1999) Shoot orientation affects vessel size, shoot hydraulic conductivity and shoot growth rate in *Vitis vinifera* L. *Plant, Cell & Environment* 22(2), 197–204.

Schultz, H.R. (2000) Climate change and viticulture: a European perspective on climatology, carbon dioxide and UV-B effects. *Australian Journal of Grape and Wine Research* 6(1), 2–12.

Schultz, H.R. (2003) Differences in hydraulic architecture account for near-isohydric and anisohydric behaviour of two field-grown *Vitis vinifera* L. cultivars during drought. *Plant, Cell and Environment* 26(8), 1393–1405.

Schwarz, M., Picazo-Bacete, J.J., Winterhalter, P. and Hermosin-Gutierrez, I. (2005) Effect of copigments and grape cultivar on the color of red wines fermented after the addition of copigments. *Journal of Agricultural and Food Chemistry* 53(21), 8372–8381.

Scienza, A., Miravalle, R., Visai, C. and Fregoni, M. (1978) Relationships between seed number, gibberellin and abscisic acid levels and ripening in Cabernet Sauvignon grape berries. *Vitis* 17, 361–368.

Scott, K.D., Ablett, E.M., Lee, L.S. and Henry, R.J. (2000) AFLP markers distinguishing an early mutant of Flame Seedless grape. *Euphytica* 113(3), 243–247.

Settle, W.H., Wilson, L.T., Flaherty, D.L. and English-Loeb, G.M. (1986) The variegated leafhopper, an increasing pest of grapes. *California Agriculture* 40, 30–32.

Seyb, A.M. (2004) Botrytis Cinerea *Inoculum Sources in the Vineyard System*. Soil Plant and Ecological Sciences Division, Lincoln University, Christchurch, New Zealand, 227 pp.

Shaulis, N.J. (1961) Associations between symptoms of potassium deficiency, plant analysis, growth and yield of Concord grapes. *American Institute of Biological Science* 8, 44–57.

Shaulis, N., Kimball, K. and Tomkins, J.P. (1953) The effect of trellis height and training systems on the growth and yield of Concord grapes under a controlled pruning severity. *Proceedings of the American Society for Horticultural Science* 62, 221–227.

Shaulis, N.J., Shepardson, E.S. and Jordan, T.D. (1967) *The Geneva Double Curtain for Vigorous Grapevines: Vine Training and Trellis Construction.* New York State Agricultural Experiment Station, Geneva, New York, 12 pp.

Shaulis, N.J., Pollock, J., Crowe, D. and Shepardson, E.S. (1973) Mechanical pruning of grapevines: progress 1968–1972. *Proceedings of the New York State Horticultural Science* 118, 61–69.

Shaw, A.B. (2001) Pelee Island and Lake Erie north shore, Ontario: a climatic analysis of Canada's warmest wine region. *Journal of Wine Research* 12(1), 19–37.

Shepardson, E.S., Studer, H.E., Shaulis, N.J. and Moyer, J.C. (1962) Mechanical grape harvesting research at Cornell III. *Journal of the American Society of Agricultural Engineers* 43(2), 66–71.

Shepardson, E.S., Markwardt, E.D., Millier, W.F. and Rehkugler, G.E. (1970) *Mechanical Harvesting of Fruits and Vegetables.* New York's Food and Life Sciences Bulletin No. 5, Cornell University Agricultural Experiment Station, Ithaca, New York, 12 pp.

Shorrocks, V.M. (1997) The occurrence and correction of boron deficiency. *Plant and Soil* 193(1), 121–148.

Shulman, Y., Nir, G., Fanberstein, L. and Lavee, S. (1983) The effect of cyanamide on the release from dormancy of grapevine buds. *Scientia Horticulturae* 19, 97–104.

Singleton, V.L. and Draper, D.E. (1964) The transfer of polyphenolic compounds from grape seeds into wines. *American Journal of Enology and Viticulture* 15(1), 34–40.

Singleton, V.L. and Esau, P. (1969) Phenolic substances in grapes and wines, and their significance. *Advances in Food Research, Supplement* 1, 1–261.

Singleton, V.L. and Trousdale, E.K. (1992) Anthocyanin–tannin interactions explaining differences in polymeric phenols between white and red wines. *American Journal of Enology and Viticulture* 43, 63–70.

Skene, K.G.M. (1968) Increase in the levels of cytokinins in bleeding sap of *Vitis vinifera* L. after CCC treatment. *Science* 159, 1477–1478.

Skene, K.G.M. and Kerridge, G.H. (1967) Effect of root temperature on cytokinin activity in root exudate of *Vitis vinifera* L. *Plant Physiology* 42, 1131–1139.

Smart, R.E. (1982) Vine manipulation to improve wine grape quality. *Proceedings of the UC Davis Grape and Wine Centennial Symposium*, Davis, California, pp. 362–375.

Smart, R.E. (1988) Shoot spacing and canopy light microclimate. *American Journal of Enology and Viticulture* 39(4), 325–333.

Smart, R.E. (1996) The rain in Spain falls . . . Who cares anymore, now we can irrigate! *Australian & New Zealand Wine Industry Journal* 11(3), 238–241.

Smart, R.E. (1999) Trees in vineyards. *Practical Winery & Vineyard* July/August, 82–84.

Smart, R.E. (2003) Portugese homoclimes in Australia. *Australian & New Zealand Wine Industry Journal* 18(1), 48–50.

Smart, R.E. and Coombe, B.G. (1999) Clone. In: Robinson, J. (ed.) *The Oxford Companion to Wine.* Oxford University Press, Oxford, UK, pp. 185.

Smart, R.E. and Robinson, M. (1991) *Sunlight into Wine: Handbook for Wine Grape Canopy Management.* Winetitles, Underdale, South Australia, 96 pp.

Smart, R.E. and Sinclair, T.R. (1976) Solar heating of grape berries and other spherical fruits. *Agricultural Meteorology* 17, 241–259.

Smart, R.E., Smith, S.M. and Winchester, R.V. (1988) Light quality and quantity effects on fruit ripening for Cabernet Sauvignon. *American Journal of Enology and Viticulture* 39, 250–258.

Smilanick, J.L., Hartsell, P., Henson, D., Fouse, D.C., Assemi, M. and Harris, C.M. (1990) Inhibitory activity of sulfur dioxide on the germination of spores of *Botrytis cinerea*. *Phytopathology* 80(2), 217–220.

Smith, B.R. and Cheng, L. (2007) Iron assimilation and carbon metabolism in 'Concord' grapevines grown at different pHs. *Journal of the American Society for Horticultural Science* 132, 473–483.

Smith, C.B., Fleming, H.K., Kardos, L.T. and Haeseler, C.W. (1972) Responses of Concord grapevines to lime and potassium. *Pennsylvania State University Agriculture Experimental Station Bulletin* 785, University Park, Pennsylvania.

Smith, L. (2002) Site selection for establishment and management of vineyards. *14th Annual Colloquium of the Spatial Information Research Centre*, University of Otago, Dunedin, New Zealand.

Sniegowski, P.D., Dombrowski, P.G. and Fingerman, E. (2002) *Saccharomyces cerevisiae* and *Saccharomyces paradoxus* coexist in a natural woodland site in North America and display different levels of reproductive isolation from European conspecifics. *FEMS Yeast Research* 1(4), 299–306.

Snyder, R.L., Tau Paw, K. and Thompson, J.F. (1992) *Passive Frost Protection of Trees and Vines*. Cooperative Extension University of California, Division of Agriculture and Natural Resources, Berkeley, California, 9 pp.

Somers, C. (1998) *The Wine Spectrum*. Winetitles, Underdale, South Australia, 136 pp.

Somers, T. (1971) The polymeric nature of wine pigments. *Phytochemistry* 10, 2175–2186.

Somers, T.C., Evans, M.E. and Cellier, K.M. (1983) Red wine quality and style: diversities of composition and adverse influences from free SO_2. *Vitis* 22, 348–356.

Sommer, K.J., Islam, M.T. and Clingeleffer, P.R. (2000) Light and temperature effects on shoot fruitfulness in *Vitis vinifera* L. cv. Sultana: influence of trellis type and grafting. *Australian Journal of Grape and Wine Research* 6, 99–108.

Souquet, J.M., Cheynier, V., Brossaud, F. and Moutounet, M. (1996) Polymeric proanthocyanidins from grape skins. *Phytochemistry* 43, 509–512.

Southy, J.M. (1992) Grapevine rootstock performance under diverse conditions in South Africa. *Rootstock Seminar: a Worldwide Perspective*, Reno, Nevada. American Society for Enology and Viticulture, Davis, California.

Spayd, S.E., Wample, R.L., Stevens, R.G., Evans, R.G. and Kawakami, A.K. (1993) Nitrogen fertilization of White Riesling in Washington: effects on petiole nutrient concentration, yield, yield components, and vegetative growth. *American Journal of Enology and Viticulture* 44(4), 378–386.

Spayd, S.E., Tarara, J.M., Mee, D.L. and Ferguson, J.C. (2002) Separation of sunlight and temperature effects on the composition of *Vitis vinifera* cv. Merlot berries. *American Journal of Enology and Viticulture* 53(3), 171–182.

Sperry, J.S., Holbrook, N.M., Zimmermann, M.H. and Tyree, M.T. (1987) Spring filling of xylem vessels in wild grapevine. *Plant Physiology* 83(2), 414–417.

Srinivasan, C. and Mullins, M.G. (1980) Effects of temperature and growth regulators on formation of anlagen, tendrils and inflorescences in *Vitis vinifera* L. *Australian Journal of Experimental Agriculture* 45, 439–446.

Srinivasan, C. and Mullins, M.G. (1981) Physiology of flowering in the grapevine – a review. *American Journal of Enology and Viticulture* 32, 47–63.

Stafne, E.T. and Carroll, B. (2006) *Rootstocks for Grape Production*. Oklahoma State University, Stillwater, Oklahoma, 4 pp.

Stapleton, J.J., Barnett, W.W., Marois, J.J. and Gubler, W.D. (1990) Leaf removal for pest management in wine grapes. *California Agriculture* 44, 15–17.

Staub, T. (1991) Fungicide resistance: practical experience with antiresistance strategies and the role of integrated use. *Annual Review of Phytopathology* 29(1), 421–442.

Staudt, G. (1982) Pollen germination and pollen tube growth *in vivo* and the dependence on temperature (m. engl. Zus.). *Vitis* 21, 205–216.

Staudt, G. (1999) Opening time of flowers and time of anthesis in grapevines, *Vitis vinifera* L. *Vitis* 38(1), 15–20.

Steudle, E. and Meshcheryakov, A.B. (1996) Hydraulic and osmotic properties of oak roots. *Journal of Experimental Botany* 47(3), 387–401.

Stoll, M., Loveys, B. and Dry, P. (2000) Hormonal changes induced by partial rootzone drying of irrigated grapevine. *Journal of Experimental Botany* 51(350), 1627–1634.

Striegler, K. and Howell, G.S. (1991) The influence of rootstock on the cold hardiness of Seyval grapevines. I. Primary and secondary effects on growth, canopy development, yield, fruit quality and cold hardiness. *Vitis* 30, 1–10.

Studer, H.E. and Olmo, H.P. (1971) The severed cane technique and its application to mechanical harvesting of raisin grapes. *Transactions of the American Society of Agricultural Engineers* 14, 38–43.

Sullivan, C.L. (2003) *Zinfandel: a History of a Grape and Its Wine*. University of California Press, Berkeley, California.

Sun, B.S., Pinto, T., Leandro, M.C., Ricardo-da-Silva, J.M. and Springer, M.I. (1999) Transfer of catechins and proanthocyanidins from solids parts of the grape cluster into wine. *American Journal of Enology and Viticulture* 50, 179–184.

Swain, T. and Bate-Smith, E.C. (1962) Flavonoid compounds. I. In: Florkin, F.M. and Mason, H.S. (eds) *Comparative Biochemistry*. Academic Press, London, pp. 755–809.

Swanepoel, J.J. and Archer, E. (1988) The ontogeny and development of *Vitis vinifera* L. cv. Chenin blanc inflorescence in relation to phenological stages. *Vitis* 27, 133–141.

Swartwout, H.G. (1925) Fruiting habit of the grape. *Proceedings of the American Society for Horticultural Science* 22, 70–74.

Szulmayer, W. (1971) From sun-drying to solar dehydration II. *Food Technology in Australia* 29(September), 494–501.

Tagliavini, M. and Rombolà, A.D. (2001) Iron deficiency and chlorosis in orchard and vineyard ecosystems. *European Journal of Agronomy* 15, 71–92.

Tai, Y., Liu, P., Zhu, B. and Li, Z. (2005) Experiment of control of grape anthracnose disease in the field. *China Fruits* 3, 31–33.

Tanton, T.W. and Crowdy, S.H. (1972) Water pathways in higher plants: II. Water pathways in roots. *Journal of Experimental Botany* 23(3), 600–618.

Taylor, L.P. and Hepler, P.K. (1997) Pollen germination and tube growth. *Annual Review of Plant Physiology and Plant Molecular Biolology* 48, 461–491.

Teissedre, P.L., Frankel, E.N., Waterhouse, A.L., Peleg, H. and German, J.B. (1996) Inhibition *of in vitro* human LDL oxidation by phenolic antioxidants from grapes and wine. *Journal of the Science of Food and Agriculture* 70, 55–61.

Thellier, M., Duval, Y. and Demarty, M. (1979) Borate exchanges of *Lemna minor* L. as studied with the help of the enriched stable isotopes and of a (n,{alpha}) nuclear reaction. *Plant Physiology* 63(2), 283–288.

Thind, S.K., Monga, P.K., Kaur, N. and Arora, J.K. (2001) Efficacy of fungicides in controlling grape anthracnose. *Journal of Mycology and Plant Pathology* 31, 211–212.

Thomas, B. and Schapel, A. (2003) Soil analysis in established vineyards – the importance of collecting the right samples. *Australian Grapegrower and Winemaker* 478, 41–43.

Thomas, C.S., Marois, J.J. and English, J.T. (1988) The effects of wind speed, temperature, and relative humidity on development of aerial mycelium and conidia of *Botrytis cinerea* on grape. *Phytopathology* 78(3), 260–265.

Thomas, C.S., Gubler, W.D., Silacci, M.W. and Miller, R. (1993) Changes in elemental sulfur residues on Pinot noir and Cabernet Sauvignon grape berries during the growing season. *American Journal of Enology and Viticulture* 44(2), 205–210.

Thompson, J., Cantwell, M., Arpaia, M.L., Kader, A., Crisosto, C. and Smilanick, J. (2001) Effect of cooling delays on fruit and vegetable quality. *Perishables Handling Quarterly*, Department of Plant Science, University of California, Davis, California, pp. 1–4.

Tipton, S., Morris, J., Main, G., Sharp, C. and McNew, R. (1999) Grape juice as an extender and sweetener for blueberry drinks. *Journal of Food Quality* 22(3), 275–285.

Tollner, E.W. and Moss, R.B. (eds) (1988) Neutron probe vs. tensiometers vs. gypsum blocks for monitoring soil moisture status. In: *Sensors and Techniques for Irrigation Management*. Center for Irrigation Technology, California State University, Fresno, California, pp. 95–112.

Tominaga, T., Baltenweck-Guyot, R., Gachons, C.P.D. and Dubourdieu, D. (2000) Contribution of volatile thiols to the aromas of white wines made from several *Vitis vinifera* grape varieties. *American Journal of Enology and Viticulture* 51(2), 178–181.

Trjapitzin, S.V. and Trjapitzin, V.A. (1999) Parasites of mealybugs (Homoptera: Pseudococcidae) on cultivated grapes in Argentina, with description of a new species of the genus *Aenasius* Walker (Hymenoptera: Encyrtidae) *Entomological Review* 79(4), 386–390.

Trought, M.C.T. (1997) The New Zealand terroir: sources of variation in fruit composition in New Zealand vineyards. *Proceedings of the Fourth International Symposium for Cool Climate Viticulture and Enology*, Rochester, New York. New York State Agricultural and Experimental Station, Geneva, New York, pp. I, 23–27.

Trought, M.C.T. and Tannock, S.J.C. (1996) Berry size and soluble solids variation within a bunch of grapes. *Proceedings of the Fourth International Symposium for Cool Climate Viticulture and Enology*, New York State Agricultural and Experimental Station, Geneva, New York, pp. V, 70–73.

Trought, M.C.T., Howell, G.S. and Cherry, N. (1999) *Practical Considerations for Reducing Frost Damage in Vineyards*. Report to New Zealand Winegrowers, Lincoln University, Christchurch, New Zealand, 44 pp.

Tukey, R.B. and Clore, W.J. (1972) Grapes – their characteristics and suitability for production in Washington. Cooperative Extension Service, College of Agriculture, Washington State University, Pullman, Washington, 12 pp.

Unwin, T. (1994) European wine sector policy and the UK wine industry. *Journal of Wine Research* 5(2), 135–146.

Uva, R.H., Neal, J.C. and DiTomaso, J.M. (1997) *Weeds of the Northeast*. Cornell University Press, Ithaca, New York.

Vaia, R. and McDaniel, M. (1996) Sensory effects of quercetin in a model wine and a Chardonnay. *Food Quality Preference* 7(3–4), 339–340.

van der Vlugt-Bergmans, C.J.B., Brandwagt, B.F., van't Klooster, J.M., Wagemakers, C.A.M. and van Kan, J.A.L. (1993) Genetic variation and segregation of DNA polymorphisms in *Botrytis cinerea*. *Mycological Research* 97(10), 1193–1200.

Van Leeuwen, C. and Seguin, G. (2006) The concept of terroir in viticulture. *Journal of Wine Research* 17(1), 1–10.

Varela, L.G., Smith, R.J. and Phillips, P.A. (2001) *Pierce's Disease*. University of California, Agriculture and Natural Resources, Oakland, California.

Velasco, R., Zharkikh, A., Troggio, M., Cartwright, D.A., Cestaro, A., Pruss, D., Pindo, M., FitzGerald, L.M., Vezzulli, S., Reid, J. *et al.* (2007) A high quality draft consensus sequence of the genome of a heterozygous grapevine variety. *PLoS ONE* 2(12), e1326.

Velicheti, R.K., Lamison, C., Brill, L.M. and Sinclair, J.B. (1993) Immunodetection of Phomopsis species in asymptomatic soybean plants. *Plant Disease* 77, 70–73.

VerCauteren, K.C., Lavelle, M.J. and Hygnstrom, S. (2006) Fences and deer-damage management: a review of designs and efficacy. *Wildlife Society Bulletin* 34(1), 191–200.

Versari, A., Parpinello, G.P., Tornielli, G.B., Ferrarini, R. and Giulivo, C. (2001) Stilbene compounds and stilbene synthase expression during ripening, wilting, and UV treatment in grape cv. Corvina. *Journal of Agricultural and Food Chemistry* 49(11), 5531–5536.

Viala, P. and Vermorel, V. (1909) Traité général de viticulture. In: *Ampelographie*. Masson, Paris.

Vidal, J., Kikkert, J., Malnoy, M., Wallace, P., Barnard, J. and Reisch, B. (2006) Evaluation of transgenic 'Chardonnay' (*Vitis vinifera*) containing magainin genes for resistance to crown gall and powdery mildew. *Transgenic Research* 15(1), 69–82.

Vrhovsek, U., Mattivi, F. and Waterhouse, A.L. (2001) Analysis of red wine phenolics: comparison of HPLC and spectrophotometric methods. *Vitis* 40, 87–91.

Wainwright, S.J. and Woolhouse, H.W. (1977) Some physiological aspects of copper and zinc tolerance in *Agrostis tenuis* Sibth.: cell elongation and membrane damage. *Journal of Experimental Botany* 28(4), 1029–1036.

Wake, C.M.F. and Fennell, A. (2000) Morphological, physiological and dormancy responses of three *Vitis* genotypes to short photoperiod. *Physiologia Plantarum* 109(2), 203–210.

Walker, A.R., Lee, E., Bogs, J., McDavid, D.A.J., Thomas, M.R. and Robinson, S.P. (2007) White grapes arose through the mutation of two similar and adjacent regulatory genes. *Plant Journal* 49(5), 772–785.

Walton, V.M. and Pringle, K.L. (2004) Vine mealybug, *Planococcus ficus* (Signoret) (Hemiptera: Pseudococcidae), a key pest in South African vineyards. A review. *South African Journal of Enology and Viticulture* 25(2), 54–62.

Wample, R.L., Shoemake, J. and Mills, L. (2000) Micrometeorology characteristics of 'grow-tubes' and their relationship to the growth and development of Cabernet Sauvignon and Chardonnay vines. *Proceedings of the Fifth International Symposium for Cool Climate Viticulture and Oenology*, Melbourne, Australia.

Wang, S., Okamoto, G., Hirano, K., Lu, J. and Zhang, C. (2001) Effects of restricted rooting volume on vine growth and berry development of Kyoho grapevines. *American Journal of Enology and Viticulture* 52(3), 248–253.

Waring, R.H. and Cleary, B.D. (1967) Plant moisture stress: evaluation by pressure bomb. *Science* 155(3767), 1248–1254.

Weaver, R.J. and McCune, S.B. (1959) Response of certain varieties of *Vitis vinifera* to gibberellin. *Hilgardia* 28, 297–350.

Weaver, R.J. and Montgomery, R. (1974) Effect of ethephon on coloration and maturation of wine grapes. *American Journal of Enology and Viticulture* 25(1), 39–41.

Wertheim, S.J. (2000) Developments in the chemical thinning of apple and pear. *Plant Growth Regulation* 31(1–2), 85–100.

Westphal, A., Browne, G.T. and Schneider, S. (2002) Evidence for biological nature of the grape replant problem in California. *Plant and Soil* 242(2), 197–203.

Westwood, M.N. (1993) *Temperate Zone Pomology. Physiology and Culture.* Timber Press, Portland, Oregon.

Whalley, W.R., Cope, R.E., Nicholl, C.J. and Whitmore, A.P. (2004) In-field calibration of a dielectric soil moisture meter designed for use in an access tube. *Soil Use Management* 20, 203–206.

Whiting, J.R. (1992) Harvesting and drying of grapes. In: Coombe, B.G. and Dry, P.R. (eds) *Viticulture Volume 2. Practices.* Winetitles, Underdale, South Australia, pp. 328–358.

Whiting, J.R. and Buchanan, G.A. (1992) Evaluation of rootstocks for phylloxera infested vineyards in Australia. *Rootstock Seminar: a Worldwide Perspective*, Reno, Nevada, American Society for Enology and Viticulture, Davis, California, pp. 15–26.

Whiting, J.R., Buchanan, G.A. and Edwards, M.E. (1987) Assessment of rootstocks for wine grape production. *Proceedings of the 6th Australian Wine Industry Technical Conference*, Adelaide, South Australia. Australian Industrial Publishers, Adelaide, South Australia, pp. 184–190.

Wicks, T., Hitch, C. and Emmett, B. (2003) Effects of temperature and application rates on the efficacy of sulphur for powdery mildew control. In: Emmett, R.W. (ed.) *Strategic Use of Sulphur in Integrated Pest and Disease Management (IPM) Programs for Grapevines.* Victoria Dept. of Primary Industries, Mildura, Victoria, Australia, pp. 86–109.

Wildenradt, H.L., Christensen, E.N., Stackler, B., Caputi, A., Jr., Slinkard, K. and Scutt, K. (1975) Volatile constituents of grape leaves. I. *Vitis vinifera* var. 'Chenin Blanc'. *American Journal of Enology and Viticulture* 26(3), 148–153.

Wilkins, M.B. (1966) Geotropism. *Annual Review of Plant Physiology* 17(1), 379–408.

Williams, C.F. (1923) Hybridization of *Vitis rotundifolia*. Inheritance of anatomical stem characters. *North Carolina Agricultural Station Technical Bulletin* 23.

Williams, C.M.J., Maier, N.A. and Bartlett, L. (2004) Effect of molybdenum foliar sprays on yield, berry size, seed formation, and petiolar nutrient composition of 'Merlot' grapevines. *Journal of Plant Nutrition* 27(11), 1891–1916.

Williams, D.W., Andris, H.L., Beede, R.H., Luvisi, D.A., Norton, M.V.K. and Williams, L.E. (1985) Validation of a model for the growth and development of the Thompson Seedless grapevine. II. Phenology. *American Journal of Enology and Viticulture* 36(4), 283–289.

Williams, L.E. (1996) 36. Grape. In: *Photoassimilate Distribution in Plants and Crops Source–Sink Relationships.* Marcel Dekker, Inc., New York, pp. 851–882.

Williams, L.E. and Araujo, F.J. (2002) Correlations among predawn leaf, mid-day leaf and mid-day stem water potential and their correlations with other measures of soil and plant water status in *Vitis vinifera*. *Journal of the American Society for Horticultural Science* 127(3), 448–454.

Williams, L.E. and Trout, T.J. (2005) Relationships among vine- and soil-based measures of water status in a Thompson Seedless vineyard in response to high-frequency drip irrigation. *American Journal of Enology and Viticulture* 56(4), 357–366.

Williams III, L. and Martinson, T.E. (2000) Colonization of New York vineyards by *Anagrus* spp. (Hymenoptera: Mymaridae): overwintering biology, within-vineyard distribution of wasps, and parasitism of grape leafhopper, *Erythroneura* spp. (Homoptera: Cicadellidae), eggs. *BioControl* 18(2), 136–146.

Williams, P.J., Strauss, C.R. and Wilson, B. (1980) New linalool derivatives in Muscat of Alexandria grapes and wines. *Phytochemistry* 19, 1137–1139.

Williamson, P.M., Sivasithamparam, K. and Cowling, W.A. (1991) Formation of subcuticular hyphae by *Phomopsis leptostromiformis* upon latent infection of narrow-leafed lupins. *Plant Disease* 75, 1023–1025.

Willocquet, L., Colombet, D., Rougier, M., Fargues, J. and Clerjeau, M. (1996) Effects of radiation, especially ultraviolet B, on conidial germination and mycelial growth of grape powdery mildew. *European Journal of Plant Pathology* 102(5), 441–449.

Willstätter, R. and Zollinger, E.H. (1915) Untersuchungen über die anthocyane: VI. Über die farbstoffe der weintraube und der heidelberre. *Liebigs Annalen* 408, 83–109.

Wilson, B., Strauss, C.R. and Williams, P.J. (1986) The distribution of free and glycosidically bound monoterpenes among skin, juice, and pulp fractions of some white grape varieties. *American Journal of Enology and Viticulture* 37(2), 107–111.

Winkler, A.J. (1926) Some responses of *Vitis vinifera* to pruning. *Hilgardia* 1, 525–543.

Winkler, A.J. (1932) Maturity tests for table grapes. *California Agricultural Experiment Station Bulletin* 529, 1–35.

Winkler, A.J. and Williams, W.O. (1935) Effect of seed development on the growth of grapes. *Proceedings of the American Society for Horticultural Science* 33, 430–434.

Winkler, A.J. and Williams, W.O. (1945) Starch and sugars of *Vitis vinifera*. *Plant Physiology* 20(3), 412–432.

Winkler, A.J., Lamouria, L.H. and Abernathy, G.H. (1957) Mechanical grape harvest – problems and progress. *American Journal of Enology and Viticulture* 8(4), 182–187.

Winkler, A.J., Cook, J.A., Kliewer, W.M. and Lider, L.A. (1974) *General Viticulture*. University of California Press, Berkeley, California.

Witzgall, P., Bengtsson, M. and Trimble, R.M. (2000) Sex pheromone of grape berry moth (Lepidoptera: Tortricidae). *Environmental Entomology* 29(3), 433–436.

Wolf, T.K. and Cook, M.K. (1992) Seasonal deacclimation patterns of three grape cultivars at constant, warm temperature. *American Journal of Enology and Viticulture* 43(2), 171–179.

Wolf, T.K. and Pool, R.M. (1987) Factors affecting exotherm detection in differential thermal analysis of grapevine dormant buds. *Journal of the American Society for Horticultural Science* 112, 520–525.

Wolf, T.K. and Warren, M.K. (1995) Shoot growth-rate and density affect bud necrosis of Riesling grapevines. *Journal of the American Society for Horticultural Science* 120(6), 989–996.

Wolf, T.K., Haeseler, C.W. and Bergman, E.L. (1983) Growth and foliar elemental composition of Seyval Blanc grapevines as affected by four nutrient solution concentrations of nitrogen, potassium and magnesium. *American Journal of Enology and Viticulture* 34(4), 271–277.

Wolpert, J.A. (1992) Rootstock use in California: history and future prospects. *Grape Rootstock Meeting Proceedings*. Oregon State University, Corvallis, Oregon, pp. 65–72.

Woodham, R.C. and Alexander, D.M. (1966) The effect of root temperature on development of small fruiting Sultana vines. *Vitis* 5, 345–350.

Xu, G., Magen, H., Tarchitzky, J. and Kafkafi, U. (1999) Advances in chloride nutrition of plants. In: Sparks, D.L. (ed.) *Advances in Agronomy*. Academic Press, New York, pp. 97–150.

Yamane, T., Jeong, S.T., Goto-Yamamoto, N., Koshita, Y. and Kobayashi, S. (2006) Effects of temperature on anthocyanin biosynthesis in grape berry skins. *American Journal of Enology and Viticulture* 57(1), 54–59.

Yilmaz, Y. and Toledo, R.T. (2004) Major flavonoids in grape seeds and skins: antioxidant capacity of catechin, epicatechin, and gallic acid. *Journal of Agricultural and Food Chemistry* 52(2), 255–260.

Yokoyama, V.Y. (1979) Effect of thrips scars on table grape quality. *Journal of the American Society for Horticultural Science* 104, 243–245.

Yun, H.K., Louime, C. and Lu, J. (2007) First report of anthracnose caused by *Elsinoe ampelina* on Muscadine grapes (*Vitis rotundifolia*) in northern Florida. *Plant Disease* 91(7), 905.

Yunusa, I.A.M., Walker, R.R. and Guy, J.R. (1997) Partitioning of seasonal evapotranspiration from a commercial furrow-irrigated Sultana vineyard. *Irrigation Science* 18(1), 45–54.

Zabadal, T.J. and Dittmer, T.W. (2003) The maintenance of fruiting potential through the winter for 'Merlot' grapevines grown in southwestern Michigan. *Small Fruits Review* 2(4), 37–44.

Zabadal, T.J., Vanee, G.R., Dittmer, T.W. and Ledebuhr, R.L. (2002) Evaluation of strategies for pruning and crop control of Concord grapevines in southwest Michigan. *American Journal of Enology and Viticulture* 53(3), 204–209.

Zazueta, F.S. and Xin, J. (1994) *Soil Moisture Sensors*. Florida Cooperative Extension Service, University of Florida, Gainesville, Florida, 11 pp.

Zoecklein, B.W. (2002) *A Review of Methode Champenoise Production*. Virginia Cooperative Extension, Arrington, Virginia, 30 pp.

Zouari, N., Romette, J.-L. and Thomas, D. (1988) A continuous-flow method for the rapid determination of sanitary quality of grape must at industrial scales. *Journal of Chemical Technology and Biotechnology* 41(3), 243–248.

Index

*Page numbers in **bold** indicate illustrations.*

(E)-2-hexanol 230
1, 1,6-trimethyl-1, 2-dihydronaphthalene (TDN) 58
abiotic stress 43
abscisic acid (ABA) on stomatal conductance 156
abscission
  calyptra 40
  fruit 43
  leaf 36
advection frost 136
adventitious root 34
Africa 198
Aliette 193
amarone style of wine 239
American hybrids, disease resistance 195
amino acids 49
  glutamate 43
ammonium 43
ampelography 25–**26**
anatomy and physiology 15–18
  above the soil 19–26
  root 15–18
animal pests 212–**213**–**214**, 216
anlage 37
antagonistic organisms on fungal growth 194
anthers 40
anthesis 25, 40, 43, Plate 8
anthocyanin 50, 51, 234
  factors affecting production 53–54
  pigments 53–54, 234
  stability of molecules 53
antioxidants 7–8, 44–45
  from pressed grapes 238
antisporulants 204–205
apoplast 49
arched cane training system 124, Plate 28
argenine 49
Argentina 234
aroma 48, 56–59
  primary compounds 57–59
  S on wine 168
Asia 198
astringency 50, 56
Australia 2, 7, 9, 10, 116, 198–199, 230
Austria 239
axillary bud 20
Azerbaijan 1
azoxystrobin fungicides 200, 204–**205**

Bacillus thuringiensis (BT) 208, 210, 218
bacterial diseases 201–202
balanced pruning 116
barrel pruner **114**, 181
basal nodes 110
basket vines 121, Plate 27
Baumé 46
beetles 210
bench grafting 92
beneficial insects and arthropods, detrimental effect of S 196
berry
  anatomy 24–25, 26
  brix:acid ratio 47, 226
  colour 33
  damage by birds 213–214
  damage by thrips 210, Plate 64
  disease defence mechanisms 193
  ethylene on colour development 225
  maturation 37
  post harvest composition 225
  ripeness and methoxypyrazines 58–59

berry – *continued*
sugars 44
table grape quality parameters 226
temperature for freezing threshold 228
see also, grape
berry development and maturation 43–59
berry growth 33–**34**
phases 33
berry size, seed numbers 42
berry thinning 159
berry-eating pests 213–214
bicarbonates, Na & K on powdery mildew 196
biocontrol
agent 117
beneficial insects and arthropods 196
diseases 194
mealybugs 209
mites 210
phylloxera 208
micophagous mites 196
organic production 218
biotic stress 43
birds 213–**214**
bitterness 56
'Black Corinth' 230, Plate 14
black rot 200, **205**
blackfoot disease 200–**201**
blind budding 110
block shape and size determination 81
bloom of berry 161, 226, **227**
Bordeaux mixture 171, 198
origin 186–187
borer insects 209
boron (B) 43, 172, **173**
botrytis bunch rot 144, 181, 187, 228, 239
Botrytis cinerea 187–195
clones 188
defence against stilbenes 193
botrytis infection **188–189**, Plates 42–43
effect of mealybug feeding 209
minimizing 190–195
'Noble Rot' 239, Plate 41
botrytis, resistance 26
box end or horizontal stay assembly **130**
branching
factors affecting 38–39
flower cluster development 38
Brazil 234
breeding and genetics 26–28
breeding from seed 27
breeding for disease resistance 13
brix on raisin yield 231
brix 28, 46, **48**, 237, 239
brix:acid ratio 47, 226
brush, berry 25, 225
brush pulling 112
bud
anatomy 18–**19**–20
distance above ground on frost damage 136
hardiness 67
lateral 110
position and fruitfulness 37
budbreak 30–31, 136, Plate 7
and flowering 31
cytokinin transport 39
effect of pruning 115
enhancing in tropical climate 69
burning
for frost control 137
for weed control 219
burying vines 68, Plate 18
bush vines **120**

'Cabernet Sauvignon' 11
calcium (Ca) 168–169
calculation
disease infection period 181
evapotranspiration 151
for balanced pruning 107–108, 116
sprayer 220–222
calibration, spray equipment 223
California 197, 201
calyptra 31
abscission zone 40
cambial layer 66
Canada 239
cane and spur pruning 109–110, 117
cane pruning, advantages and disadvantages **110**–111
cane **19**
cross section 20–**21**
fruit and dormant 108
maturation 35–36
pruning 109–**110**, Plate 28
selection for cordon 113
weight for counting 117
cannon, bird scaring 215–**216**
canopy
air movement and relative humidity 191, 195
climate 60
density, planning ahead 118
exposure and row orientation 81–82
height and row spacing 87

openness and spray penetration 220
qualifying change in 146–147
shaded 146, Plate 38
canopy management 139–147
benefits 147
downy mildew control 198
goals and tools 139–140
mechanization **144**, 178–179
on harvest composition 49
root pruning 145
things to avoid 146
cap 32, Plate 8
capacitance measurement 152–153
capfall 40, Plate 8
carbaryl 209
carbohydrate
pruning for balance 106
status on flower clusters 39
carbohydrate partitioning **63**–64
carbohydrate reserve 42–43
on flowering 31
carbohydrate translocation 169
carotenoids 58
'Cardinal' 6
'Catawba' 196
cell division 172
rate 44–45
Chardonnay 136, 196–**197**, Plate 6
chemotaxonomy 49–51
Chile 2, 6–7, 10, 196, **208**, 225, 230, 234
chilling requirement 36
China 7, 9, 10, 197, 225, 230, 239
chloride toxicity 168, Plate 34
chlorine (Cl) 172–173
chlorophyll 169–170
chloroplast 21–22
chlorosis 166, 168–172, Plates 20, 39
citric acid 44
cleft graft 93
cleistothecia, powdery mildew 195, Plate 45
climate change 72–73
climate on organic production 218
climate
for icewines 239
on downy mildew growth 197–198
on pest damage 185
on site selection 76–77
quantification 60–62
requirements for growth 65
three levels of 60–62
tropical vine growth 36
clones 14–15
identification 15
selection of characteristics 15
single gene mutation 15
white vs. red grapes 15
cluster compactness 159
on botrytis infection 190
cluster infection, powdery mildew 195
cluster thinning 159–**160**
colchicine and polyploidy 27
cold damage, managing 67–68
cold hardiness 4, 65–68
cold injury and crown gall development 201
cold press method 235
cold resistance 35–36
cold temperature probabilities **66**
cold-damaged vines 66–67, Plates 16–17
colour 47–48
berry 225
berry kins and juice 52
effect of metal ions 54
effect of quercetin 54
effect of sun exposure 53–54
effect of temperature 53
sun dried raisins 232
compound bud 18–**19**, 20
'Concord' grapes 7, 116, 233–234
planting in USA 234
condensed tannins 55
consumer
demand on organic production 218
influence on cultivar plantings 13
preferences, aroma 59
content vs concentration 47–**48**
control
black rot infection 200
borer insects 209
crown gall 202
eutypa 200
fanleaf degeneration 203
fungal diseases 204–205
grape berry moth 209–210
leafhoppers 208–209
mammals and birds 214–215
nematode 212
phomopsis 199
phylloxera 208
postharvest diseases 228–230
viral diseases 205
cool climate grape growing 32
cool climate viticulture, summer temperature and frost 66
cool climate
carbohydrate reserve on fruit set 43
harvest and leaf-fall 36
post harvest root growth 35
temperature on crop load 39

cooling, fresh grapes 228–229
co-pigmentation 54–**55**
copper (Cu) 170–171, 218
cordon pruning 109
cordon 112
  wrapping too tight on wire 112–**113**
count nodes 109
cover crops 82–**83**, 132–**133–134**, 140
  advantages 132
  disadvantages 132–134
  on botrytis infection 191
  P application 167
  selecting 133
covercrop management 217
crop load
  on dry weight **63**
  on fruit quality **158**
crown gall 201–**202**
crushed glass for mulch 219
crusher/de-stemmer **238**
cultivar choice, frost prevention 136
cultivar identification, ampelography 26
cultivar on seed number 41
cultivars
  breeding for disease resistance 13
  choice for commercial crop **14**, 70
  complexity and common names 11
  consumer influence on planting 13
  databases 11–12
  disease tolerant 186
  DNA fingerprinting 12
  for juice 234
  for raisin production 230
  genetic similarities **12**
  response to girdling 156
  rootstocks 13
  selecting cold hardy 67
  susceptibility to botrytis 189–190
cultivation 5, **86**, **120**, **132**
  weed management 176–**177**–178
cultivation, effect on frost 138, **139**
cultural practices
  for disease control 204
  grape berry moth control 210
  leafhopper control 208
  minimizing botrytis infection 190–191
  powdery mildew control 195
cuticle 25
cuttings
  for propagation 91–92
  root growth 34
  storing 92
Cylindrocarpon spp. 200
cytokinins (CKs) 27, 39–40

dagger nematodes 212
Daktulosphaira vitifoliae, *see* phylloxera
databases
  cultivar and nomenclature 11–12
  historical climate information 63
day length on dormancy 36
debudding 99–100
deer, defoliation by **213**
deficit irrigation
  on ABA transport 156
  on shoot growth 156
  regulated deficit irrigation (RDI) 155–156
degree of polymerization (DP) 55
'Dolcetto' 136
demethylation inhibitors (DMIs) 196
diagonal stay end assembly **129**
dielectric constant of soil 152
digital imaging for canopy data 147
dioecious vine 24
disbudding 99, 100, **102**
disease forecasting 194–195
disease resistance
  breeding 13
  cultivars and rootstock 189–190
  induced 193
  phytoalexins 192–193
  preformed phenolic compounds 191
diseases
  fungal 187–202
  introduction to new areas 185–186
  post harvest 228
disorders 206
dithianon 199
divided canopy systems **102**, 124–**125**–126
DNA fingerprinting 12, 50, 203
dormancy 29–30, 36–37, 63
dormant cane 108
  survival 67
downy mildew 197–198, Plates 47–48
drainage 83
dried grapes 230–233
dried on vine (DOV) raisin production 127, 231
droplet size, spray 221
dry matter production, calculation for pruning 108

early bunchstem necrosis 43, Plate 15
ecodormancy 36
economic threshold 185
economics of site selection 81
Egypt 2

'Einset Seedless' 41, Plate 3
elderberry juice in wines 51
electrical conductivity sensors 152
'Emperor' 6
end assemblies **128**–131
causes of failure **130–131**, Plate 40
determining type 131
End Point Principle (EPP) 110–**111**
endodormancy 36–37
Endopiza viteana 209
engustation 33
environmental monitoring, mechanization 181–182
environmental/climatic influences on growth 59–62
enzyme-linked immunosorbent assay (ELISA) 203
enzymes
fungal 191, 193
for fruit formation 171
for photosynthesis 170
for reduction/oxidation 172
ϵ-viniferin 192
ethylene 225
Europe 197–200
European Cooperative Programme for Plant Genetic Resources 11–12
European grape berry moth 210
European Network for Grapevine Genetic Resources Conservation and Characterization 11
eutypa die back 117, 166, 198, 199–200, Plate 50
Eutypa lata 199
Euvitis 2–4
evapotranspiration 151

false K deficiency 168
false spider mites 210
fanleaf degeneration (GFLV), symptoms 203
fanleaf virus (GFLV) 212
fans
for frost control 136–**137**
for reducing cold damage 67
fatty acids 44
favanols 54–55
fences 215
fermentation 5–6, 49
of pressed grape skins 238
fertilization 140
fertilizer programme 165
fertilizer, mono-potassium phosphate on powdery mildew 196
field grafting 92–93
fixed growth 20
'Flame Seedless' 76, **157**
flavan-3-ols, presence in tissues 52
flavonols 50–51
presence in tissues 52–54
flavour
development 33
foxy 4, 56–57
grappa 238
human detection of compounds 59
kerosene 58
muscat 230
quantification 28, 57
flood irrigation 148, **149**
flower and berries 24–26
flower cluster **160**, Plate 2
anatomy 25
initiation 32–33
primordia 19
thinning 158–159
flower clusters, relationship to tendrils 37–38, Plate 12
flower development and anthesis 40
flower initiation, fruit development and berry maturation 37
flowering 31–33
date, predicting 40
duration 40
flowers per cluster 40
foliage wires 113, 143, 181
fosetyl-aluminum 193–194
foxy flavour 4, 56–57
France, introduction of new pests 186
free growth 20
Free Volatile Terpenes (FVT) 58
freezing threshold of berries 228
'French Colombard' 7
French hybrids 13
frost damage Plates 19–20, 32
and botrytis Plate 42
and cover crop 134, 138
green tissue 69
frost
active and passive control 136–138, Plates 33–35
fans 136–**137–138**
management 134–139
tolerance 69
types of event 136
fructose 44
fruit cane 108
fruit maturity 46–47
factors affecting 48–49

fruit removal, young vines 101
fruit set 156, 172
  factors influencing 42–43
  predicting 32
  success rate 32
fruit thinning 156, 158
fruit weight to pruning weight ratio 107–109
fruit weight, calculation for pruning 108
fruit
  black rot infection 200
  development 37
  formation, effect of Zn 171
  growth affected by girdling 156
  leaf area needed to ripen 108
  on root growth 35
  sun exposure on maturity 143
fruitfulness 37
  spur pruning 112
fruiting potential 37
fruiting wire 122
fruiting zone 112–113
fungal diseases, control 204–205
fungicides
  black rot control 200, **206**
  downy mildew control 198
  organic 218–219
  post infection 204
  powdery mildew control 195–196
  soil 212

gallic acid equivalents (GAE) 52
garlic paste 69
gas chromatography 28
gas chromatography-mass spectrophotometry 57
Geneva Double Curtain (GDC) 124–**125**, 136, 142, 175, 179
geographic distribution
  downy mildew 197
  eutypa 199
  phomopsis 198
  Pierce's disease 201
Georgia 1
Germany 239
'Gewürztraminer' 188
gibberellic acid (GA) 39, 159–**160**
girdling 156, **157**
Global Information System (GIS) 81
Global Positioning Systems (GPS) 80
glucose 44
glycosylation on aroma 57
goals of pesticide application 219
golden raisins, production techniques 232
grafting 88, 92–**93**, 186
grape berry moth 209–210
grape composition 44–**45**
  canopy management 49
  effect of oil sprays 196
  primary metabolites 44
  secondary metabolites 49–**51**–56
grape production, by country **8–9**
grape products
  fermented 5–6
  health-related 7–8
  international markets 9–10
  juice 7–8
  raisins 7
grape, post harvest utilization 225, 230, 233, 235
grapevine leafroll virus (GLR) 203, Plate 52
grapevine:
  as perennial plant 63–64
  botanical classifications 2–4
  climate requirements 65
  cold hardiness 4
  forced evergreen production 36
  history of cultivation 4
  nutrients in prunings 164
  origins 1–2
  see also, vine
grappa 238
grazing pests 213
Greece 7, 230
Greek Vitis Database 11
green cuttings 93–94
grey mould 187
growing degree days (GDD) 60–62
  calculations **61**–62
growing season 60
growth patterns
  root 34–35
  tendrils 35
growth phases, berry 33, 44, 49
growth
  balancing vegetative and reproductive 64
  coordinating for tropical climate 69
  crop load and leaf removal on dry weight 64
  effect of climate change 72–73
  effect of UV radiation 73
  effect of vine density 85–87
  environmental/climatic influences 59–62
  heat and light requirements 70
  influence of rootstock 89
  projection 182
  soil requirements 71

terroir 70–72
using vine shelters 98
water use requirements 70–71
guidelines for pruning 107–109
Guignardia bidwellii 200

hail 206, Plate 56
harvesting 159–162
by hand **161**, 226, 231, 236
by machine 161–**162**, 180–**181**
juice grapes 235
method on phenolics extracted 56
method on wine quality 159–161
raisins 231–232
sugar content on date 47–48
table grapes 226
wine grapes 235–237
haustoria 195
health-related products 7–8
heat stress 206, Plate 57
hedge-type system 178
hedging 143
mechanization **144**, 181
hen and chicken 172
herbicides
for young vines 99
toxicity symptoms 206, Plates 53–55
weed control 217
high performance liquid chromatography (HPLC) 47
hillers 176
honey dew 208
horizontal shaking system 175
hot foam 178
hot press method 234
human detection of flavour compounds 59
hydrogen cyanamide and dormancy 36, 69
hydroponics 163

icewine 239
indole acetic acid (IAA) 171
infection period, botrytis 189
inflorescence necrosis 43, Plate 15
inflorescence, flower differentiation 40
insect pests 206–210, **211**
insect resistance 23–24
insect vectors 185, 202
insecticides
leafhopper control 208
organic 218
integrated pest management 209
international trade of grape juice 234
internode **19**, 20
influencing shoot length 20
length 113
inversion layer 136, 137
Iraq 1
iron (Fe) 170
irrigation 70–71, 77, 94, **95**, 140, 148–156
deficit 156
for frost control 137–138
methods and equipment 148–150
monitoring soil moisture 151–153
on diseases 204
rootstock selection 89
scheduling 154–155
'Isabella,' volatiles on botrytis growth 230
isotherm procedure, predicting bud hardiness 67
'Italia Pirovana' 173
Italy 6, 225, 239

Japan 197
JMS Stylet-oil 196–**197**
juice 7–8, 234–235
juice/preserves 233–235
juvenile state 27

Kazakhstan 10
kerosene character of wine 58

labour
hand harvesting 226
hand thinning and harvesting 156, 158–159, 161
hand weed control 176
Point Quadrat data collection 147
removal of immature clusters 236
repetitive strain injuries 127
requirements 112, 124
shoot topping 143
training for safety 222
lasers, for straight rows 176
lateral buds 110
lateral shoots 143
Latitude Temperature Index (LTI) 62
layering 91
leaf
age on disease resistance 191–192
ampelographic measurements **26**

leaf – *continued*
  anatomy on insect resistance 23–24, 26
  area needed to ripen fruit 108–**109**
  cross section **23**
  downy mildew symptoms 197, Plate 47
  leafhopper feeding injury 208, Plate 61
  number requirement per shoot for fruit 143
  phylloxera symptoms 207, Plate 60
  powdery mildew symptoms 195, Plates 44–45
  water loss for monitoring water status 154
leaf blade 20
leaf petiole analysis 140
leaf primordia 19
leaf removal 143–145, Plate 36
  amount optimum for wine grape 144
  damage as a result of 144–145, Plate 37
  mechanization **179**
  on botrytis infection 144, 191
  on dry weight 64
  optimum timing 144–145
  using pulsed air 179
leaf-fall 36
leafhoppers 201, 208–209
  feeding symptoms 208, Plate 61
lenticels 25, Plate 6
light
  on branching 39
  on fruit set 42
lignification 35
  seed 41
lignin, active response to infection 192
Lobesia botrana 210
Lyre trellis **125**

macroclimate 48
macronutrients 166
magnesium (Mg) 169
  effect of high K 169
  'Malaga' 6
malic acid 44
malvin 53
management
  for uniform vineyard 117
  young vines 81–100
mancozeb 170, 200
manganese (Mn) 170
Marlborough Sauvignon blanc style 58–59
marc 238
marginal chlorosis 168
Market News Raisin Update 230
market research for site selection 75
materials other than grapes (MOG) 180, **233**
maturity 156
  fruit 46–47
  wine grapes 235
mealybugs 209, Plate 63
  as vectors 203
mean temperature of the warmest month (MTWM) 62
measurements of fruit maturity 46–47
mechanical harvester
  early development 175
  for crop reduction 179–180
mechanical harvesting 161–**162**, **181**, 231
  juice grapes 235
  on botrytis inoculum 194
  raisins 180
  wine grapes 180, 235–236
mechanical injury, on botrytis infection 189
mechanical pruning, viability 181
mechanical weed control 217, 219
mechanization 98
  canopy management 178
  cluster cutting 175
  environmental monitoring 181–182
  harvesting 175
  limitations 176
  pesticide application 175
  sampling harvested wine grapes **237**
  trellis design 119, 122, 124, 184
  weed control **86**, 176–178
Meloidogyne spp. 212
mesoclimate 48, 60, 77
metal ions 54
methoxypyrazines 57–58
methyl anthranilate 57
methyl bromide 212
Mexico 234
micophagous mites 196
microclimate 60
micronutrients 170–173
micropropagation 94
micropyle 41
microsatellite DNA markers 12
microsprinkler irrigation 138, 148, Plate 35
microwave 178
milk and whey for powdery mildew control 196
millerandage 172
minerals 44–45
minimal pruning 122–**123**
  and mechanical harvesting 180
mites 210, **211**, **212**

molecular biology and breeding 27
molybdenum (Mo) 172
monoecious vine 24
Movable Free Cordon System 183
mowers 176
  under vine **178**
mulchers, hydraulic 176, **177**
mulches 134, 219, Plates 25, 31
  benefits to vines 134–**135**
Munson T-type trellis **127**
Muscadelle 136
Muscadine 2–3, 13, 234
Muscadinia cvs. 228
'Muscat of Alexandria' 230
mussel shells for mulch 134, 219, Plate 31
must 238
mutations
  berry colour 15, Plate 5
  seedless cultivars 41
myclobutanil 204–**205**
mycorrhizae **17**–18

necrosis, leaf tissue 206
nematodes 82, 208, 212
  as vector 203
netting 214–**215**
neutron probe, accuracy and cost 153
New York 116
New Zealand 9, 12, 36
nitrate reductase 172
nitrogen (N) 166–167
  for yeast 49
  on disease severity 197
  timing of application 166–**167**
'Noble Rot' 239, Plate 41
node counting 40
nodes **19**
  cane pruning 110
  count and non-count 109–110
  down and up facing on cordon 112
  pruning 106–109
non-count cane **19**
non-count shoots **108**, 140–141
norisoprenoids 57–58
North America 197–199
nucleic acids 44
nursery stock
  crown gall-free 202
  heat stress 206
  virus-free 203
nutrient
  analysis 164–165
  application, timing 164
  deficiencies and excesses, symptoms 164, 166
  deficiencies, on fruit set 43
  remedies 165–173
  supplements 8

Oats, as a cover crop **133**
Oechsle 46
organic
  pesticides 218
  postharvest disease control 230
  production 218–219
organophosphates 209
Oxford Companion to Wine 15
ozone gas 230

packing house cooling 229
packing, table grapes 226, **228**
panicle inflorescence 24
parasitic wasps 209, 218
parthenocarpy 41–42
Partial Rootzone Drying (PRD) 155–156
Passito wines 239
pathogenesis-related proteins (PRPs) 192
Pendelbogen training system 124
Penicillium 228
'Perlette' 6, **42**
periderm 35, Plate 10
permanent wood 42
personal protection apparatus (PPA) 222
Peru 2
pest control, second year vines 103
pest damage 185
pesticides
  application equipment 222–223
  methods of application 219–222
  organic 218
  resistance 217
petiole sampling 165
pH 47, 78, 168
phases of growth, berry 33, 44, 49
phenolics 7–8, 44–45, 161, 191, 197
  on wine colour 54
  structure and metabolism 50–**51**
phenology 29–34
phenylalanine 50
pheromone, for insect control 209, 210, 218
phloem 20
Phomopsis viticola 198
phomopsis 198–199, Plate 49

phosphorus (P) 167
application on K supply 167
photoassimilates 44
photosynthesis 21–24, 39, 156
developing shoot 31
effect of oil sprays 196
enzymes 170
factors affecting 22–23
phylloxera 4, 13, 26, 186, 206, **207**, 208, Plate 58
aggressive strain 88–89
control 208
spreading by man 208
vine symptoms 207, Plates 59–60
phytoalexins 192–193
phytotoxicity
spray materials 221
see also, toxicity
piceid 192
Pierce's disease 185–186, 201
pigments 53–54, 58, 234
'Pinot gris' 102
'Pinot noir' 134, **160,** 188
plant growth regulators 39–40, 41, 42, 88
table grape production 159, 160
plant tissue analysis 165
plant water status, pressure chamber for monitoring 153–**154**
planting 94–96
records 96
tools **95, 96**
use of lasers 176
Plasmopara viticola 197
ploidy 27
Point Quadrat method 147
pollen 172
count 32
growth **32,** 40–41
sacs 40
pollination 31–32
polymerase chain reaction (PCR) 203
polymers of tannins 55
polythene for weed control 97
pomace 8
potassium (K) 168–**169**
deficiency, soil pH and Mg 168
excess and juice pH 168–169
Potential Volatile Terpenes (PVT) 58
powdery mildew 186, 195–197, Plates 44–46
control 168
introduction into France 186–187
Pratylenchus spp. 212
precipitation assays 47
pre-cooling, post harvest 227–228
predatory mites 210
pre-plant soil fungicides 212
pressure chamber for plant water status 153–**154**
primary grape aroma compounds 57–59
primary metabolites 44
primordia
flower cluster 19
leaf 19
primordium 37
proanthocyanidins 55
processing, grape products 232, 234, **236**–237
production by country 8–10
proline 49
propagation 34, 91–93, 202, Plates 23–24
layering 91, Plate 22
material and disease dissemination 199
protein synthesis 168
proteins 44
pruning weight 107–109
pruning 36, **106–107**
aims 105–106
balanced **106**–109
basic principles 117
count and non-count nodes 109
delaying 118, 138
dormant **19,** 105
effect on growth and crop 107–108
for reducing cold damage 68
for young vines 99
mechanical 181–182
mechanization 179, 181–182
minimal 122
on budbreak 115
paint for larger cuts 117
second year vines 102–103
root 145
shoots 140, 143
summer 105
tips to smooth process 113
training for decision making 111–112
types 109–113
unpruned vines **107**
pterostilbene 192
pulsed air for leaf removal 179
pyrethroids 208
pyrethrum 218

quality
indicators 27–**28**
manipulation of secondary metabolites 50

quantification, fruit 46–47
table grapes 226
wine 134
quercetin 50, 54

rabbits 213
radiation frost 136
rainfall on downy mildew infections 198
raisins 7, 230–**233**
cultivars 230
drying 230–232, Plates 4, 65
for wine making 239
mechanical harvesting 180,
sorting and storage 233
techniques for drying 231–232
trellis systems 127
world production 230
recioto style of wine 239
record keeping, disease trouble spots 195
regulated deficit irrigation (RDI) 155–156
replacement spurs 113
replanting 101
disorders 212
issues 79
resveratrol 192–**193**–194
retrofited trellis 126
Rhizopus rot 228
'Ribier' 6
'Riesling' 136, Plate 41
ripping soil **145**
structure and function 15–**16**, 18
'Rish Baba' 6
robotic pruner 181
root borer 209, Plate 62
root cap 15
root growth
depth 34
seed vs. cutting 34
root hairs 16
root-knot nematodes 212
root placement on blackfoot disease 200–**201**
root pruning **145**
root tip 15–**16**
root
and mycorrhizae **17**–18
growth patterns 34–35
effect of high Cu 171
killing with systemic herbicide 212
nematode pests 212
normal turnover 19
phylloxera injury **207**
rooting of cuttings 92
rootstock 24, 88–91
and scion 140
and water table 35
choice 89–91
cultivars 13
classification by root angles 34–35
iron-chlorosis resistant 170
nematode resistance 203
on diseases 204
phylloxera resistant 186
resistance to botrytis 190
root growth patterns 34–35
selection 77, **90**
selection for phylloxera resistance 88–89
selection for reducing risk 91
rotenone 218
row orientation 140
determination 81–82
on botrytis infection 190
row spacing **86–87**,121
RuBISCO 21, 72, 166
'Ruby Seedless' 6
ryania 218

sabadilla 218
Saccharomyces cerevisae 1
sampling, mechanical arm **237**
sap 29, 40–41
sap ball **24**
sap flow meter **155**
scaring devices, bird damage control 214, **216**
scion 89
rooting 68
sclerotia 188
Scott-Henry trellising system **119**, 124–**126**, 142
scouting 211
second season vine care 100–103
second set 33
secondary metabolites 49–56
on bitterness and aroma 52
seed 25
number per berry 41
on berry size 42
remnants 41, **42**
seedlessness 13, 41–42
on growth 33
seeds 25
Selective Inverted Sumps **138**
self pollination 24

self- or stake-supported trellis 119–121
  advantages and disadvantages 121
sensory analysis 28
shoot development and growth 20–21, 26
shoot positioning **142**–143
shoot thinning **141**, Plate 30
  timing 140–142
  on botrytis infection 191
shoot topping 143–**144**
shoot **31**
  bifuricated **38**
  cross section **21**
  horizontal on canopy density **142**
  growth and leaf area **108**
  non-count **108**, 140–141
  position on growth 110–111, 112
  rate of growth 30
shot berries 171–172
sigmoid growth curve 33–**34**
single gene mutation 15
single wire trellis 121–122
sinker roots 34
site
  climate 60
  planning 81–87
  preparation 82–83
site selection 61–62, 69–72, 75–81, 140
  economic and conflict issues 79–80
  factors affecting 76
  on botrytis infection 190–191
  temperature moderating influences 77
slope and aspect of land **79, 80**–81, **82**
smudge pots 137
soil moisture, methods of monitoring 151–153
soil preparation, preplant 82–83
soil testing, preplant 164–165
soil type 60
soil 71
  accumulation of Cu through sprays 198
  analysis, nutrients 164
  Ca on high Na 169
  cultivation for frost control 138, **139**
  depth 78, Plate 21
  moisture monitoring 151, **152**, 153
  on site selection 77–79
  pH 78
  pH on Mg deficiency 169, 170
  preplant fungicides 212
  sand content on phylloxera 207
  temperature and pH on chlorotic symptoms 170
  vineyard floor management 131
  water holding capacity 78
sooty mould 208
sorting wine grapes 237
South Africa 6, 7, 10, 199, 230
South America 200
spacing, vine 85–**86**–**87**
  and canopy height 87
  between-row 85–87
  within-row 86–87
Spain 10, 181
sparkling wine 236
  production 161
specific gravity 46
spectrophotometry 47
spider mites 210
spores, means of spreading 188–189
spray coverage
  evaluating 220
  minimizing problems 221–222
spray drift reduction 221–222
sprayer
  calculation 220
  calibrating spray output **223**
sprayers 219–**220**, **222**–**223**
spreader roots 34
spring fever 168
spur death 112
spur pruning **112**–113
  advantages and disadvantages 112–113
staked vines 119, **121**
stenospermocarpic 41
sterile offsprings from crosses 41
stilbene-synthase 193
stomatal conductance 154
stomates 22–**23**, 189, 198
storage
  fresh grapes 229
  raisins 233
stress 43
stuck fermentation 49
styles of wines 58–59, 239
sucrose 44
sugars
  equivalent measures **46**
  in berry 44
  tracking for harvest 47–**48**
sulphites
  bleaching of wine 54
  golden raisins 232
sulphur dioxide
  corrosiveness 229
  human sensitivity 229
  juice treatment 235
  treatment for storage 229

sulphur (S) 168, 186, 195–196, 205, 218
effect on predatory mites 210
'Sultana' 6–7, 11, **115**, 230, 234, Plate 3
summer pruning 143
sun drying on raisin colour 232
Sylvoz training system 124, Plate 29
symplast 49
symptoms
blackfoot disease 200–**201**
borer injury 209, Plate 62
crown gall 201–**202**
fanleaf degeneration 203
herbicide toxicity 206, Plates 53–55
leafhopper injury 208, Plate 61
leafroll virus 203, Plate 52
nutritional deficiencies 164–**171**–173
'Syrah' 11
Syria 2
systemic acquired resistance (SAR) 193
table grapes 159, 225–230
astringency from tannins 56
hand harvesting 226
harvesting and packing 161
industry and production 6
major cultivars 6
markets 9–10
packing 226–**228**–229
quality parameters 226
seedless 41
thrips damage 210
trellis systems 127
world production 225

table wines 161
tannins 50, 54–56, 191
definition 54
polymerization 55–56
precipitating proteins 55
taproot 34
targeted irrigation **150**
tartaric acid 44
T-budding 93
teinturier type grapes 52, Plate 11
temperature
for budbreak 30
for powdery mildew growth 195
GDD 61–62
on anthocyanin 53
on botrytis growth 187
on branching 39
on dormant survival 67
on flower number 40
on fruit set 32, 42
on monoterpenes 58
on post harvest diseases 228
tannin production 53
winter minimum for site **66**
tendrils 20, 25, Plate 9
growth pattern 35
relationship to flower clusters 37–38
tensiometer 152
terpenes 57–58
terroir 71–72
The International Grape Genome Program 12
thinning
flower cluster
fruit 156, 158–159
shoot 140–**141**–142, 191, Plates 30, 36
thiols 168
'Thompson Seedless' 6, 11, **109**
thrips 210
tieback end assembly **128**-129
tiller 176
time-domain reflectometry (TDR) 152–153
titratable acidity (TA) 47, 143
'Tokay' 6
tolerance, diseases 186
topoclimate 60
toxicity
B 172, **173**
Cl 168, 173
copper 198
herbicide 206
potash 168
sulphur 196
see also, phytotoxicity
training and trellising 118–131
training system 124
basket vine 121, Plate 27
bush vines **120**
changing 118
cordon 112
overhead 126–127, Plate 30
young vines **121**
training, first year 96–**97**–98, **100**, **101**
transporting wine grapes 236
T-trellis **127**
trellis design for mechanization 183
trellis system on mechanization 178
trellis 5, 96
damaged **120**
divided canopy **102,** 124–125
end assemblies 128–131
fixing broken posts 113
foliage wire 143
hedge type 122–124

trellis – *continued*
- overhead systems 126–127
- shoot positioning 142–143
- second year vines 102–103
- types of 87, 119–**122**–128

trellising, goals 118–119
Trichoderma spp. 194
trichomes 196
tropical climate 68–69
- vine growth 36

Trunk Proximity Principle 111
trunk 19, **30**
- damage by animals **214**
- debudding 99–100, **102**
- multiple replacement 67–68

turgor pressure, on capfall 40
Turkey 1, 7, 213, 225, 230
twin furrow irrigation 148

Umbrella Kniffen training system 124, Plate 28
Uncinula necator or *Erysiphe necator* 195
upright growth habit 121, 124
Uruguay 138
USA 6, 7, 10, 225, 230, 234
user friendliness, trellis systems 127
UV radiation 73
- on disease severity 197
- on flavonol production 52
- on resveratrol production by berries **193**

veraison 33, 43, 49
vertical shoot positioning (VSP) 118, 122, **123**, 124, 136, 142
vesicular-arbuscular mycorrhizae (VAM) 170, 172
vine
- balance and shoot thinning 142
- capacity 63, 108, 115, 117
- decline 185, 212
- density on vigor 86
- establishment 96–103
- N status 165
- planting density 85–87, 140
- shelters 98, Plate 26
- size 118
- spacing 85–87, 110
- training, young vines **100, 101**
- vigorous **119**
- *see* also, grapevine

vine weevil 210, **211**
vineyard
- cleanliness on diseases 194
- floor management 131–135
- recording disease trouble spots 195
- uniformity and mechanization success 175

vineyard management
- canopy 139–147
- frost prevention 134–139
- pruning 105–118
- use of phenology 29
- young vines 81–100

vineyard manager, setting threshold for irrigation 155
viral diseases 202–203
- control 205
- methods of detection 202–203, Plate 52
- mode of transmission 202

Virginia Dare 13
vitamins 44–45
Vitis amurensis 2
Vitis labrusca 2, 4, 22, 230
- argenine 49
- berry composition 45
- S toxicity 196

Vitis riparia **3**, Plate 1
Vitis vinifera 1, 2, 203, 228, 234
- berry composition 45
- day length on dormancy 36
- disease susceptibility 195, 218
- pest susceptibility 186

water availability
- for site 83
- pre-veraison stress **148**
- site selection 77
- stress on vines 156–**157**

water holding capacity (WHC) 78
water status monitoring 182
water table, rootstock selection 35
water use efficiency (WUE) 70–71
wax, epicuticular **227**
weather station 182–**183**
- predicting infection 194–195

weather
- dampness and phomopsis infection 198–199
- humidity and eutypa infection 199
- on fruit set 31 32
- on spraying 221–222
- rain on (DoV) grapes 231–232

weed control 217
- first year 96–97

organic production 219
preplant 82
use of heat 178
weed management, mechanization **132,** 176–**177**–178
weed spray timing 182
weeds 216–217
wild grapevines, on insect and disease 209
wind and photosynthesis 22–23
wind protection 83–85
windbreaks
benefits 83–**84**
drawbacks 85
types **84**–85
wine grapes
removal of immature clusters 236
mechanical harvesting 180
sampling for quality 237
wine making 49
basket press **236**
crusher/de-stemmer 237, **238**
influence of terroir 72
method on phenolics release 56
wine markets 9–10
wine quality
effect of harvesting method 159–161
effect of mussel shell mulch 134
wine styles 58–59, 239
wine **35,** 235–239
bitterness and aroma 52
chemotaxonomy 51
colour enhancement 54
kerosene character 58
metal ions on colour 54
off flavour with S residue on berries 196
red, processing 237–238
religious 234
sulphite bleaching 54
white, fermenting juice 236
wipers, herbicide 219
within-row spacing 86–87
worker training
cane pruning 111–112
safety 222
world production, grape juice 234

Xiphinema index 203
Xiphinema spp. 212
Xylella fastidiosa 201
xylem 20
xylem exudate 29, **30**

yeast assimilable nitrogen (YAN) 49
yeast, wine 1, 49
addition for wine 236
yield component **115**
young vine care 99–103

'Zante Currant' 230, Plate 14
zinc (Zn) 43, **171**–172
'Zinfandel', as table grape 6